Theo Siegrist
X-Ray Structure Analysis

Also of Interest

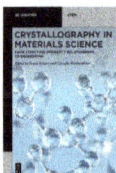

Crystallography in Materials Science
From Structure-Property Relationships to Engineering
Schorr, Weidenthaler (Eds.), 2021
ISBN 978-3-11-067485-9, e-ISBN 978-3-11-067491-0

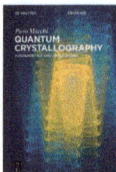

Quantum Crystallography
Fundamentals and Applications
Macchi, 2022
ISBN 978-3-11-060710-9, e-ISBN 978-3-11-060712-3

Complementary Bonding Analysis
Grabowsky (Ed.), 2021
e-ISBN 978-3-11-066007-4

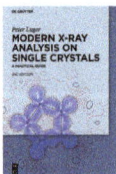

Modern X-Ray Analysis on Single Crystals
A Practical Guide
Luger, 2014
ISBN 978-3-11030823-5, e-ISBN 978-3-11-030828-0

Theo Siegrist

X-Ray Structure Analysis

—

DE GRUYTER

Author
Prof. Theo Siegrist
Florida State University
FAMU-FSU College of Engineering
2525 Pottsdamer Street
32310 Tallahassee, FL
USA
siegrist@eng.famu.fsu.edu

ISBN 978-3-11-061070-3
e-ISBN (PDF) 978-3-11-061083-3
e-ISBN (EPUB) 978-3-11-061092-5

Library of Congress Control Number: 2021943339

Bibliographic information published by the Deutsche Nationalbibliothek
The Deutsche Nationalbibliothek lists this publication in the Deutsche Nationalbibliografie;
detailed bibliographic data are available on the Internet at http://dnb.dnb.de.

© 2022 Walter de Gruyter GmbH, Berlin/Boston
Cover image: Theo Siegrist, CrystalMaker®
Typesetting: VTeX UAB, Lithuania
Printing and binding: CPI books GmbH, Leck

www.degruyter.com

Preface

Crystallographic techniques are used in many scientific disciplines, condensed matter physics, chemistry, mineralogy, materials science and engineering, each area with a different emphasis on the applications of crystallographic techniques. In many text, the techniques are briefly discussed based on Bragg's law, which is derived in a traditional manner. The ingenious devices invented over the last century to explore the reciprocal space and accurately measure integrated X-ray intensities, intensity profiles and intensity positions, are often used as black box systems, with limited understanding of the devices. In this text, an overview of basics of crystallography is presented, with brief discussions of current methods.

The text is the result of the lectures in *Structure Determination* given at the FAMU-FSU College of Engineering from 2013 to 2020. The lectures were designed for graduate students in physics, chemistry, geology, material science, and engineering. The different levels of preparations of the students necessitated the inclusion of a number of introductory chapters, covering geometry, waves and symmetry. Chapters introducing single crystal and powder diffractometry cover implementations of different types of systems. The text builds on the excellent textbooks of A. Guinier, *X-ray Diffraction in Crystals, Imperfect Crystals and Amorphous Bodies* (Dover Publications, Inc, New York, 1994), and D. Schwarzenbach, *Cristallographie* (Presses polytechniques et universitaires romandes, Lausanne, 1993).

While this text is necessarily incomplete, the basic topics are covered in enough detail that the reader will grasp the power that diffraction techniques provide for determining and visualizing atomic arrangements in periodic structures. Individual chapters were designed for inclusion in other courses, and thus, repetitions in the discussions have purpose. The text also reflects the personal preferences of the author and includes discussion of data taken in the author's laboratory.

This text would not be possible without support from my wife and family, and my colleagues. I thank the many students who have taken this courses and who helped to improve the chapters and pointing out errors and areas where explanations needed to be expanded. Thanks go to the manufacturers of diffraction equipment who graciously provided images of their diffraction equipment; images were provided by Bruker (https://www.bruker.com/en.html), Rigaku (https://www.rigaku.com/products/xrd), Stoe (https://www.stoe.com) and Huber (https://www.xhuber.com). Thanks go to Dr. M. L. Steigerwald at Columbia University for providing the sample of $Co_2(CO)_6(PEt_3)_2$ used as an example. Permission to reproduce symmetry entries from the *International Tables for Crystallography, Volume A* (https://it.iucr.org/A) is gratefully acknowledged.

The text was formatted using LaTEX (https://www.latex-project.org), with figures drawn with "tikz" and "pgf" (https://github.com/pgf-tikz/pgf). The tikz code for the Wulff net was provided by Dr. G. Nolze, Federal Institute for Materials Research and Testing, Berlin (Germany). Structure drawings and images were generated using

https://doi.org/10.1515/9783110610833-201

CrystalMaker®: a crystal and molecular structures program for Mac and Windows (CrystalMaker Software Ltd, Oxford, England (www.crystalmaker.com)). Molecular structure images were generated with the program Mercury – Crystal Structure Visualisation, Exploration and Analysis Made Easy (Cambridge Crystallographic Data Centre, https://www.ccdc.cam.ac.uk/Community/csd-community/freemercury).

Tallahassee, 2021

Contents

1 Introduction

To characterize a crystalline structure at the atomic scale, knowledge of the type and positions of all the atoms in a given volume is desired. With the atomic positions known, the nature of the bonding in a material can be determined and structure-property relationships derived. Modern electronic structure codes based on density functional theory can give a detailed description of the behavior of the electrons in a solid. As input, they use the structural information: the unit cell size and the atomic positions [1]. For a full description of the structure of a material, a set of coordinates for each atom can be written. The question then becomes: How many atomic coordinates are needed to describe a given volume of a material? With each atom occupying a distinct volume, how many atoms are found in $1\,cm^3$? With the size known, the number of sets of coordinates needed for such an enumeration can be estimated. Since atoms are always in motion, even at the absolute temperature of $0\,K$, a full description not only needs 3 coordinates x, y, z, but also coordinates that describe the motion, the momentum coordinates p_x, p_y, p_z or the average deviation from the equilibrium position.

Estimate the size of an atom?

The volume of a spherical atom can be estimated from the density of the material and the atomic mass of its atoms. As an example, silver (Ag), has an atomic mass of about $107\,g/mol$, and a density of $10.5\,g/cm^3$. Therefore, $1\,cm^3$ of Ag contains roughly 0.1 moles of Ag atoms, about 6×10^{22} atoms. Assuming dense sphere packing with an atomic packing factor of 0.74 gives $1.43\,\text{Å}$ for the radius of an Ag atom. Obviously, the number of atoms even in $1\,mm^3$ volume is exceedingly large and does not lend itself to listing all the coordinates. Even if this would be attempted for a volume of $1\,mm^3$, a 6×10^{19} sets of coordinates are required, a data set obviously too large to handle. A different way is therefore needed. A random packing of spheres is a possibility, but the lattice energy of such an assembly will be relatively high due to a large number of irregular bonds. An example of a planar model illustrating sphere packing, an "amorphous" sphere arrangement is shown in Figure 1.1. The arrangement does not have long range order, but locally, some order may exist.

For spherical atoms, with increasing order, the number of nearest neighbor atoms will increase and eventually reach 12, with all bonds equal in length and strength, an arrangement with lower energy and lower entropy. Any assembly of atoms will thus have a strong tendency to form regions where atoms are regularly arranged over distances that are many bond lengths. Such regular areas will be periodic, with a unique repeating volume containing a much smaller and manageable number of atoms. Listing the coordinates of all the atoms in this unique volume is now possible, adding the requirement that this volume is representative of the total assembly of atoms. Figure 1.2 shows a planar model of sphere packing, where order has developed over a

https://doi.org/10.1515/9783110610833-001

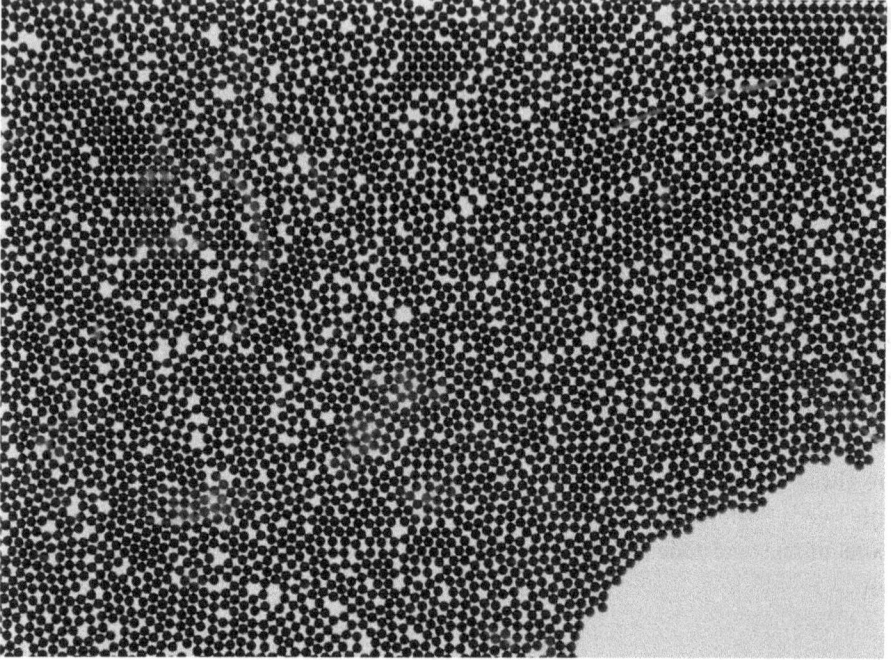

Figure 1.1: Planar sphere model: Amorphous sphere arrangement.

Figure 1.2: Planar sphere model: Crystalline sphere arrangement. Point defects (missing spheres) and grain boundaries can be seen.

long range. In addition, point defects (missing spheres) and "grain boundaries" exist as well, where the dense sphere packing is interrupted.

Long range order is found for most metals, minerals, ceramics and small molecules, whereas polymers and biomolecules tend to have a higher propensity to form amorphous or partially ordered assemblies. With the problem reduced to the determination of the atomic coordinates in a small representative volume of space, it is necessary to consider the best way to image the atoms in this volume of space.

Imaging atoms

Using a regular light microscope, the maximum lens-based magnification is of the order of 2000×, making it possible to image dimensions of the order of a fraction of 10^{-6} m. With the wavelength in the middle of the visible spectrum, about 500 nm, dimensions down to 250 nm can in principle still be resolved. However, this dimension is about a factor of 2500 larger than the radius of a silver atom. To image atoms, it is thus necessary to use light that has a wavelength comparable to the dimension of an atom. This also means that this light is about 2500 times more energetic than the visible light in the middle of the visible spectrum, putting the wavelength into the X-ray region.

The interactions of the light with matter, or more generally, the interaction of electromagnetic waves with a wavelength of the order of an atomic radius, is therefore at the heart of structure determinations. To produce an image, the scattering and diffraction of the light from matter needs to be understood, so that a transfer function can be constructed that describes the scattering body. This is necessary since optical devices such as lenses are not easily constructed for wavelengths in the X-ray region. In addition, the scattering can be either elastic, with the wavelength (energy) of the radiation unchanged, or inelastic, with the wavelength increased or reduced. Often, the energy change is negative, so that an impinging electromagnetic wave will emerge from them scattering process with a longer wavelength (lower energy).

Imaging of atoms can also be achieved by using the quantum mechanical wave nature of particles with appropriate wavelengths, and recording their scattered/ diffracted intensities. Particles used for probing matter include electrons and neutrons at suitable energies. As electrons are the lightest elementary particles, their wave nature is readily accessible, and the Davisson and Germer experiment was the first demonstration of electron diffraction from a periodic atom arrangement [2]. What makes electrons special is that focusing optical elements can be constructed for electrons, using electric and magnetic fields. High magnifications are thus possible in a *Transmission Electron Microscope* (TEM). However, the interaction of an electron beam with the electrons in a sample is strong, such that only thin samples can be imaged, as the incident electrons are strongly absorbed in thicker samples. Since higher particle energies give better penetration, an electron microscope uses acceleration voltages of the order of 80 to 300 keV. The lower value is mostly used for organic materials to reduce radiation induced damage, whereas the higher value is used for inorganic samples.

Neutrons of thermal energies, of the order of 25 meV or less, are used for diffraction and scattering experiments, with a wavelength of the order of 1 to 2 Å. In contrast to electromagnetic radiation and electrons, which both interact with the electrons in the sample, neutrons interact with the atomic nuclei, which, to first order, are point-like objects. As with electromagnetic radiation, the interactions of electrons and neutrons with matter can be elastic (no energy transfer), or inelastic (with energy transfer). Inelastic scattering is used in many areas to probe the dynamics and excitations in matter. In this text, inelastic processes will not be considered. The restriction to elastic scattering allows treating the scattering process using semiclassical methods, and Fourier formalism will give an elegant solution for scattering effects from periodic systems.

Arrangements of atoms will have order and arrange into periodic structures. The ideal atomic arrangement, being periodic, will have to fill space, and due to high symmetry and periodicity, obey symmetry relationships. In a real crystal, defects, impurities, grain boundaries and atomic motion disturb the perfect symmetry (see Figure 1.2). A small number of point defects and grain boundaries, however, do not severely affect the periodicity of a structure, and the time scale of atomic motion is different from the time scale of the scattering process. Perfect symmetry is thus a good assumption in many cases, and approximations have been developed to deal with atomic motion, grain boundaries, defects, etc.

Model assumptions

To arrive at a model of the scattering of an array of atoms, the following assumptions are made:

- The interaction of an electromagnetic wave with an array of atoms is modeled, using quantum mechanics to determine the scattering power of an individual atom.
- The scattering power of the individual atoms is then added up under consideration of the phase differences that occur for a given direction. The radiation is therefore considered coherent over a finite distance. For a periodic array, symmetry considerations are used to arrive at a compact formulation of the resulting sums.
- The observation of diffracted intensities is carried out in the "far field," far away from the atoms array by a distance of many wavelengths of the radiation.
- The dynamic displacements of the scattering atoms due to thermal energies (phonons) is considered random and Gaussian.
- Detectors register intensities, not wave amplitudes and phases. Therefore, the phase information is not accessible (phase problem).

For X-ray scattering and diffraction, the model that will be developed can account for the observed intensities to a very high degree, approaching 95 % fidelity or better. This renders diffraction methods quantitative in their measurement of electron density and interatomic distances in a periodic solid. It is even possible to obtain valence electron density information in the case of low Z atoms, such as Li, Be, B, C, N, O, F. For heavier atoms, the core electrons will dominate the scattering power, but valence electron densities have been obtained for light transition element compounds by carefully measuring the diffracted intensities to a high precision. Consequently, structure determinations have provided key information on the atomic arrangements, electron densities in solids and the bonding between atoms. Since the electronic bonding determines physical and chemical properties, the determination of the structure, based on the models developed, gives crucial information of the physical and chemical behavior. More recently, crystallography has provided insights into the biological activity of proteins, where methods have been developed to acquire data from small, weakly scattering crystals with very large unit cells. However, the text will not cover this area of research.

Bibliography

[1] D. J. Singh and L. Nordstrom. Planewaves, Pseudopotentials and the LAPW Method. Springer, 2006. https://doi.org/10.1007/978-0-387-29684-5. ISBN: 978-0-387-29684-5.
[2] C. J. Davisson and L. H. Germer. Reflection of Electrons by a Crystal of Nickel. Proc. Natl. Acad. Sci., 14(4):317–322, 1928. https://doi.org/10.1073/pnas.14.4.317.

2 Geometry and coordinate systems

2.1 Coordinate systems and geometrical relationships

Describing objects in three-dimensional space (object space) requires a coordinate system so that positions can be unambiguously specified. Different types of coordinate systems are used, depending on the particulars of the space and objects that are described. In addition to the position information, a description of planes is needed as well. In the following, positions (vectors) and planes (defined by plane normal directions) are discussed for a general coordinate systems.

2.1.1 Vectors

The coordinate system in 3 dimensions needs to be defined. Three non-coplanar vectors $\mathbf{a}, \mathbf{b}, \mathbf{c}$, the base vectors, are given. These vectors are generally of different length, as well as nonorthogonal. A general point P in space is defined by a set of coordinates u, v, w, and the vector from the origin to the point P is given as $\mathbf{r} = u\mathbf{a} + v\mathbf{b} + w\mathbf{c}$. In Figure 2.1, a coordinate system with general angles and lengths of vectors $\mathbf{a}\,\mathbf{b}\,\mathbf{c}$ is shown.

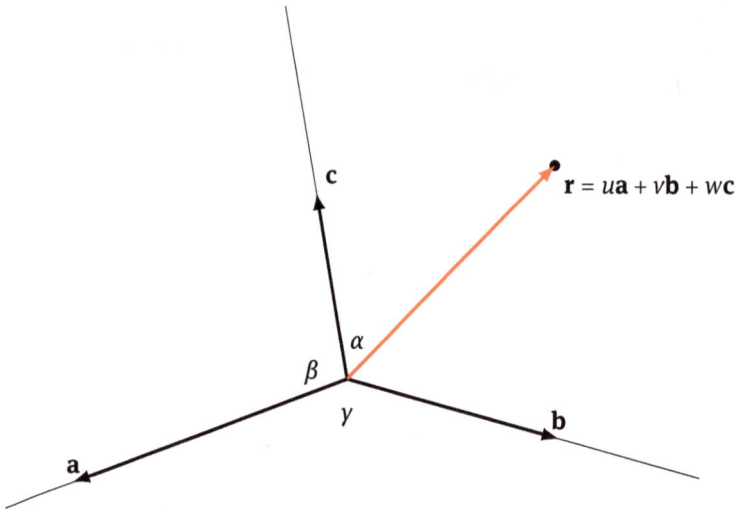

Figure 2.1: Coordinate system: points and vectors.

The following convention applies: The coordinate system is right-handed, meaning that using the fingers of the right hand, $\mathbf{a}$ is along the thumb, $\mathbf{b}$ along the index finger and $\mathbf{c}$ along the middle finger. Furthermore, the angle between $\mathbf{a}$ and $\mathbf{b}$ is γ, the angle between $\mathbf{b}$ and $\mathbf{c}$ is α and the angle between $\mathbf{a}$ and $\mathbf{c}$ the angle β. A general point

https://doi.org/10.1515/9783110610833-002

therefore has the coordinates u, v, w, and is represented by the vector $\mathbf{r} = u\mathbf{a} + v\mathbf{b} + w\mathbf{c}$.

2.1.2 Planes

The equation of a general plane is $hu + kv + lw = 1$, the same form as for an orthonormal (unitary) coordinate system. For coordinates $v = w = 0$, one obtains $u = 1/h$; h is therefore the reciprocal value of the segment from the origin O to the point A where the plane intersects the $\mathbf{a}$-axis. If $|\mathbf{a}|$ is given in meters, then the length of the segment OA is therefore a/h meters (see Figure 2.2).

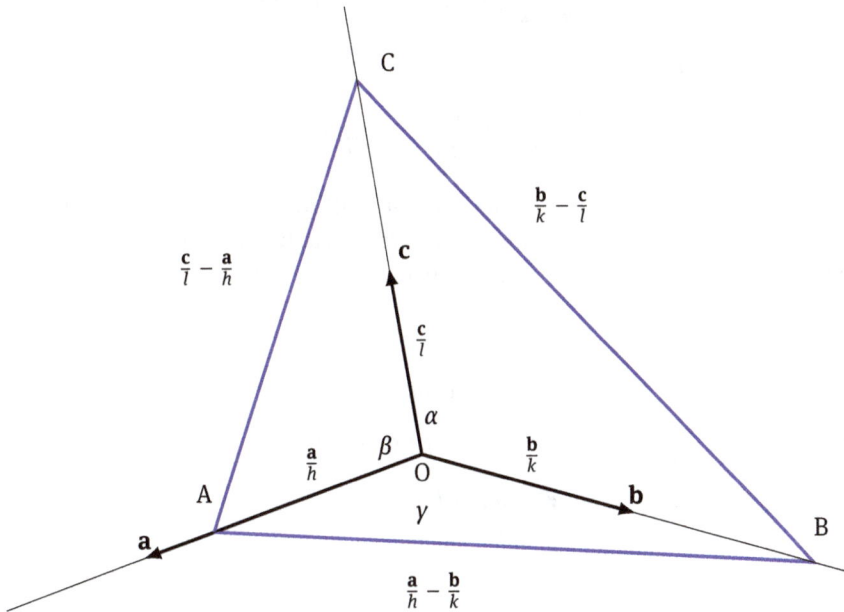

Figure 2.2: Coordinate system: planes.

The plane is therefore defined by three noncollinear points, or by two nonparallel vectors. For the plane $hu + kv + lw = 1$, the plane normal oriented toward the origin O is determined by the vector product

$$\mathbf{N} = \text{sign}(hkl)\left[\left(\frac{\mathbf{a}}{h} - \frac{\mathbf{b}}{k}\right) \times \left(\frac{\mathbf{b}}{k} - \frac{\mathbf{c}}{l}\right)\right] = \frac{1}{|hkl|}\left[h(\mathbf{b} \times \mathbf{c}) + k(\mathbf{c} \times \mathbf{a}) + l(\mathbf{a} \times \mathbf{b})\right] \quad (2.1)$$

The plane is therefore defined by the plane normal $\mathbf{N}$ and all vectors perpendicular to the plane normal direction, and the distance d from the origin (equation (2.1)). For an orthogonal and isometric system, the cross products can be handled easily, and the plane normal vector is given by the triple hkl, the index of the plane, which in this special case is parallel to a vector $[hkl]$.

2.2 Reciprocal coordinate systems

Every coordinate system has a uniquely defined reciprocal coordinate system that is represented by all the plane normal vectors. Thus, the length and direction of the plane normal vector has to be calculated. This uses the expression (2.1) and the fact that a trigonal pyramid is formed by $OABC$. The volume of this pyramid is one-third of the product of the area of the triangle ABC times the height d:

$$V = \frac{1}{3}\left[\frac{1}{2}\left\|\left(\frac{\mathbf{a}}{h} - \frac{\mathbf{b}}{k}\right) \times \left(\frac{\mathbf{b}}{k} - \frac{\mathbf{c}}{l}\right)\right\|\right] \times d = \frac{1}{6}d\|\mathbf{N}\|$$

Therefore, the volume is

$$V = \frac{1}{6}\left\|\left(\frac{\mathbf{a}}{h} \times \frac{\mathbf{b}}{k}\right)\frac{\mathbf{c}}{l}\right\| = \frac{1}{6|hkl|}(\mathbf{a\,b\,c})$$

where $(\mathbf{a\,b\,c}) = \mathbf{a} \cdot (\mathbf{b} \times \mathbf{c}) = \mathbf{b} \cdot (\mathbf{c} \times \mathbf{a}) = \mathbf{c} \cdot (\mathbf{a} \times \mathbf{b})$ (the triple product is equal to the volume of a parallelepiped spanned by the vectors $\mathbf{a}$, $\mathbf{b}$ and $\mathbf{c}$). The unit of $\|\mathbf{N}\|$ is in $(\text{length})^2$, related to the cross products of two vectors. Therefore, the plane normal vector is

$$\|\mathbf{N}\| = \frac{1}{d}\frac{(\mathbf{abc})}{|hkl|}$$

The vector $\mathbf{r}^* = |hkl|\mathbf{N}/(\mathbf{abc})$ has the following properties:

$$\mathbf{r}^* = h\mathbf{a}^* + k\mathbf{b}^* + l\mathbf{c}^*;$$

$$\mathbf{a}^* = \frac{(\mathbf{b} \times \mathbf{c})}{(\mathbf{abc})}$$

$$\mathbf{b}^* = \frac{(\mathbf{c} \times \mathbf{a})}{(\mathbf{abc})}$$

$$\mathbf{c}^* = \frac{(\mathbf{a} \times \mathbf{b})}{(\mathbf{abc})}$$

$$\|\mathbf{r}^*\| = \frac{1}{d}$$

The vector $\mathbf{r}^*$ is perpendicular to the plane $hu + kv + lw = 1$ and is oriented from the origin O toward the plane, and has a length of $\frac{1}{d}$. The three vectors $\mathbf{a}^*$, $\mathbf{b}^*$, $\mathbf{c}^*$ define the *reciprocal lattice* that is related to $\mathbf{a}$, $\mathbf{b}$, $\mathbf{c}$. The reciprocal vectors $\mathbf{a}^*$, $\mathbf{b}^*$, $\mathbf{c}^*$ are in general not parallel to the vectors $\mathbf{a}$, $\mathbf{b}$, $\mathbf{c}$, unless the angles $\alpha = \beta = \gamma = 90°$ and $\mathbf{a} = \mathbf{b} = \mathbf{c}$. It is further observed that

$$\mathbf{a}^* \cdot \mathbf{a} = \mathbf{b}^* \cdot \mathbf{b} = \mathbf{c}^* \cdot \mathbf{c} = 1$$

and

$$\mathbf{a}^* \cdot \mathbf{b} = \mathbf{a}^* \cdot \mathbf{c} = \mathbf{b}^* \cdot \mathbf{a} = \mathbf{b}^* \cdot \mathbf{c} = \mathbf{c}^* \cdot \mathbf{a} = \mathbf{c}^* \cdot \mathbf{b} = 0$$

These expressions are simplified if the base lattice vectors are labeled $\mathbf{a}_1$, $\mathbf{a}_2$, $\mathbf{a}_3$, and correspondingly the reciprocal vectors $\mathbf{a}_1^*$, $\mathbf{a}_2^*$, $\mathbf{a}_3^*$. The above expressions is then written as

$$\mathbf{a}_i \cdot \mathbf{a}_j^* = \delta_{ij}$$

The equation for a plane hkl simplifies to

$$hu + kv + lw = 1 = \mathbf{r} \cdot \mathbf{r}^*$$

with the projection of the vector $\mathbf{r}$ onto the plane normal $\mathbf{r}^*$ equal d. The following relations between the direct and reciprocal lattices are derived:

$$\mathbf{a}^* \cdot \mathbf{b}^* = (\mathbf{b} \times \mathbf{c}) \cdot (\mathbf{c} \times \mathbf{a})/(\mathbf{abc})^2$$
$$= [(\mathbf{b} \cdot \mathbf{c})(\mathbf{a} \cdot \mathbf{c}) - c^2(\mathbf{a} \cdot \mathbf{b})]/(\mathbf{abc})^2$$
$$\mathbf{a}^* \times \mathbf{b}^* = [(\mathbf{b} \times \mathbf{c}) \times (\mathbf{c} \times \mathbf{a})]/(\mathbf{abc})^2$$
$$c(\mathbf{abc}) = \|(\mathbf{b} \times \mathbf{c}) \times (\mathbf{c} \times \mathbf{a})\| = abc^2 \sin\alpha \sin\beta \sin\gamma^*$$
$$(\mathbf{abc}) = abc \sin\alpha \sin\beta \sin\gamma^*$$
$$= abc \sin\alpha \sin\beta^* \sin\gamma$$
$$= abc \sin\alpha^* \sin\beta \sin\gamma$$
$$(\mathbf{abc})^2 = a^2\|\mathbf{b} \times \mathbf{c}\|^2 - \|\mathbf{a} \times (\mathbf{b} \times \mathbf{c})\|^2$$
$$= a^2 b^2 c^2 (1 - \cos^2\alpha - \cos^2\beta - \cos^2\gamma + 2\cos\alpha \cos\beta \cos\gamma)$$
$$(\mathbf{a}^* \mathbf{b}^* \mathbf{c}^*) = (\mathbf{abc})^{-1}$$
$$\cos\gamma^* = (\cos\alpha \cos\beta - \cos\gamma)/(\sin\alpha \sin\beta)$$
$$\cos\alpha^* = (\cos\beta \cos\gamma - \cos\alpha)/(\sin\beta \sin\gamma)$$
$$\cos\beta^* = (\cos\alpha \cos\gamma - \cos\beta)/(\sin\alpha \sin\gamma)$$
$$a^* = (a \sin\beta \sin\gamma^*)^{-1} = (a \sin\beta^* \sin\gamma)^{-1}$$
$$b^* = (b \sin\gamma \sin\alpha^*)^{-1} = (b \sin\gamma^* \sin\alpha)^{-1}$$
$$c^* = (c \sin\alpha \sin\beta^*)^{-1} = (c \sin\alpha^* \sin\beta)^{-1}$$

2.3 Metric tensor

The length of the vector $\mathbf{r} = u\mathbf{a} + v\mathbf{b} + w\mathbf{c}$ is obtained by developing the terms of the expression

$$\|\mathbf{r}\|^2 = u^2 a^2 + v^2 b^2 + w^2 c^2 + 2uv(\mathbf{a} \cdot \mathbf{b}) + 2uw(\mathbf{a} \cdot \mathbf{c}) + 2vw(\mathbf{b} \cdot \mathbf{c})$$

This is written more conveniently using matrix notation

$$\|\mathbf{r}\|^2 = (u \quad v \quad w) \begin{pmatrix} a^2 & \mathbf{a} \cdot \mathbf{b} & \mathbf{a} \cdot \mathbf{c} \\ \mathbf{a} \cdot \mathbf{b} & b^2 & \mathbf{b} \cdot \mathbf{c} \\ \mathbf{a} \cdot \mathbf{c} & \mathbf{b} \cdot \mathbf{c} & c^2 \end{pmatrix} \begin{pmatrix} u \\ v \\ w \end{pmatrix} = \mathbf{u}^T \mathbf{M} \mathbf{u} \qquad (2.2)$$

Here, $\mathbf{u}^T$ is the "line" vector (uvw), transposed from the column vector $\mathbf{u}$. The determinant of the metric tensor is

$$|\mathbf{M}| = (\mathbf{abc})^2 \qquad (2.3)$$

The same holds for the reciprocal vector $\mathbf{r}^* = h\mathbf{a}^* + k\mathbf{b}^* + l\mathbf{c}^*$, where

$$\|\mathbf{r}^*\|^2 = h^2 a^{*2} + k^2 b^{*2} + l^2 c^{*2} + 2hk(\mathbf{a}^* \cdot \mathbf{b}^*) + 2hl(\mathbf{a}^* \cdot \mathbf{c}^*) + 2kl(\mathbf{b}^* \cdot \mathbf{c}^*)$$

Again, this is elegantly expressed using matrix notation

$$\|\mathbf{r}^*\|^2 = (h \quad k \quad l) \begin{pmatrix} a^{*2} & \mathbf{a}^* \cdot \mathbf{b}^* & \mathbf{a}^* \cdot \mathbf{c}^* \\ \mathbf{a}^* \cdot \mathbf{b}^* & b^{*2} & \mathbf{b}^* \cdot \mathbf{c}^* \\ \mathbf{a}^* \cdot \mathbf{c}^* & \mathbf{b}^* \cdot \mathbf{c}^* & c^{*2} \end{pmatrix} \begin{pmatrix} h \\ k \\ l \end{pmatrix} = \mathbf{h}^T \mathbf{M}^* \mathbf{h} \qquad (2.4)$$

with $\mathbf{h}^T$ the "line" vector (hkl), the transposed column vector $\mathbf{h}$. The reciprocal metric tensor is the inverse of the direct space metric tensor:

$$\mathbf{M}^* = \mathbf{M}^{-1} \qquad (2.5)$$

In the special case of an unitary coordinate system (orthogonal isometric), the metric tensor is the unitary matrix $\mathbf{M}_{ij} = \mathbf{M}_{ij}^* = \delta_{ij}$.

The general dot product of two direct space vectors is

$$\mathbf{r}_1 \cdot \mathbf{r}_2 = \mathbf{u}_1^T \mathbf{M} \mathbf{u}_2$$

and correspondingly

$$\mathbf{r}_1^* \cdot \mathbf{r}_2^* = \mathbf{h}_1^T \mathbf{M}^* \mathbf{h}_2$$

Additionally, the dot product between a direct and reciprocal vector is

$$\mathbf{r}_1 \cdot \mathbf{r}_2^* = \mathbf{u}_1^T \mathbf{h}_2$$

The *vector product* of two vectors $\mathbf{r}_1$ and $\mathbf{r}_2$ divided by $(\mathbf{abc})$ is a member of the reciprocal space

$$(\mathbf{r}_1 \times \mathbf{r}_2)/(\mathbf{abc}) = (v_1 w_2 - v_2 w_1)\mathbf{a}^* + (w_1 u_2 - w_2 u_1)\mathbf{b}^* + (u_1 v_2 - u_2 v_1)\mathbf{c}^* = \mathbf{r}_{hkl}^*$$

In analogy, the *vector product* of two reciprocal vectors $\mathbf{r}_1^*$ and $\mathbf{r}_2^*$ divided by $(\mathbf{a}^*\mathbf{b}^*\mathbf{c}^*)$ is a member of the direct space

$$(\mathbf{r}_1^* \times \mathbf{r}_2^*)/(\mathbf{a}^*\mathbf{b}^*\mathbf{c}^*) = (k_1 l_2 - k_2 l_1)\mathbf{a} + (l_1 h_2 - l_2 h_1)\mathbf{b} + (h_1 k_2 - h_2 k_1)\mathbf{c} = \mathbf{r}_{uvw}$$

The representations in direct and reciprocal lattice are unique, the two lattices have a defined relationship and the two representations are equivalent. In direct space, the coordinates represent points in space, whereas in the reciprocal lattice, the (reciprocal) coordinates represent directions or planes that are perpendicular to the directions, which are given in units of inverse distance $((\text{length})^{-1})$. For the description of planes, such as faces of a mineral specimen, the reciprocal lattice representation is favored. Furthermore, the combination of direct and reciprocal space will allow a concise description of wave phenomena, since the propagation vector of a wave defines a direction and the perpendicular wave front is defined as a plane.

2.4 Coordinate transformation

If the coordinate system is transformed from $\mathbf{a}_i$ to $\mathbf{a}_i'$, the direct coordinates (uvw) and the reciprocal coordinates (hkl) do not transform in the same manner. The coordinate transformation is given by (3×3) matrices C_a and C_{a*} for the direct and reciprocal coordinate systems, C_u and C_h for the coordinates. Coordinate systems transform according to

$$\begin{pmatrix} \mathbf{a}' \\ \mathbf{b}' \\ \mathbf{c}' \end{pmatrix} = C_a \begin{pmatrix} \mathbf{a} \\ \mathbf{b} \\ \mathbf{c} \end{pmatrix}; \quad \begin{pmatrix} \mathbf{a}^{*\prime} \\ \mathbf{b}^{*\prime} \\ \mathbf{c}^{*\prime} \end{pmatrix} = C_{a*} \begin{pmatrix} \mathbf{a}^* \\ \mathbf{b}^* \\ \mathbf{c}^* \end{pmatrix};$$

and coordinates transform according to

$$\begin{pmatrix} u' \\ v' \\ w' \end{pmatrix} = C_u \begin{pmatrix} u \\ v \\ w \end{pmatrix}; \quad \begin{pmatrix} h' \\ k' \\ l' \end{pmatrix} = C_h \begin{pmatrix} h \\ k \\ l \end{pmatrix}$$

Since the vectors $\mathbf{r}$ and $\mathbf{r}^*$, as well as the scalar product (dot product) $\mathbf{r}\cdot\mathbf{r}^*$ are invariant in respect to a coordinate transformation, the following relationships are derived:

$$\mathbf{r} = (uvw)\begin{pmatrix} \mathbf{a} \\ \mathbf{b} \\ \mathbf{c} \end{pmatrix} = (u'v'w')\begin{pmatrix} \mathbf{a}' \\ \mathbf{b}' \\ \mathbf{c}' \end{pmatrix} = (uvw)C_u^T C_a \begin{pmatrix} \mathbf{a} \\ \mathbf{b} \\ \mathbf{c} \end{pmatrix} \qquad (2.6)$$

$$\mathbf{r}^* = (hkl)\begin{pmatrix} \mathbf{a}^* \\ \mathbf{b}^* \\ \mathbf{c}^* \end{pmatrix} = (h'k'l')\begin{pmatrix} \mathbf{a}^{*\prime} \\ \mathbf{b}^{*\prime} \\ \mathbf{c}^{*\prime} \end{pmatrix} = (hkl)C_h^T C_{a*} \begin{pmatrix} \mathbf{a}^* \\ \mathbf{b}^* \\ \mathbf{c}^* \end{pmatrix} \qquad (2.7)$$

$$\mathbf{r}^* \cdot \mathbf{r} = (hkl) \begin{pmatrix} u \\ v \\ w \end{pmatrix} = (h'k'l') \begin{pmatrix} u' \\ v' \\ w' \end{pmatrix} = (hkl)C_h^T C_u \begin{pmatrix} u \\ v \\ w \end{pmatrix} \tag{2.8}$$

$$\mathbf{r} \cdot \mathbf{r}^* = (uvw) \begin{pmatrix} h \\ k \\ l \end{pmatrix} = (u'v'w') \begin{pmatrix} h' \\ k' \\ l' \end{pmatrix} = (uvw)C_u^T C_h \begin{pmatrix} h \\ k \\ l \end{pmatrix} \tag{2.9}$$

It follows that

$$C_a = C_h = (C_{a^*}^T)^{-1}$$
$$C_{a^*} = C_u = (C_a^T)^{-1}$$

The direct base vectors and the reciprocal coordinates (hkl) transform covariantly!
The reciprocal base vectors and the direct coordinates (uvw) transform contravariantly!
This distinction needs to be kept in mind when dealing with geometrical operations, such as rotations, mirror operations, etc. If the operation acts on the coordinate, then the transformation is contravariant. If the operation acts on the direct base vectors, e. g., the coordinate system, the transformation is covariant.

Construction of the reciprocal lattice
Draw an oblique coordinate system in the plane. Assume that the c-axis is perpendicular to the plane. Construct the reciprocal lattice according to the above prescription. With the c-axis perpendicular (out of the plane), the reciprocal $\mathbf{a}^*$ has to be perpendicular to the plane spanned by the b- and c-axes. The length is in reciprocal units. The same holds for $\mathbf{b}^*$, which is perpendicular to the c- and a-axes. The length ratio $(a/b) = (b^*/a^*)$ and the angle y transforms to $y^* = \pi - y$. A solution is given in Figure 2.3.

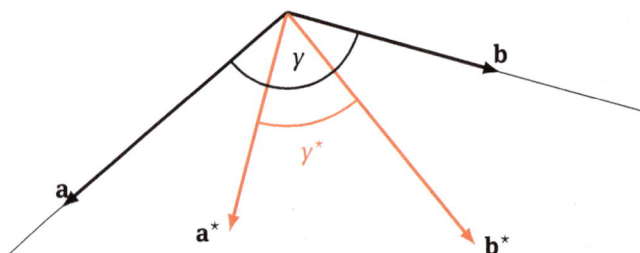

Figure 2.3: Direct and reciprocal vectors. The c-axis is perpendicular to the (ab)-plane.

2.5 Application to crystal forms

The form of a crystal, its *habit*, is defined by the arrangement of planes (crystal faces) and the edges and points where the planes intersect. In mineralogy, it was recognized very early that the same minerals always showed identical angles between like faces. This law was first formulated for natural quartz crystals in 1669 by Nils Steensen (Nicolaus Steno). In particular, the following observations are formulated:

-   the angle between two faces does not change when a crystal grows; the angle therefore does not depend on the distance to a given point
-   the angles between corresponding faces of two individual crystals of the same mineral are equal (at equal temperature and pressure)
-   under defined physical conditions, the angles between faces are therefore characteristic for a particular crystalline mineral

These statements indicate that only the directions of the edges and plane normals are important. It is possible to measure angles between planes to a high precision by using a goniometer and a parallel beam light source (autocollimator), to within a second of arc (about 0.0003°). Furthermore, mineralogical studies determined that the faces of a crystal do not form an arbitrary polyhedron. It is thus possible to choose a coordinate system for the analytical description of the crystal form. In general, the coordinate system does not need to be unitary (isometric and orthogonal). One chooses three noncoplanar edges to define the directions of the axes $\mathbf{a}, \mathbf{b}$ and $\mathbf{c}$. The ratio between the lengths of these axes $a : b : c$ is determined by a fourth edge that is by definition $\mathbf{a} + \mathbf{b} + \mathbf{c}$. Therefore, the equation of a plane is again $hu + kv + lw = \text{const}$; with the ratios $h : k : l$ defining the orientation of the plane. By analogy, the direction of an edge is given by the ratio $u : v : w$. In practice, only the orientation of faces needs to be determined.

The procedure is as follows:

-   choose three intersecting faces to define $\mathbf{a}, \mathbf{b}$ and $\mathbf{c}$
-   and a fourth face to determine the ratio $h : k : l$.

All the other faces are referenced in this coordinate system, with the faces and edges satisfying the *law of rational indexes*:

-   the ratio $h : k : l$ of all the faces, and the ratio $u : v : w$ of all the edges is a rational number
-   The coordinates h, k and l of all the faces, as well as the coordinates u, v and w of all the edges of a crystal are small integer numbers.

These numbers are usually found in the interval from −10 to 10. The indexes h, k, l and u, v, w are called **Miller indexes** of the faces and directions (edges). The following nomenclature is used: for **faces** (planes), the indexes are written between *parenthesis*: (hkl) (no commas), with negative numbers written as $(\bar{h}\bar{k}\bar{l})$, for example $(3\bar{2}\bar{4})$. All faces $(hk0)$ are parallel to $\mathbf{c}$, all faces $(h0l)$ are parallel to $\mathbf{b}$ and all faces $(0kl)$ are parallel to $\mathbf{a}$. The coefficients (hkl) define reciprocal vectors $\mathbf{r}^* = h\mathbf{a}^* + k\mathbf{b}^* + l\mathbf{c}^*$. Similarly, the indexes of edges are written between *brackets*: $[uvw]$, and they define a direct space vector, $\mathbf{r} = u\mathbf{a} + v\mathbf{b} + w\mathbf{c}$.

If the indexes of two faces are known, then the index of the edge direction between the faces is calculated by using the previously defined relation (2.9). By analogy, the index of a face defined by two edges $[u_1v_1w_1]$ and $[u_2v_2w_2]$ is obtained using equation (2.9). For example, the intersections of the (111) and the (100) planes is the $[0\bar{1}1]$ direction, the intersection of the (010) and the (001) planes is the [100] direction.

2.5.1 Zone axis

The description of a crystal using the plane normal directions allows a concise description of the *habit* of a crystal specimen. It is also possible to use the edges, by replacing an edge by the plane that contains the edge. Parallel faces belong to the same zone. If two faces intersect which depends on the particular habit of the crystal, this direction is parallel to an edge. The zone axis defines therefore an existing physical edge, or an edge that may exist. The associated plane normal directions form the *zone planes*. The description of the crystal in geometric terms is therefore done in the reciprocal space, and is uniquely associated with the direct space coordinate system.

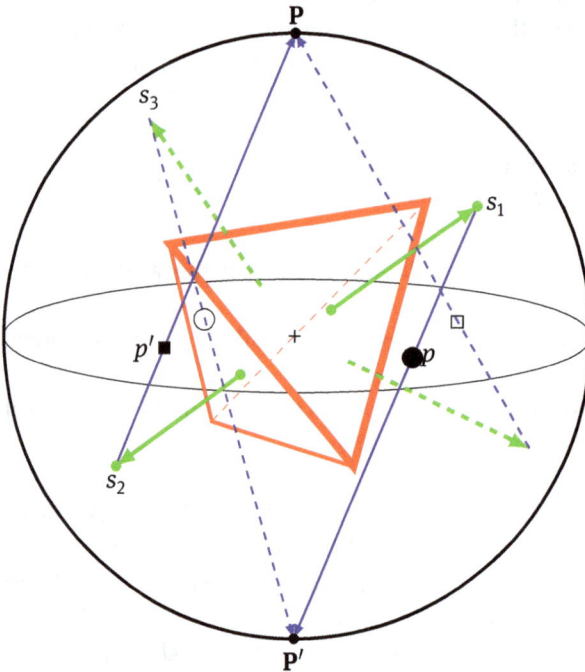

Figure 2.4: Stereographic projection.

2.5.2 Stereographic projection

A method needs to be developed to represent the faces of a 3-dimensional crystal in the 2-dimensional plane. For this, the stereographic projection will be used, together with the *Wulff Net*.

Figure 2.4 displays the stereographic projections of a body. The crystal is placed at the center of a sphere. Then all of the intersections s of the plane normal directions with the sphere constitute the spherical projection. The intersection points are then projected onto the equatorial plane via the pole P for points below the equator and P' for intersection points above the equator.

Projections from pole P (intersection above the equatorial plane) are shown in round symbols, while projections from below the equatorial plane (pole P') are shown in square symbols. The solid in the center can be oriented in any arbitrary direction, but symmetry will be apparent if a unique symmetry direction is chosen. The stere-

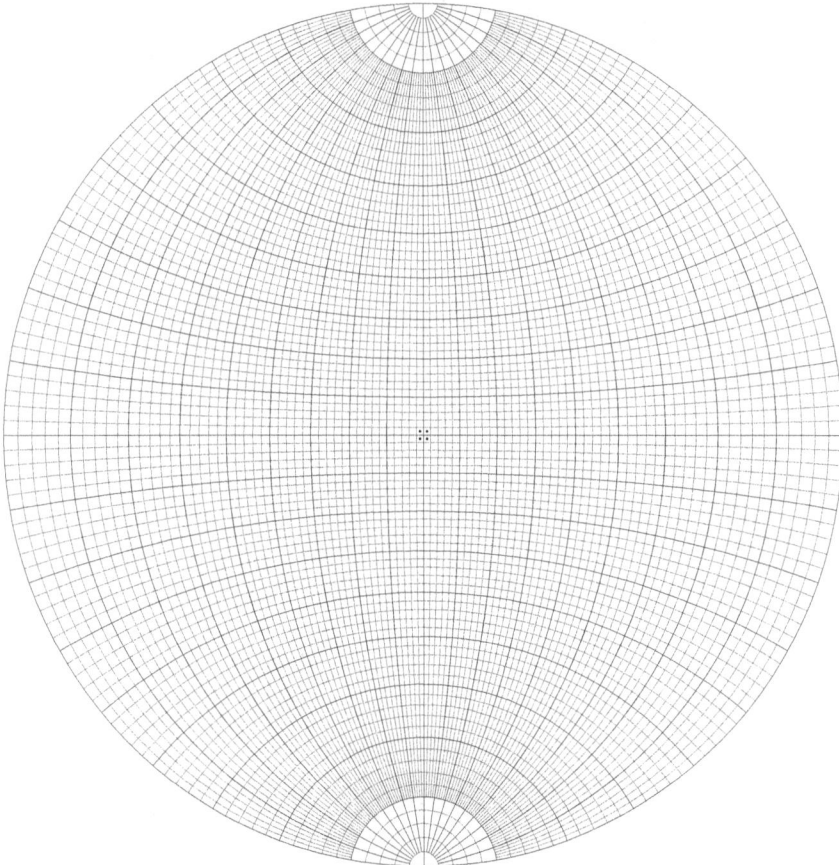

Figure 2.5: Wulff net: The projection is angle preserving.

ographic projection produces a projection of a spherical coordinate system onto a plane, with latitude and longitude coordinates. Depending on the view of the stereographic projection that is needed, different plots can be generated. If the stereographic projection of a hemisphere is required, then a projection with the equator in the center is used. This projection is called the *Wulff net* and is shown in Figure 2.5.[1] This pattern has both the poles P and P' on the circumference, and the equator is a straight line in between the poles. In addition, the latitude and longitude lines intersect at right angles, as they do on the surface of a sphere. This is a consequence of the stereographic projection which is angle preserving. The Wulff net was widely used, since its graphic pattern performs the calculations for the stereographic projection. Today, stereographic projections can be performed using computer programs, and visualization with real time rotations have been implemented.

---

1 tikz code of the Wulff net courtesy of G. Nolze, Bundesanstalt für Materialforschung, Berlin.

3 Waves

Waves are general physical phenomena that are found under many different conditions. In wave phenomena, energy is transported from point A to point B without involving moving mass from A to B. It is therefore illuminating to look at the derivation of the wave equation in the case of a fixed string, and review solutions of this differential equation. For a string under tension, it is assumed that the only relevant forces are the tensions acting on an element of the string, and that gravity and other forces can be neglected. Any motion of the string will therefore be confined to a single plane, so that a (x, y) coordinate system is sufficient. The string will be uniform with a mass density $\mu(x) = \mu = $ constant [g/cm]. In Figure 3.1, a piece of the string in a position (x, y) with the tension forces acting on it is displayed. It is assumed that the string only moves in the y-direction, and that therefore the horizontal forces are equal. The amplitude along the y-axis in position x and time t is given as $u(x, t)$.

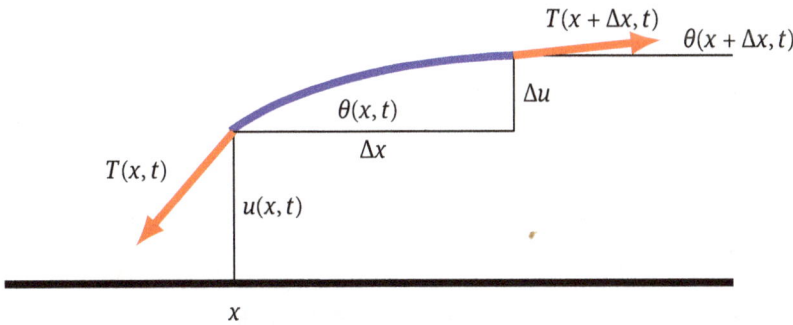

Figure 3.1: Forces acting on a string element.

Here, the basic notation in Figure 3.1 describes:
- $u(x, t)$ = vertical displacement amplitude of the string from the x axis at position x and time t
- $\theta(x, t)$ = angle between the string and a horizontal line at position x and time t
- $T(x, t)$ = tension in the string at position x and time t
- $\mu(x)$ = mass density of the string at position x, will be constant

The forces that act on the tiny string element are thus
- tension pulling to the right and left, acting at an angle, with their horizontal components canceling
- various external forces, assumed to act vertically (set to zero)

https://doi.org/10.1515/9783110610833-003

The motion of the string element is governed by the forces acting on it. Using Newton's second law,

$$F(x,t) = \mu(x)\frac{\partial^2 u(x,t)}{\partial t^2}$$

and the balance of forces, the following equation is obtained:

$$\mu(x)\sqrt{\Delta x^2 + \Delta u^2}\frac{\partial^2 u(x,t)}{\partial t^2} = T(x + \Delta x,t)\sin\theta(x + \Delta x,t) - T(x,t)\sin\theta(x,t) + F(x,t)\Delta x$$

Simplifying this equation by dividing by Δx and taking the limit for $\Delta x \longrightarrow 0$, assuming that only small vibrations (small $u(x,t)$) are considered ($\theta(x,t) \ll 1$ for all x and t) and the tension $T(x,t) = $ const, it is realized that the horizontal forces balance due to the fact that $T(x,t)\cos\theta(x,t)$ for small θ cannot produce a horizontal acceleration of the string. Also, $\cos\theta = 1 - \theta^2$ for small θ, and $\sin\theta = \theta = \frac{\Delta u}{\Delta x} = \frac{du}{dx}$. Furthermore, the transverse components of the tension (in direction of $u(x,t)$) differ by a value that is of first order in $\frac{du}{dx}$. The equation then becomes

$$T(x + \Delta x)\theta(x + \Delta x) - T(x)\theta(x) = T\left(\frac{\Delta u(x + \Delta x,t)}{\Delta x} - \frac{\Delta u(x,t)}{\Delta x}\right)$$

For an infinitesimal Δx, this is a difference of a difference, and can be expressed as the second derivative $\frac{d^2 u(x,t)}{dx^2}$. The expression, under the assumption that the tension T is constant, is then written as

$$\mu(x)dx\frac{d^2 u(x,t)}{dt^2} = T \times (\theta(x + dx) - \theta(x)) + F(x,t)dx = T\frac{d^2 u(x,t)}{dx^2}dx + F(x,t)dx$$

The term $F(x,t)$ describing external forces on the string is taken as small versus the tension $T(x)$ (the string is tight, and thus gravity is not playing a role). Thus the term $F(x,t)dx$ can be dropped. Under these assumptions, one arrives at a differential equation that describes the vertical motion $u(x,t)$ of the string:

$$\frac{\partial^2 u(x,t)}{\partial t^2} = \frac{T}{\mu}\frac{\partial^2 u(x,t)}{\partial x^2} = c^2\frac{\partial^2 u(x,t)}{\partial x^2} \tag{3.1}$$

where $c^2 = T/\mu$. This differential equation links the second derivative of $u(x,t)$ in space (coordinate x) to the second derivative of $u(x,t)$ in time. It can be solved by the separation of variables, expressing $u(x,t) = X(x)T(t)$, with functions $X(x)$ depending on the spatial coordinate x only and $T(t)$ on time t only. With this ansatz, the wave equation (3.1) becomes

$$X(x)\ddot{T}(t) = c^2\ddot{X}(x)T(t) \tag{3.2}$$

with $\ddot{T}(t)$ the second derivative in t, and $\ddot{X}(x)$ the second derivatice in x. By dividing this equation (3.2) by $X(x)T(t)$, the following equation is obtained:

$$\frac{\ddot{T}(t)}{T(t)} = c^2\frac{\ddot{X}(x)}{X(x)} \tag{3.3}$$

The right side of equation (3.3) is independent of time t, therefore, its left side is also independent of t. Since the left side of equation (3.3) is independent of dimension x, the right side has to be independent of x as well. Therefore, both sides have to be constant, resulting in

$$\frac{\ddot{X}(x)}{X(x)} = \frac{1}{c^2}\frac{\ddot{T}(t)}{T(t)} = k \tag{3.4}$$

and individual equations for space x and time t are obtained

$$\ddot{X}(x) - kX(x) = 0$$
$$\ddot{T}(t) - c^2 kT(t) = 0$$

The two ordinary differential equations based on (3.4) can be solved in the usual way. Furthermore, boundary conditions need to be defined. The general solution according to d'Alambert is given by

$$u(x,t) = f(x - ct) + g(x + ct) \tag{3.5}$$

describing two waves, the first traveling to the right (in direction $+x$), the second traveling to the left (in direction $-x$). A convenient way for this solution uses the complex notation

$$u(x,t) = Ae^{i(kx-\omega t)} + Be^{i(kx+\omega t)}; \quad \frac{\omega}{k} = c \tag{3.6}$$

For a particular solution of equation (3.6), the boundary conditions need to specified.

3.1 Example: solution for a string fixed at both ends

The string with length ℓ is attached at the origin ($x = 0$), and the end of the string is at $x = \ell$. The general solution for this condition is

$$u(x,t) = U(x)f(t) = \left(C\cos\left(\frac{\omega x}{c}\right) + D\sin\left(\frac{\omega x}{c}\right)\right)(A\cos(\omega t) + B\sin(\omega t))$$

The following boundary conditions are defined:

$$u(0,t) = 0 \quad \text{for all } t > 0$$
$$u(\ell,t) = 0 \quad \text{for all } t > 0$$
$$u(x,0) = f(x) \quad \text{for all } 0 < x < \ell$$
$$\dot{u}(x,0) = g(x) \quad \text{for all } 0 < x < \ell$$

With the boundary conditions in $x = 0$ and $x = \ell$, the general solution becomes

$$u(x,t) = \sum_{n=1}^{\infty} b_n(t) \sin\left(\frac{n\pi x}{\ell}\right) \quad \text{with } n = \text{integer}$$

The odd expansion satisfies the first two boundary conditions, and the coefficients b_n have to be determined from the functions $f(x)$ and $g(x)$. If the first two boundary conditions are considered, the solutions can be evaluated in more detail. First, it is obvious that $C = 0$ is required. Second, $u(\ell, 0) = 0$ imposes $D \sin(\frac{\omega}{c}\ell) = 0$. This is possible if $D = 0$, but this solution is trivial, and does not describe any oscillations. Therefore, $D \neq 0$ and $\sin(\frac{\omega}{c}\ell) = 0$. This restricts the angle values to $\omega = n\pi\frac{\ell}{c}$, with n an integer number. Therefore, the allowed values for ω are $\omega = n\pi\frac{\ell}{c}$, and $U(x)$ is written as

$$U_n(x) = D_n \sin\left(\frac{n\pi x}{\ell}\right)$$

These are the eigenmodes of the string. They also impose a value for the time dependent part of the solution:

$$f_n(t) = A_n \cos\left(\frac{n\pi c t}{\ell}\right) + B_n \sin\left(\frac{n\pi c t}{\ell}\right)$$

The constants D, A, B need to be determined from the unused boundary conditions. Figure 3.2 shows qualitative solutions to the wave equation.

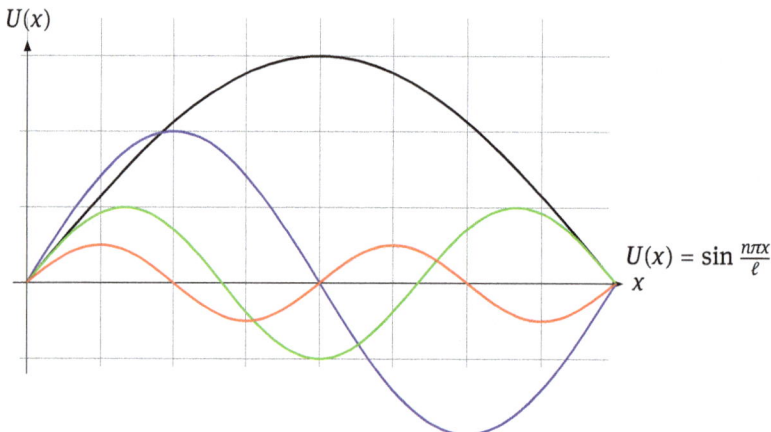

Figure 3.2: Solution to the wave equation.

The general solutions to the wave equation consists of an oscillatory time dependent part, as well as an oscillatory spatially dependent part. The velocity of the wave $c = \sqrt{T/\mu}$ depends on the tension T as well as on the mass distribution along the string μ.

For instance, high tension increases c, and with it, the frequency ω, whereas higher mass lowers c, and correspondingly, the frequency ω. This is familiar in the tuning of a string instrument, where the string tension is changed to adjust the pitch. For lower pitch, heavier gauge strings are used, often loaded with an additional winding, increasing the mass per unit length.

For an observer at a given position x, the string element will oscillate up and down. If the observer is moving along the position $+x$, the amplitude of the oscillation will change, and "nodes" where $U(x) = 0$ may be observed. If only a particular frequency allowed by the above condition is excited, then a single harmonic (or natural frequency) is observed, with $\omega_n = \frac{n\pi c}{l}$, depending on the length l of the string, and also depending on c, which includes the string tension and the string mass density.

If the solution is expressed as

$$u(x,t) = Ae^{i(\pm kx - \omega t)} \quad \text{with} \quad \frac{\omega}{k} = c \qquad (3.7)$$

with c the velocity of the traveling wave, it is seen that a linear relationship between k and ω is obtained. Furthermore, the wavelength (spatial oscillation) is given by $\lambda = 2\pi/k$, and the time period of the oscillation at any given point is $\tau = 2\pi/\omega$. A general form of the solution, with the direction dependence generalized in three dimensions (bold symbols indicate vectors), may therefore be written as

$$u(\mathbf{x},t) = C_1 \cos(\mathbf{k}\mathbf{x} + \omega t) + C_2 \sin(\mathbf{k}\mathbf{x} + \omega t) + C_3 \cos(\mathbf{k}\mathbf{x} - \omega t) + C_4 \sin(\mathbf{k}\mathbf{x} - \omega t) \quad (3.8)$$

or

$$u(\mathbf{x},t) = D_1 \cos \mathbf{k}\mathbf{x} \cos \omega t + D_2 \sin \mathbf{k}\mathbf{x} \sin \omega t + D_3 \sin \mathbf{k}\mathbf{x} \cos \omega t + D_4 \cos \mathbf{k}\mathbf{x} \sin \omega t \quad (3.9)$$

where the $u(\mathbf{x})$ part is generalized in three dimensions. Since the wave equation is linear, the most general solution is a sum of such expressions that contain different pairs of $\mathbf{k}$ and ω, with $\omega/k = c$.

A solution of the type $f(x - ct)$ satisfies the wave equation. This is easily seen if this solution is applied as $u(x,t) = f(x - ct)$, provided that $c = \omega/k$. If the function $f(z)$ with $z = (x - ct)$ is written as a Fourier series

$$f(z) = \int_{-\infty}^{\infty} C(k)e^{ikz}\, dk \qquad (3.10)$$

then the function becomes

$$f(x - ct) = \int_{-\infty}^{\infty} C(k)e^{ik(x-ct)}\, dk = \int_{-\infty}^{\infty} C(k)e^{i(kx-wt)}\, dk \qquad (3.11)$$

This shows that wave phenomena can be described using Fourier analysis. Any arbitrary function can be expressed as a sum (integral) of sine (cosine) waves. If the propagation of a sine wave is known, the propagation of any complicated waveform consisting of a superposition of sine (cosine) waves is known as well.

3.2 Electromagnetic waves

Electromagnetic waves are general solutions to the Maxwell equations for a space without charges. It means that electromagnetic waves can propagate in vacuum and are not tied to the presence of matter or charges. This is in contrast to acoustic waves that are coupled to a medium, either a solid, liquid or gas. In the case of air, the speed of sound at 1 atm and 25 °C is of the order of 300 m/s, with the acoustic wave a longitudinal pressure wave.

Maxwell's equation for a charge-free space in vacuum are written:

$$\nabla \cdot \mathbf{E} = 0$$

$$\nabla \times \mathbf{E} = -\frac{\partial B}{\partial t}$$

$$\nabla \cdot \mathbf{B} = 0$$

$$\nabla \times \mathbf{B} = \mu_0 \varepsilon_0 \frac{\partial E}{\partial t}$$

Taking the curl of the curl equations, the following relationships are obtained:

$$\nabla \times (\nabla \times \mathbf{E}) = -\frac{\partial}{\partial t} \nabla \times \mathbf{B} = -\mu_0 \varepsilon_0 \frac{\partial^2 E}{\partial t^2}$$

$$\nabla \times (\nabla \times \mathbf{B}) = \mu_0 \varepsilon_0 \frac{\partial}{\partial t} \nabla \times \mathbf{E} = -\mu_0 \varepsilon_0 \frac{\partial^2 B}{\partial t^2}$$

With the vector identity,

$$\nabla \times (\nabla \times \mathbf{V}) = \nabla(\nabla \cdot \mathbf{V}) - \nabla^2 \mathbf{V},$$

where $\mathbf{V}$ is a vector function of space, two conditions are obtained:

$$\nabla \cdot \mathbf{E} = 0$$

$$\nabla \cdot \mathbf{B} = 0$$

The first term on the right side vanishes and two wave equations, one for $\mathbf{E}$ and $\mathbf{B}$ each are formed:

$$\frac{\partial^2 E}{\partial t^2} - c_0^2 \cdot \nabla^2 \mathbf{E} = 0 \tag{3.12}$$

$$\frac{\partial^2 B}{\partial t^2} - c_0^2 \cdot \nabla^2 \mathbf{B} = 0 \tag{3.13}$$

with

$$c_0 = \frac{1}{\sqrt{\mu_0 \varepsilon_0}} = 2.99892 \times 10^8 \text{ m/s}$$

the speed of light in free space. The electric and magnetic field vector are mutually perpendicular and are phase shifted, resulting in alternating electric and magnetic fields. The propagation vector (Poynting vector) is perpendicular to both the electric and magnetic field vectors, in direction of the wave vector **k**. It is therefore sufficient to consider either the electric or the magnetic wave equation to discuss the wave propagation of light. To describe the interaction of an electromagnetic wave with matter, it is convenient to use the electric field vector alone, as the electrons and nuclei in matter will accelerate along an electric field, but will get deflected in a direction perpendicular to the magnetic field.

The propagation vector **k** of the wave defines a planar wave front that is perpendicular to **k**. In a plane perpendicular to **k**, the phases of the electric and magnetic vectors of the wave are constant. The reciprocal space coordinates that were defined by plane normal directions are therefore the natural coordinates for the **k** vectors that give the direction of the wave propagation.

A general wave may be described by

$$\Psi(\mathbf{x}, t) = A e^{i(\mathbf{k}\mathbf{x} - \omega t)} = A e^{2\pi i(\mathbf{s}\mathbf{r} - vt)} \tag{3.14}$$

with $2\pi\mathbf{s} = \mathbf{k}$ and $2\pi v = \omega$. This is a plane wave, with defined direction vector **k** or **s**, and a wave front that has infinite extension perpendicular to the propagation direction.

If an electromagnetic plane wave interacts with matter, part of the wave gets scattered. The wave interacts with the negative charges (electrons) and positive charges (nuclei) present in matter. The electrons have the smallest mass and, therefore, accelerate the most in the electric field of the wave. Nuclei also accelerate, but due to their much larger masses, the acceleration is significantly smaller, and consequently, do not produce a large scattering effect. In the case of an incident electron beam, the impinging electrons interact with the electrons and nuclei in the material and get scattered via Coulomb interactions. Neutrons, which do not carry charge, interact with nuclei, and may further include nuclear reactions. For a magnetic material with long range magnetic order, the neutron spin will interact with the magnetic spins present in the material, making neutrons excellent probes to study the spin structures in solids.

If a wave is scattered by a piece of matter, the resulting intensity distribution is usually observed far from the scattering body. For a charge density distribution, every infinitesimal charge volume will be driven by the external electrical field of the wave, and will emit a spherical wave (dipole radiation). Observing all these spherical waves far from the scattering body, they can be treated as plane waves with different **k**-vectors, and the waves are added up under consideration of their relative phase differences. The resulting pattern is thus the interference pattern of the scattering body. If there is a lens or other device that could reassemble the scattered waves, then it is possible to get a magnified image of the scattering body. Unfortunately, lenses for X-rays and neutrons are difficult to achieve, so it is not generally easy to produce an

X-ray image with magnification as it is possible for visible light. However, X-ray and neutron mirrors exist and can be used in imaging applications. In contrast, electrons are charged particles, and electrostatic and magnetic lenses can be constructed. Such lenses are used in a transmission electron microscope, and images with high magnification and high resolution are possible.

3.3 Addition and subtraction of waves: interference of plane waves

The mathematical description of a plane wave describes the excitation $\Psi(\mathbf{r}, t)$ at position $\mathbf{r}$ (with $\mathbf{r}$ a position vector in real space) at time t:

$$\Psi(\mathbf{r}, t) = Ae^{i(\mathbf{kr}-\omega t+\phi)} = Ae^{2\pi i(\mathbf{sr}-vt+\phi')} \tag{3.15}$$

where ϕ is the phase at the origin, $\omega = 2\pi v$ the frequency, $\mathbf{k}$ the wave vector (or $2\pi \mathbf{s} = \mathbf{k}$), and $\|\mathbf{k}\| = 2\pi/\lambda$, and λ the wavelength. In physics $\mathbf{k}$ and ω are preferred, whereas crystallography uses $\mathbf{s}$ and v. If two plane waves with the same wave vector interfere, the different amplitudes as well as the relative phase need to be considered. Therefore, the superposition of two plane waves gives the following relation (using $\cos(\alpha + \beta) = \cos \alpha \cos \beta - \sin \alpha \sin \beta$):

$$\Psi = \Psi_1 + \Psi_2 = A_1 \cos(\mathbf{kr} - \omega t + \phi_1) + A_2 \cos(\mathbf{kr} - \omega t + \phi_2)$$
$$= A \cos(\mathbf{kr} - \omega t + \phi)$$
$$A^2 = A_1^2 + A_2^2 + 2A_1A_2 \cos(\phi_1 - \phi_2)$$
$$\tan \phi = \frac{A_1 \sin \phi_1 + A_2 \sin \phi_2}{A_1 \cos \phi_1 + A_2 \cos \phi_2}$$

These equations represent the summation of two vectors with different azimuthal angles, which can be represented very elegantly using complex functions:

$$\Psi(\mathbf{k}, t) = Ae^{i(\mathbf{kr}-\omega t+\phi)}$$
$$= Ae^{2\pi i(\mathbf{sr}-vt+\phi')}$$
$$A = A_1 e^{i\phi_1} + A_2 e^{i\phi_2}$$
$$= A_1 e^{2\pi i\phi_1'} + A_2 e^{2\pi i\phi_2'}$$

As an example, two waves with different amplitude and different phase are added. Note that the wavelength remains the same, but the resulting new wave now has a different amplitude and phase. In Figure 3.3, the resulting wave is shown in red.

If the two waves are out-of-phase (phase angle 180° or π), then the amplitudes subtract directly (see Figure 3.4). The interference of the two waves is thus destructive.

$U(x)$

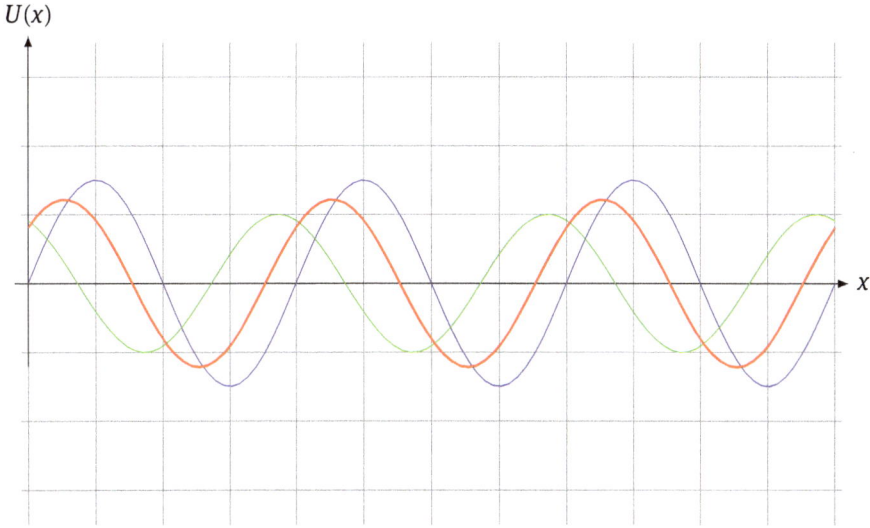

Figure 3.3: Addition of two waves with different amplitudes and different phases: The red curve is the sum of the green and blue curves.

$U(x)$

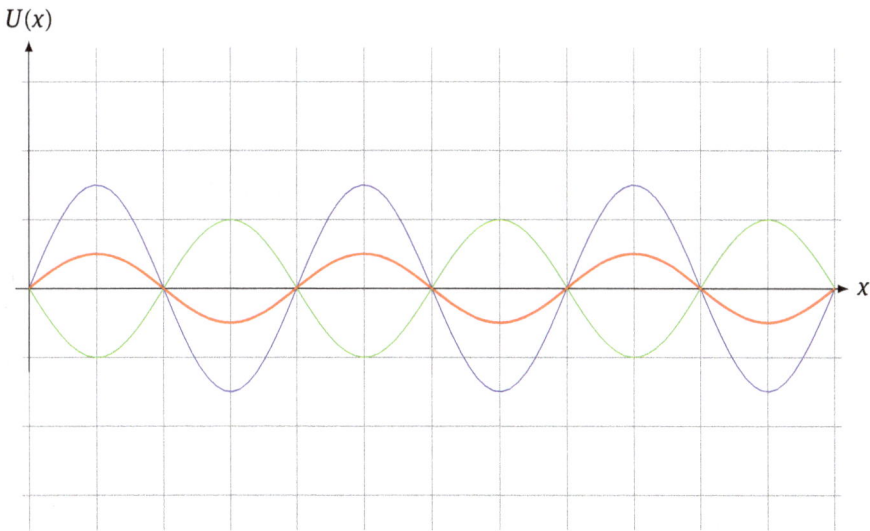

Figure 3.4: Wave addition (out-of-phase): The red curve is the sum of the blue and green curves.

Mathematically, the exponential form of the plane waves is well adapted to handle both amplitude and phase, making the notation compact. Depending on the problem, however, the explicit form with trigonometric functions, or the exponential form will be used. Additionally, the factor 2π needs to be accounted for, either explicitly, or implicitly included in the variables $\omega = 2\pi v$ and $k = \frac{2\pi}{\lambda}$.

3.3.1 Fourier transformation

The discussion of the Fourier transformation will follow the book by A. Guinier [3].

A periodic function $f(x)$ with a period of a is assumed. The Fourier theorem states that $f(x)$ can be expressed as a double series of trigonometric functions with arguments of type $(2\pi n x/a)$, with n a positive integer number.

$$f(x) = A_0 + 2 \sum_{n=1}^{n=\infty} A_n \cos \frac{2\pi n x}{a} + 2 \sum_{n=1}^{n=\infty} B_n \sin \frac{2\pi n x}{a}$$

The coefficients A_n and B_n can be obtained by using the following identities that define orthogonality:

$$\int_{-a/2}^{a/2} \cos \frac{2\pi n x}{a} \cos \frac{2\pi m x}{a} \, dx = \delta_{nm} \times \frac{a}{2} \tag{3.16}$$

$$\int_{-a/2}^{a/2} \cos \frac{2\pi n x}{a} \sin \frac{2\pi m x}{a} \, dx = 0 \tag{3.17}$$

$$\int_{-a/2}^{a/2} \sin \frac{2\pi n x}{a} \sin \frac{2\pi m x}{a} \, dx = \delta_{nm} \times \frac{a}{2} \tag{3.18}$$

Since $f(x)$ is expressed as a sum of trigonometric functions, the coefficients A_m are determined by multiplying both sides of the equation by $\cos(2\pi m x/a)$ followed by integration from $-a/2$ to $a/2$. The following equation for A_m is obtained:

$$\int_{-a/2}^{a/2} f(x) \cos \frac{2\pi m x}{a} \, dx = A_m a$$

The coefficients B_m are obtained in the same way

$$\int_{-a/2}^{a/2} f(x) \sin \frac{2\pi m x}{a} \, dx = B_m a$$

and finally

$$\int_{-a/2}^{a/2} f(x) \, dx = A_0 a$$

The function $f(x)$ can be written in a more symmetrical form

$$f(x) = \sum_{-\infty}^{+\infty} A_n \cos \frac{2\pi n x}{a} + \sum_{-\infty}^{+\infty} B_n \sin \frac{2\pi n x}{a} \tag{3.19}$$

with n an integer number and the coefficients A_n and B_n determined above. Furthermore, the coefficients have the following relationships: $A_n = A_{-n}$, $B_0 = 0$ and $B_n = -B_{-n}$.

If $f(x) = f(-x)$, then $f(x)$ is an even function and, therefore, the coefficients $B_n = 0$. Conversely, of $f(x) = -f(-x)$, then $f(x)$ is an odd function, and all the coefficients $A_n = 0$. Any general function $g(x)$ that is neither odd nor even, can be decomposed into an even and an odd function so that

$$g(x) = f_1(x) + f_2(x) = \frac{g(x) + g(-x)}{2} + \frac{g(x) - g(-x)}{2}$$

with the coefficients A_n relating to $f_1(x)$, while the coefficients B_n relate to $f_2(x)$. A compact formulation for $f(x)$ uses imaginary numbers and the exponential function

$$f(x) = \sum_{-\infty}^{+\infty} C_n e^{2\pi i \frac{nx}{a}}$$

with

$$C_n = A_n - iB_n$$

Writing this function out explicitly, the following expression is found:

$$f(x) = \sum \left(A_n \cos \frac{2\pi nx}{a} + B_n \sin \frac{2\pi nx}{a} \right) + i \sum \left(A_n \sin \frac{2\pi nx}{a} - B_n \cos \frac{2\pi nx}{a} \right)$$

The imaginary part of the sum has to be zero, since $A_n = A_{-n}$ and $B_n = -B_n$, with the coefficients C_n determined via

$$C_n = A_n - iB_n = \frac{1}{a} \int_{-\frac{a}{2}}^{\frac{a}{2}} f(x) e^{-2\pi i \frac{nx}{a}} \, dx$$

Note that the signs in the exponential functions are opposite for $f(x)$ and C_n, with the choice made by convention. Using complex variables makes the Fourier transform formalism very compact, but it has to be kept in mind that these exponential functions represent trigonometric functions.

3.4 The Fourier integral

In the previous section, functions that are periodic in an interval a were expressed as a sum of their Fourier components. If the interval a is expanded toward infinity, then the sum will become an integral. In the following, a derivation of the Fourier integral is presented. A periodic function $g(x)$ with period a is assumed, which is based on a general function $f(x)$ that is continuous in the interval $-\frac{a}{2}$ to $\frac{a}{2}$. The function $g(x)$ in this interval is expressed as a Fourier series, but $g(x)$ will have a problem at the boundary, since it is not required that $f(-\frac{a}{2}) = f(\frac{a}{2})$.

The following function is defined for the interval $\frac{-a}{2}$ to $\frac{a}{2}$:

$$F(s) = \int_{-\frac{a}{2}}^{\frac{a}{2}} f(x)e^{-2\pi i s x}\, dx,$$

which does have values aC_n with $s = \frac{n}{a}$ and n an integer number, and the prefactor a to C_n a normalization factor. For an arbitrary fixed position x_0, the following function of s can be written:

$$Z(s) = F(s)e^{2\pi i s x_0}\, dx$$

In the case when $s = \frac{n}{a}$, it is possible to write

$$Z\left(\frac{n}{a}\right) = aC_n e^{2\pi i \frac{n x_0}{a}}.$$

The sum

$$\sum_{n=-\infty}^{n=+\infty} \frac{1}{a} Z\left(\frac{n}{a}\right)$$

must be equal to

$$\frac{1}{a}\sum_{-\infty}^{+\infty}[aC_n e^{2\pi i \frac{n x_0}{a}}] = f(x_0)$$

and approximates the integral

$$\int_{+\infty}^{-\infty} Z(s)\, ds$$

that is calculated by summing all the intervals with a width of $\Delta = 1/2$. It now follows that

$$f(x_0) \approx \int_{-\infty}^{+\infty} Z(s)\, ds = \int_{-\infty}^{+\infty} F(s)e^{2\pi i s x_0}\, ds$$

If the interval a grows to infinity, the above expression is no longer approximate, and it becomes valid for any x_0, and the following relationships are obtained:

$$f(x) = \int_{-\infty}^{+\infty} F(s)e^{2\pi i s x}\, ds \qquad (3.20)$$

$$F(s) = \int\limits_{-\infty}^{+\infty} f(x)e^{-2\pi i s x}\, dx \qquad (3.21)$$

This derivation is not rigorous, but it shows the relationship between $f(x)$ and $F(s)$.

The function $F(s)$ is the *Fourier transform* of the function $f(x)$, and vice versa. As mentioned earlier, even and odd functions in x can be defined. Furthermore, if $f(x)$ is a real function, then $F(s) = F^*(-s)$. Setting $F(s) = X(s) - iY(s)$, with $X(s)$ an even function and $Y(s)$ an odd function, the Fourier transform can be written using trigonometric functions

$$X(s) = \int\limits_{-\infty}^{+\infty} f(x) \cos(2\pi s x)\, dx$$

$$Y(s) = \int\limits_{-\infty}^{+\infty} f(x) \sin(2\pi s x)\, dx$$

$$f(x) = \int\limits_{-\infty}^{+\infty} \left[X(s) \cos(2\pi s x) + Y(s) \sin(2\pi s x) \right] ds \qquad (3.22)$$

If the function $f(x)$ is real and even, then $F(s)$ is real, and $Y(s) = 0$; if $f(x)$ is odd, then $X(s) = 0$ and $F(s)$ is a complex function. For implementation in computer programs, equations (3.22) are preferred.

If the Fourier transform is for a one-dimensional variable, then both x and s are linear, with x in direct space and s in reciprocal space. The Fourier transformation therefore relates a given function in direct space to a corresponding function in reciprocal space. In analogy, a Fourier transform of a time dependent function expresses the function in frequency space, which has units of reciprocals of time. Thus, if $f(x)$ is periodic, with period a, the transform gives discrete values in reciprocal space of period $1/a$, with, in general, complex values $F(s)$. Furthermore, if $f(x)$ is not periodic (equivalent $a \to \infty$), the Fourier transform $F(s)$ is a continuous function of s that is complex for the general case.

3.4.1 Examples

The Fourier transform of a simple function can be calculated:

Delta function (Kronecker δ) at position $x = 0$: $f(x) = \delta(x)$:

$$F(s) = \int f(x)e^{-2\pi i s x}\, dx = e^0 = 1$$

The delta function is located at $x = 0$, and the Fourier transform is real and constant.

Delta function (Kronecker δ) at position $a : f(x) = \delta(x - a)$:

$$F(s) = \int f(x)e^{-2\pi i s x}\, dx = \int \delta(x - a)e^{-2\pi i s x}\, dx = e^{2\pi i a s}$$

The delta function is placed at position a, and the Fourier transform acquires a phase factor.

Similarly, two symmetrical delta functions $f(x) = \delta(x - a) + \delta(x + a)$ result in

$$F(s) = \int f(x)e^{-2\pi isx}\, dx = \int \left(\delta(x - a) + \delta(x + a)\right)e^{-2\pi isx}\, dx = e^{2\pi ias} + e^{-2\pi ias} = 2\cos(2\pi as)$$

In time varying functions, the function $f(x)$ is a function of time t, $f(t)$. The Fourier transform is used to describe periodic functions in time, with the reciprocal of the time period the frequency. The Fourier transform thus switches between the time domain and the frequency domain, with $2\pi i\nu t = i\omega t$ the relevant variables. Applying the Fourier transform therefore allows decomposition of any time-varying signal into its respective frequency components. The frequency ν is the inverse of the time interval t, and the unit for frequency, the Hertz, is in inverse seconds. The angular frequency, $\omega = 2\pi\nu$, needs to be distinguished from ν, and the factor of 2π needs to be carried. In physics, ω is preferred, whereas in engineering, ν is generally used.

In the case of a space dimension, this analogy is not as "easily" grasped, since the variables s or k are now of inverse dimension, with $2\pi sa = ka$. Therefore, one expression is "spatial frequency" when referring to s or k. Furthermore, extension to 3 dimensions in space is straightforward, where the expression kr is replaced by $\mathbf{kr} = 2\pi\mathbf{sr}$, with $\mathbf{k}$, $\mathbf{s}$ and $\mathbf{r}$ now 3-dimensional vectors, and the product $\mathbf{kr}$ or $2\pi\mathbf{sr}$ the dot-product of two vectors. The vector product of two vectors $\mathbf{kx} = $ const shows that the vector $\mathbf{x}$ in direct space has an associated vector $\mathbf{k}$ in reciprocal space, and the geometry formalism developed earlier applies. The function $f(\mathbf{x})$ is therefore expressed by plane waves in direction $\mathbf{s}$.

3.5 Point functions in object space

A function of three spatial variables will now be considered. The variables are x, y, z, and they represent a vector $\mathbf{x}$ in a coordinate system given by three non-coplanar vectors $\mathbf{a}, \mathbf{b}, \mathbf{c}$. The function $f(\mathbf{x})$ is related to the point M of object (direct) space, defined by the vector from the origin O to M, $\mathbf{x} = \overline{\mathbf{OM}}$.

The function $f(\mathbf{x})$ shall be triply periodic in a lattice that is generated by the vectors $\mathbf{a}, \mathbf{b}, \mathbf{c}$, with periodicity along the three vectors. This function is expanded into a Fourier series

$$f(\mathbf{x}) = f(x, y, z) = \sum_h \sum_k \sum_l C_{hkl} \exp\left[2\pi i\left(\frac{hx}{a} + \frac{ky}{b} + \frac{lz}{c}\right)\right] dx\, dy\, dz$$

with the coefficients C_{hkl} defined by a set of three integer numbers.

The coefficients C_{hkl} are calculated in the same way as for a single variable. The following integral will be calculated

$$C_{h'k'l'} = \int_{V_c} f(\mathbf{x}) \exp\left[-2\pi i\left(h'\frac{x}{a} + k'\frac{y}{b} + l'\frac{z}{c}\right)\right] dv$$

in the unit cell volume V_c, and h', k', l' integer. Setting $X = \frac{x}{a}, Y = \frac{y}{b}, Z = \frac{z}{c}$ as fractional coordinates, the volume element is given by

$$dv = V_c \, dX \, dY \, dZ$$

The integral can now be evaluated as

$$\int_{V_c} f(\mathbf{x}) \exp[-2\pi i(h'X + k'Y + l'Z)] \, dv$$

$$= V_c \sum_h \sum_k \sum_l C_{hkl} \int_0^1 \exp[2\pi i(h - h')X] \, dX \times \int_0^1 \exp[2\pi i(k - k')Y] \, dY$$

$$\times \int_0^1 \exp[2\pi i(l - l')Z] \, dZ$$

The integral

$$\int_0^1 \exp(2\pi i m x) \, dx = \frac{1}{2\pi i m}[1 - \exp(2\pi i m)]$$

is zero unless $m = 0$, where the integral is 1. The terms in the summations all disappear with the exception of the hkl term, resulting in

$$C_{hkl} = \frac{1}{V_c} \int_{V_c} f(\mathbf{x}) e^{-2\pi i(hX+kY+lZ)} \, dv$$

The expression $(hX + kY + lZ)$ is a vector product of two vectors, where one is a direct space vector (X, Y, Z), and the other one a reciprocal space vector (h, k, l). The Fourier transform of the direct lattice is therefore the reciprocal lattice, and vice versa. The two vectors are defined by $\mathbf{x} = X\mathbf{a} + Y\mathbf{b} + Z\mathbf{z}$ and $\mathbf{r}_{hkl} = h\mathbf{a}^* + k\mathbf{b}^* + l\mathbf{c}^*$. Thus, the vector product is written as

$$\mathbf{x}\mathbf{r}_{hkl}^* = hX + kY + lZ$$

The Fourier series for the periodic function $f(\mathbf{x})$ now becomes

$$f(\mathbf{x}) = \sum_{hkl} C_{hkl} e^{2\pi i \mathbf{x}\mathbf{r}_{hkl}^*} \tag{3.23}$$

with C_{hkl} the coefficient related to the node (hkl) of the reciprocal lattice. With the vector $\mathbf{s}$ equal to $\mathbf{r}_{hkl}^*$, the function $F(s)$ in reciprocal space is equal to $V_c C_{hkl}$. As V_c is extended to infinity, the summations can be expressed as integrals:

$$f(\mathbf{x}) = \int F(\mathbf{s})e^{2\pi i s x}\, dv_s \tag{3.24}$$

and

$$F(\mathbf{s}) = \int f(\mathbf{x})e^{-2\pi i s x}\, dv_x \tag{3.25}$$

The properties that follow can be easily demonstrated
- if $f(\mathbf{x})$ is real, then $F(-\mathbf{s}) = F^*(\mathbf{s})$
- if $f(\mathbf{x})$ is real and $f(\mathbf{x}) = f(-\mathbf{x})$ (even), then $F(\mathbf{s})$ is real

Example: sawtooth function

The Fourier transform of a sawtooth function will be calculated, using the formalism with the complex exponential notation. The sawtooth function $f(x) = Ax$ in the interval from $-\frac{a}{2}$ to $\frac{a}{2}$ represents a periodic function with A the amplitude. To calculate the Fourier coefficients C_n, the following integral is evaluated:

$$C_n = \frac{1}{a}\int\limits_{-\frac{a}{2}}^{\frac{a}{2}} Axe^{-2\pi inx/a}\, dx \tag{3.26}$$

The integral with the general form of $\int xe^{-qx}\, dx$ is obtained by integration by parts (or from an integration table):

$$\int xe^{-qx}\, dx = -\frac{x}{q}e^{-qx} - \frac{1}{q^2}e^{-qx}$$

The special case of $q = 0$ is treated separately, where the integral becomes

$$\int x\, dx = \frac{x^2}{2}$$

Since $q = 2\pi in/a$, the coefficients C_n are obtained as

$$C_n = -\frac{A}{a}\left(\frac{xa}{2\pi in}e^{-2\pi inx/a}\Big|_{-a/2}^{a/2} + \left(\frac{a}{2\pi in}\right)^2 e^{-2\pi inx/a}\Big|_{-a/2}^{a/2}\right)$$

The second term vanishes since the limit terms are equal. This leaves the first term as

$$C_n = -\frac{A}{a}\cdot\frac{(a/2)a}{2\pi in}\left(e^{-\pi in} + e^{\pi in}\right)$$

The sum of the exponentials depends on n as $2(-1)^n$. The final result is written as

$$C_n = (-1)^n\frac{iAa}{2\pi n} \quad \text{(for } n \neq 0)$$

Since the sawtooth function is odd, $C_0 = 0$, the function $f(x)$ is expressed as

$$f(x) = \sum_{n \neq 0} (-1)^n \frac{iAa}{2\pi n} e^{2\pi inx/a}$$

with the sum running from $-\infty$ to $+\infty$, skipping 0. Expressing the function in terms of sine and cosine functions gives

$$f(x) = \sum_{n \neq 0} (-1)^n \frac{iAa}{2\pi n} \left(\cos\left(\frac{2\pi nx}{a}\right) + i \sin\left(\frac{2\pi nx}{a}\right) \right)$$

Since $(-1)^n/n$ is an odd function in n, all the even terms, the cosines, cancel out. The odd terms, the sines, are grouped together, and using $i^2(-1)^n = (-1)^{n+1}$, the following equation is obtained:

$$f(x) = \frac{Aa}{\pi} \sum_{n=1}^{\infty} (-1)^{n+1} \frac{1}{n} \sin\left(\frac{2\pi nx}{a}\right)$$

This is the Fourier series for the sawtooth function. The same results is obtained if the explicit sine and cosine functions are used from the start. It should be noted that the series includes all harmonic frequencies, with diminishing amplitudes $\propto \frac{1}{n}$. In Figure 3.5, the Fourier series with one, two, three and ten terms in the summation are shown, with ten terms giving a good approximation of the actual sawtooth function.

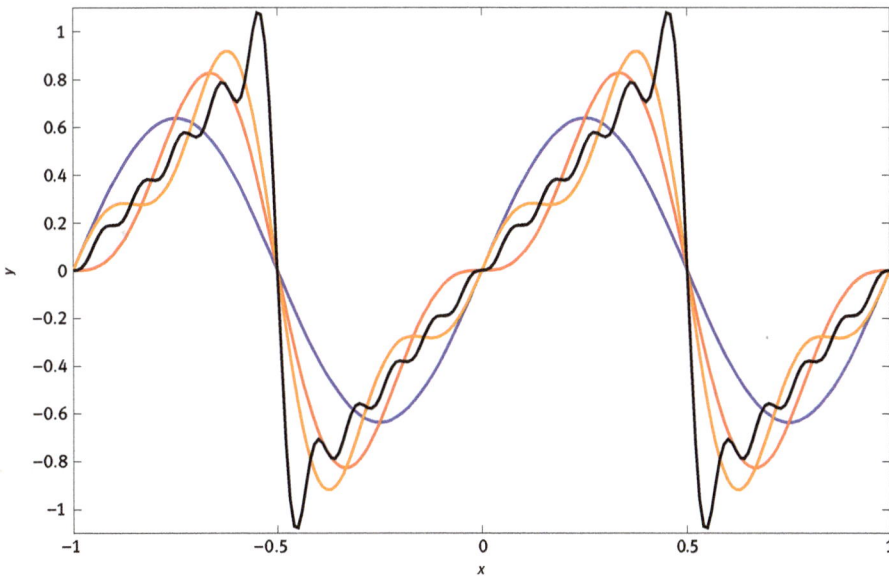

Figure 3.5: Fourier components of the sawtooth function: blue: 1 term, red: 2 terms, orange: 3 terms, black: 10 terms.

An accurate description of the jump from +1 to −1 requires a larger number of terms in the Fourier summation and, therefore, high resolution requires the inclusion of sufficient number of terms in the Fourier summation.

Bibliography

[3] A. Guinier. X-ray Diffraction in Crystals, Imperfect Crystals and Amorphous Bodies. Dover Publications, Inc., Mineola, N. Y., 1994. ISBN: 0-486-68011-8.

4 Symmetry

4.1 Symmetry of finite objects

The symmetry chapter follows the discussion given by D. Schwarzenbach in "Cristal-lographie" [4]. Symmetry is found in natural objects in many places. Examples include plants and flowers, animals, snowflakes and minerals, to mention just a few. The symmetry might not be perfect, for instance, a human face may not be perfectly mirror-symmetric, but approximate symmetries can be identified in many instances.[1] For example, in chemistry, molecules are described in terms of their symmetry, which symmetry operation will map the molecule onto itself. The symmetry elements utilized are rotations and mirror planes, plus improper rotations where a rotation is followed by mirroring about a plane perpendicular to the rotation axis. Since the molecule is a discrete unit, it can have a simple or a complicated symmetry. For instance, the C_{60} molecule, where 60 carbon atoms are arranged in a pattern that is also found on a soccer ball (Fussball), forms a highly symmetric arrangement with a large number of symmetry elements.

In crystallography, the equilibrium shape of crystalline materials is classified as to the presence of symmetries. The symmetry operations used are rotations, mirrors, inversion (centrosymmetry, $(x, y, z) \rightarrow (-x, -y, -z)$) and improper rotations, the roto-inversion where a rotation is followed by an inversion. In contrast to chemistry that utilizes rotations, mirrors and combined rotations followed by mirroring (roto-reflection), the roto-inversions are used as improper rotations in crystallography. The two descriptions are equivalent, but different notation have been developed, which will be introduced below.

4.1.1 Definitions

The following definitions will be used:

Symmetry element An axis, plane or a point that "anchors" the symmetry operation.

For example, a rotation by a given angle is around a rotation axis. Similarly, a mirror operation is in respect to a plane.

Symmetry operations A symmetry operation applied to a body will result in a new state indistinguishable from the original state; thus, the inherent symmetry of the body is becoming apparent. For instance, an equilateral triangular prism transforms into itself upon rotation by 120° around an axis that is perpendicular to the basal plane

---

1 Symmetry online resources: Bilbao Crystallography Server: http://www.cryst.ehu.es; Crystallography at the EPFL: escher.epfl.ch

https://doi.org/10.1515/9783110610833-004

(the symmetry element). Repeating this symmetry operation generates again a trans-formation of the prism into itself, while a third application of the symmetry operation returns to the initial state.

4.1.2 Point groups

Single symmetry elements may be present, but combinations of different symmetry elements can produce a set of interacting symmetry operations (symmetry elements). The appearance of new symmetry elements when two distinct symmetry elements are combined can be verified easily. This set of interacting symmetry elements is referred to as a *point group*. In particular, a point group is defined as a set of compatible symmetry elements that all have at least one point in common; therefore this point is invariant under all possible symmetry operations of the set. It is convenient to use this point as the origin of a coordinate system. The coordinate axes, if possible, are then chosen along high symmetry directions. If the symmetry operations leave a line or a plane unchanged, it is necessary to arbitrarily choose the origin of the coordinate system on the line or the plane, respectively. In this way, a symmetry-adapted coordinate system is set up, and it allows to describe the symmetry operations using matrix notation.

4.1.3 Symmetry operations

The definition of *symmetry* describes the invariance of an object or a structure in rela-tion to certain operations. A *geometric symmetry operation* is an application of space onto itself, and this operation transforms an object into itself. The *representation* of a geometric symmetry operation in 3-dimensional space can be described by a matrix $\mathbb{R}$ acting on a vector plus a translation $\mathbf{t}$, relating the point (vector) $\mathbf{x}' = (x_1', x_2', x_3')$ to the original point (vector) $\mathbf{x} = (x_1, x_2, x_3)$:

$$\begin{pmatrix} x_1' \\ x_2' \\ x_3' \end{pmatrix} = \begin{pmatrix} r_{11} & r_{12} & r_{13} \\ r_{21} & r_{22} & r_{23} \\ r_{31} & r_{32} & r_{33} \end{pmatrix} \times \begin{pmatrix} x_1 \\ x_2 \\ x_3 \end{pmatrix} + \begin{pmatrix} t_1 \\ t_2 \\ t_3 \end{pmatrix}$$

$$\mathbf{x}' \quad = \quad \mathbb{R} \quad \times \quad \mathbf{x} \quad + \quad \mathbf{t}$$

The combined operation is abbreviated by the symbol $(\mathbb{R}, \mathbf{t})$. This operation does not change the dimensions of the object and, therefore, conserves the metric and is there-fore an *affine transformation*. Conservation of the metric means that the vector $|\mathbf{x}'| = |\mathbf{x}|$. The matrix $\mathbb{R}$ transforms $\mathbf{x}$ into $\mathbf{x}''$ and any general $\mathbf{r}$ into $\mathbf{r}''$ and is *independent* of the origin choice of the coordinate system. In contrast, the vector $\mathbf{t}$ depends on the origin choice. The repeated application of the symmetry operation $(\mathbb{R}, \mathbf{t})$, for instance

$(\mathbb{R}, \mathbf{t})^2 = (\mathbb{R}^2, \mathbb{R}\mathbf{t} + \mathbf{t})$, $(\mathbb{R}, \mathbf{t})^3 = (\mathbb{R}^3, \mathbb{R}^2\mathbf{t} + \mathbb{R}\mathbf{t} + \mathbf{t}), \ldots$, produces a symmetric structure that is invariant in respect to $(\mathbb{R}, \mathbf{t})$ and all powers $(\mathbb{R}, \mathbf{t})^n$.

4.1.4 Groups

If a structure is invariant in respect to two different symmetry operations $(\mathbb{P}, \mathbf{t}_P)$ and $(\mathbb{Q}, \mathbf{t}_Q)$, it is invariant in respect to a successive application of these two operations, which is a *product* of the two operations. If the operation $(\mathbb{P}, \mathbf{t}_P)$ is applied first, followed by $(\mathbb{Q}, \mathbf{t}_Q)$, the vector $\mathbf{x}$ is transformed to $\mathbf{x}'' = \mathbb{Q}\mathbf{x}' + \mathbf{t}_Q = \mathbb{Q}\mathbb{P}\mathbf{x} + \mathbb{Q}\mathbf{t}_P + \mathbf{t}_Q$. The multiplication of the matrices $\mathbb{Q}$ and $\mathbb{P}$ is executed from the *right to the left*: $\mathbb{P}$ is applied first, followed by $\mathbb{Q}$:

$$(\mathbb{Q}, \mathbf{t}_Q)(\mathbb{P}, \mathbf{t}_P) = (\mathbb{Q}\mathbb{P}, \mathbb{Q}\mathbf{t}_P + \mathbf{t}_Q)$$
$$(\mathbb{P}, \mathbf{t}_P)(\mathbb{Q}, \mathbf{t}_Q) = (\mathbb{P}\mathbb{Q}, \mathbb{P}\mathbf{t}_Q + \mathbf{t}_P) \tag{4.1}$$

The multiplication of two operations is generally not commutative.

The *identity operation* $(\mathbb{E}, \mathbf{0})$ is a symmetry operation existing for all objects conceivable where $\mathbb{E}$ is the identity represented by the unit matrix and the vector $\mathbf{0}$ with zero length. If $(\mathbb{P}, \mathbf{t}_P)$ is a symmetry operation, then the *inverse operation*, $(\mathbb{P}, \mathbf{t}_P)^{-1}$ with the property that $(\mathbb{P}, \mathbf{t}_P) \times (\mathbb{P}, \mathbf{t}_P)^{-1} = (\mathbb{P}, \mathbf{t}_P)^{-1} \times (\mathbb{P}, \mathbf{t}_P) = (\mathbb{E}, \mathbf{0})$ is also a symmetry operation. This leads to the following relationships:

$$(\mathbb{P}, \mathbf{t}_P)^{-1} = (\mathbb{P}^{-1}, -\mathbb{P}^{-1}\mathbf{t}_P)$$

The multiplication product of symmetry operations is *associative*, resulting in

$$[(\mathbb{R}, \mathbf{t}_R)(\mathbb{Q}, \mathbf{t}_Q)](\mathbb{P}, \mathbf{t}_P) = (\mathbb{R}, \mathbf{t}_R)[(\mathbb{Q}, \mathbf{t}_Q)(\mathbb{P}, \mathbf{t}_P)]$$
$$= (\mathbb{R}\mathbb{Q}\mathbb{P}, \mathbb{R}\mathbb{Q}\mathbf{t}_P + \mathbb{R}\mathbf{t}_Q + \mathbf{t}_R) \tag{4.2}$$

These properties ensure that the symmetry operations of an object constitute a mathematical group.

The group formed by an operation $(\mathbb{R}, \mathbf{t}_R)$ and its powers $(\mathbb{R}, \mathbf{t}_R)^2$, $(\mathbb{R}, \mathbf{t}_R)^3$, $\ldots$, $(\mathbb{R}, \mathbf{t}_R)^n$ is a *cyclic group,* and if all operations of a group commute such that $(\mathbb{R}_j\mathbf{t}_j)(\mathbb{R}_i\mathbf{t}_i) = (\mathbb{R}_i\mathbf{t}_i)(\mathbb{R}_j\mathbf{t}_j)$, then the group is *abelian.*

If a group consists of symmetry operations, the matrices $\mathbb{R}$ and the vectors $\mathbf{t}$ constitute a *representation of the group* that depends on the choice of the coordinate system and its origin. Therefore, an infinite number of representations are possible.

4.1.5 Rotations, roto-reflections and roto-inversions

A symmetry operation $(\mathbb{R}, \mathbf{t})$ ensures that the absolute magnitude of symmetry related vectors $\mathbf{r}$ and $\mathbf{r}'$ are equal. For a unitary coordinate system, the following condition holds:

$$\|\mathbf{r}'\|^2 = \mathbf{r}^T \mathbb{R}^T \mathbb{R} \mathbf{r} = \|\mathbf{r}\|^2$$

This condition ensures that $\mathbb{R}^T \mathbb{R} = \mathbb{E}$, with $\mathbb{E}$ the identity matrix, and $\mathbb{R}^T$ the transposed matrix of $\mathbb{R}$. The matrix $\mathbb{R}$ is thus an *orthogonal matrix*, with eigenvalues of $e^{i\phi}$, $e^{-i\phi}$, ± 1, where ϕ is given by

$$\cos \phi = [\text{trace}(\mathbb{R}) \pm 1]/2.$$

The values for ϕ are in the range of $0 \leq \phi \leq 2\pi r$ with $r = m/n$ a rational number. The matrix $\mathbb{R}$ will therefore resemble a matrix $\mathbb{U}$ with

$$\mathbb{U}(\phi) = \begin{pmatrix} \cos \phi & -\sin \phi & 0 \\ \sin \phi & \cos \phi & 0 \\ 0 & 0 & \pm 1 \end{pmatrix}, \quad \text{and} \quad \phi = \frac{m}{n} 2\pi, \; m \text{ and } n \text{ integer and smallest}$$

(4.3)

A matrix $\mathbb{X}$ must exist such that $\mathbb{U}(\phi) = \mathbb{X}^{-1} \mathbb{R} \mathbb{X}$. Furthermore, it can be shown that $\mathbb{U}^2(\phi) = \mathbb{U}(2\phi)$ and $\mathbb{U}^{-1}(\phi) = \mathbb{U}^T(\phi) = \mathbb{U}(-\phi)$, and that an integer number p exists such that $p < n$ and $pm/n(\text{mod } 1) = 1/n$, where $\mathbb{U}^p(\phi) = \mathbb{U}(\phi')$ with $\phi' = 2\pi/n$. Therefore, if $\mathbb{U}(\phi)$ is a symmetry operation, then $\mathbb{U}(\phi')$ is also a symmetry operation. This leads to the following types of symmetry operations: *Rotations, roto-reflections,* and *roto-inversions* (see 4.1). These symmetry operations can be *represented* by the following matrices, assuming that the rotation axis coincides with the coordinate z-axis:

$$\text{rotation } \mathbb{A}_n \quad \text{matrix similar to} \quad \begin{pmatrix} \cos \phi & -\sin \phi & 0 \\ \sin \phi & \cos \phi & 0 \\ 0 & 0 & +1 \end{pmatrix} \quad \phi = \frac{2\pi}{n}; \quad (4.4)$$

$$\text{roto-reflection } \mathbb{S}_n \quad \text{matrix similar to} \quad \begin{pmatrix} \cos \phi & -\sin \phi & 0 \\ \sin \phi & \cos \phi & 0 \\ 0 & 0 & -1 \end{pmatrix} \quad \phi = \frac{2\pi}{n}; \quad (4.5)$$

$$\text{roto-inversion } \mathbb{I}_n \quad \text{matrix similar to} \quad \begin{pmatrix} -\cos \phi & \sin \phi & 0 \\ -\sin \phi & -\cos \phi & 0 \\ 0 & 0 & -1 \end{pmatrix} \quad \phi = \frac{2\pi}{n}; \quad (4.6)$$

The **rotation** $\mathbb{A}_n$ (see (4.4) and Figure 4.1) around an axis transforms a left-hand into a left-hand, and a right-hand into a right-hand: the rotation therefore preserves the *chirality* of an object. The rotation is called an *operation of the first kind*. The determinant of any matrix representing an operation of the first kind is therefore $\|\mathbb{A}_n\| = +1$.

The **roto-reflection** $\mathbb{S}_n$ (see (4.5) and Figure 4.1) around an axis is a rotation followed by a reflection by a plane perpendicular to the rotation axis. This combined operation is neither a true reflection nor a true rotation. It transforms a left-hand into a right-hand, and vice versa, and thus switches chirality, and is called an *operation*

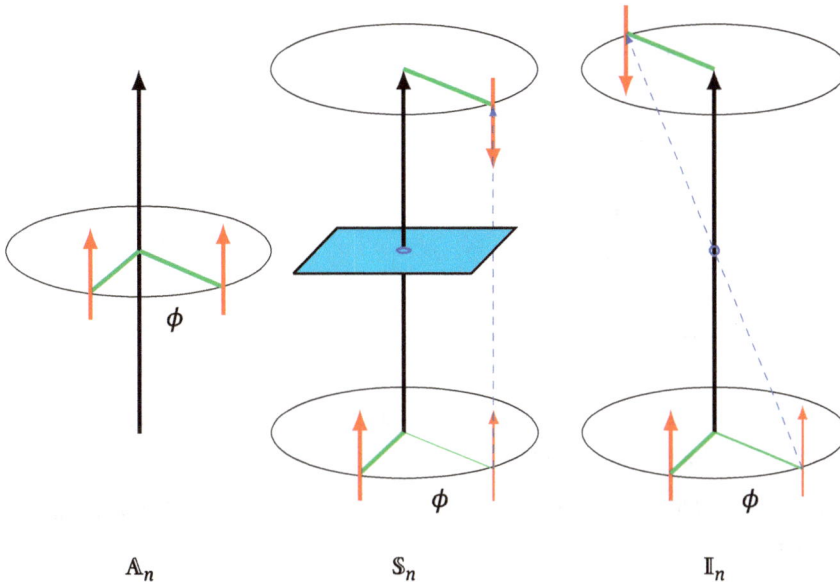

Figure 4.1: Rotation $\mathbb{A}_n$, roto-reflection $\mathbb{S}_n$ (with mirror plane) and roto-inversion $\mathbb{I}_n$.

of the second kind. The determinant of any matrix representing an operation of the second kind is $\|\mathbb{S}_n\| = -1$. The roto-reflection $\mathbb{S}_0$ ($\phi = 0$) is a reflection by a symmetry plane, a *mirror* reflection. Additionally, the roto-reflection $\mathbb{S}_2$ ($\phi = \pi$) is the *inversion*, relating ($\mathbf{r} \rightarrow -\mathbf{r}$).

The **roto-inversion** $\mathbb{I}_n$ (see (4.6) and Figure 4.1) around an axis is a rotation by the angle ϕ followed by an inversion through a point situated on the rotation axis. The roto-inversion is also an operation of the second kind, neither a pure inversion nor a pure rotation. Each roto-inversion is equivalent to a roto-reflection and vice versa:

$$\mathbb{I}(\phi) = \mathbb{S}(\pi + \phi); \quad \mathbb{S}(\phi) = \mathbb{I}(\pi + \phi) \tag{4.7}$$

Operations of the second kind are represented by either roto-inversions or roto-reflections. Either one of these sets of symmetry operations can be used in a description of the symmetry, and both systems of symmetry operations are in use: the *Schoenflies* notation based on rotations and roto-reflections and the *Hermann–Mauguin* notation (*international system*) based on rotations and roto-inversions. Table 4.1 lists the equivalent roto-inversion and roto-reflections. The *Schoenflies* notation is used in chemistry and physics to describe the symmetry of molecules and finite objects, whereas the *Hermann–Mauguin* notation is preferred to describe the symmetry of lattices and crystal shapes in crystallography.

For groups of *finite order*, a finite number n exists so that $\mathbb{A}_n^n = \mathbb{E}$ (for both n even or odd), $\mathbb{S}_n^n = \mathbb{I}_n^n = \mathbb{E}$ (for n even), $\mathbb{S}_n^{2n} = \mathbb{I}_n^{2n} = \mathbb{E}$ (for n odd), with $\mathbb{E}$ the identity operation. These groups are formed by the operations $\mathbb{A}_n$ and $\mathbb{I}_n$ (or $\mathbb{S}_n$) and their powers

Table 4.1: Equivalence of roto-inversions and roto-reflections.

$I_1 = S_2 =$ inversion ($\bar{1}$)	$S_1 = I_2 =$ mirror
$I_2 = S_1 =$ mirror m	$S_2 = I_1 =$ inversion ($\bar{1}$)
$I_3 = S_6^{-1} = S_6^5$	$S_3 = I_6^{-1} = I_6^5$
$I_4 = S_4^{-1} = S_4^3$	$S_4 = I_4^{-1} = I_4^3$
$I_5 = S_{10}^{-3} = S_{10}^7$	$S_5 = I_{10}^{-3} = I_{10}^7$
$I_6 = S_3^{-1} = S_3^2$	$S_6 = I_3^{-1} = I_3^2$
$I_7 = S_{14}^{-5} = S_{14}^9$	$S_7 = I_{14}^{-5} = I_{14}^7$
$I_8 = S_8^{-3} = S_8^5$	$S_8 = I_8^{-3} = I_8^5$

(repeated applications). At least one point in space is invariant under application of all the operations $\mathbb{A}_n$ and $\mathbb{I}_n$ (or $\mathbb{S}_n$):

The groups formed by the rotations and roto-inversion (or roto-reflections) are called point groups.

It is important to distinguish between a symmetry operation and its *representation* by a matrix (see equations (4.4), (4.5) and (4.6)) since the matrix elements depend on the particular coordinate system that is chosen.

4.1.6 Translation

A translation is represented by the operation $(\mathbb{E}, \mathbf{T}_{uvw})$, where $\mathbb{E}$ represent the identity operation (unity matrix), and $\mathbf{T}_{uvw} = u\mathbf{a} + v\mathbf{b} + w\mathbf{c}$ a vector in three dimensions with the coordinate systems spanned by three non-coplanar vectors $\mathbf{a}$, $\mathbf{b}$ and $\mathbf{c}$, and (uvw) a triple of integer numbers. The symmetry of the translations allows arranging a large number of identical atoms or molecules in a way that all are strictly equivalent in an infinitely large crystal (or at least large in relation to the shortest translations). The order of translations is not critical, as the translation operations are commutative, the order of the translation does not affect the outcome. Atoms or molecules at the surface of the crystal are mutually equivalent if they form a 2-dimensional translation lattice that is parallel to a certain plane of the 3-dimensional lattice. An example of a 1-dimensional translation lattice formed by a stretched chain is shown in Figure 4.2, where all chain members are of equal length, but rotated by 90° from each other. In this example, the translation period is two times the length of a chain link.

The group formed by all translations $\mathbf{T}_{uvw} (-\infty \leq u, v, w, \leq \infty)$ is of infinite order and abelian.

The existence of a translation lattice implies long-range order. However, it is not necessary to postulate the existence of long-range forces acting over long distances.

Figure 4.2: Stretched chain (links are perpendicular to each other).

For example, in the stretched chain (Figure 4.2), each link interacts only with its two neighbors left and right, but the chain itself is periodic with specific orientations of the individual links.

4.2 Symmetry elements

4.2.1 Fixed points, rotation axes and reflection planes

There are symmetry operations $(\mathbb{R}, \mathbf{t})$ that leave some points in space unchanged, and such points may lend themselves as possible locations for the origin of the coordinate system. Does a preferred origin exist, and if yes, how should the origin location be chosen? It has to be kept in mind that the vector $\mathbf{t}$ depends on the origin choice of the coordinate system.

An origin shift by the vector $\mathbf{v}$ transforms the affine (symmetry) transformation $\mathbf{x}' = \mathbb{R}\mathbf{x} + \mathbf{t}$ to $\mathbf{x}' - \mathbf{v} = \mathbb{R}(\mathbf{x} - \mathbf{v}) + \mathbf{t}_v = (\mathbb{R}\mathbf{x} + \mathbf{t}) - (\mathbb{R}\mathbf{v} + \mathbf{t}) + \mathbf{t}_v$. Therefore, the affine (symmetry) operation $(\mathbb{R}, \mathbf{t})$ becomes $\mathbb{R}, \mathbf{t}_v$, with the translation:

$$\mathbf{t}_v = (\mathbb{R} - \mathbb{E})\mathbf{v} + \mathbf{t} \qquad (4.8)$$

A fixed point of an affine transformation is a point that is transformed onto itself, with $\mathbf{x}' = \mathbb{R}\mathbf{x} + \mathbf{t} = \mathbf{x} = \mathbb{E}\mathbf{x}$; therefore it is seen that $(\mathbb{R} - \mathbb{E})\mathbf{x} = -\mathbf{t}$.

There are four different cases to distinguish:

- The matrix $(\mathbb{R} - \mathbb{E})$ is invertible, the inverse $(\mathbb{R} - \mathbb{E})^{-1}$ exists and, therefore three nonzero eigenvalues exist. By displacing the origin to a fixed point, $\mathbf{v} = \mathbf{x} = -(\mathbb{R} - \mathbb{E})^{-1}\mathbf{t}$, the translation vector "annihilates" to $\mathbf{t}_v = \mathbf{0}$, and the affine transformation becomes linear $(\mathbb{R}, \mathbf{0})$. This is the case for all roto-inversions $\mathbb{I}_n$ represented by the matrices given earlier ((4.4), (4.5), (4.6)), with the exception of $n = 2$, representing the reflection by a plane (or by all roto-reflections $\mathbb{S}_n$ with the exception of $n = 1$). For example, the operation $\mathbb{I}_1$ followed by a translation that transforms the point $\mathbf{x}$ with coordinates (x_1, x_2, x_3) to a point $\mathbf{x}'$ with coordinates $(\frac{1}{2} - x_1, \frac{1}{2} - x_2, \frac{1}{2} - x_3)$ has a fixed point at $(\frac{1}{4}, \frac{1}{4}, \frac{1}{4})$. This fixed point associated with the operation $\mathbb{I}_n$ is called *center of inversion* or *symmetry center*. Placing the origin of the coordinate system at the inversion center is convenient and often the best choice. In this way, the vector $\mathbf{x}$ is related to the vector $-\mathbf{x}$. In contrast, the plane reflection, $\mathbb{I}_2 = \mathbb{S}_1$ does not possess a unique fixed point.
- The matrix $(\mathbb{R} - \mathbb{E})$ possesses one eigenvalue zero, and two nonzero eigenvalues; the matrix is therefore not invertible and $(\mathbb{R} - \mathbb{E})^{-1}$ does not exist. This is the case

for all rotations $\mathbb{A}_n$ represented by matrices of the form given in equation (4.4), with the exception if the identity operation given by $n = 1$. The eigenvector $\mathbf{x}_0$ corresponds to the zero eigenvalue and is the *rotation axis*, which is invariant in respect to $\mathbb{R}$, $\mathbb{R}\mathbf{x}_0 = \mathbf{x}_0$. A displacement of the origin along the rotation axis does not change the value of the translation vector $\mathbf{t}$ of the affine operation, whereas a displacement perpendicular to the rotation axis modifies the vector $\mathbf{t}$. The origin of the coordinate system can be placed on the rotation axis, with a translation vector with the nonzero component parallel to the rotation axis (for instance, $\mathbf{t} = (0, 0, t_3)$ for a rotation axis parallel to $\mathbf{x}_3$). If the translation vector is zero, $\mathbf{t} = \mathbf{0}$, then each point located on the rotation axis becomes a fixed point.

- The matrix $(\mathbb{R} - \mathbb{E})$ possesses two zero eigenvalues and one nonzero eigenvalue. This is the case for a reflection by a plane, $\mathbb{S}_1 = \mathbb{I}_2$, which is represented by a matrix similar to

$$\begin{pmatrix} 1 & 0 & 0 \\ 0 & 1 & 0 \\ 0 & 0 & -1 \end{pmatrix} \tag{4.9}$$

a reflection (mirror) plane perpendicular to $\mathbf{x}_3$. An eigenvector corresponding to the nonzero eigenvalue is normal to the *reflection plane*. The affine operation $(\mathbb{I}_2, \mathbf{t})$ is therefore invariant in respect to a displacement of the coordinate system origin that is located in the plane. The vector $\mathbf{t}$ has two nonzero components, both parallel to the mirror plane. In a 2-dimensional example, the reflection along a line $\mathbf{x}_2$ transforms x_1 into $-x_1$, and $\mathbf{t}$ is parallel to the reflection line. In the case $\mathbf{t} = \mathbf{0}$, each point in the reflection (mirror) plane is fixed.

- The matrix $(\mathbb{R} - \mathbb{E})$ has three zero eigenvalues, therefore, $\mathbb{R} = \mathbb{E}$, the identity operation. This operation is a pure translation and has no fixed points or preferred origin location.

A *symmetry element* is the ensemble of the fixed points of a symmetry operation. For the fixed points of the symmetry operation $\mathbb{I}_n$ and $\mathbb{S}_n$, addition of the corresponding rotation axis $\mathbb{A}_n$ or the plane perpendicular to it, is needed. For the symmetry operations $\mathbb{A}_n$, these are the rotation axes, for the operation $\mathbb{I}_n$, the rotation axis and the inversion, and for the operation $\mathbb{S}_n$, the rotation axis and reflection plane. The center of symmetry corresponds to $\mathbb{I}_1$, the reflection plane to $\mathbb{I}_2$ (mirror). The symmetry elements are further used to visualize the graphic representation of a symmetry group, and the respective symbols are shown in Tables 4.2 and 4.4.

The symmetry group is an assembly of symmetry operations, whereas a *symmetry element* is a geometric location. It is possible to represent symmetry operations of the second kind by either roto-reflections or roto-inversions, but there are *symmetry elements* that *cannot* be built from a combination of rotations, inversions and plane reflections. These are the roto-inversions $\bar{4}$, $\bar{8}$, etc. All the axes $\bar{n}$ produce cyclic groups and, therefore, the symbols $\bar{3}$ and $\bar{6}$ are preferred over the combinations of rotations, reflections and inversions. Examples are given in Tables 4.2 and 4.3, and Figure 4.3 depicts roto-inversions.

Table 4.2: Symbols of symmetry elements (m = reflection plane).

Symmetry elements	Symbol
rotation axes	1, 2, 3, ..., x
roto-inversions	$\bar{1}, \bar{2} = m, \bar{3}, \ldots, \bar{x}$
roto-reflections	$\tilde{1} = m, \tilde{2}, \tilde{3}, \ldots, \tilde{x}$

Table 4.3: Equivalence of roto-inversion and roto-reflection axes.

$\bar{1} = \tilde{2}$	center of symmetry (inversion center)
$\bar{2} = \tilde{1} = m$	reflection plane (mirror)
$\bar{3} = \tilde{6}$	ternary axis in combination with a center of symmetry
$\bar{4} = \tilde{4}$	
$\bar{6} = \tilde{3}$	ternary axis in combination with a reflection plane
$\bar{n}, n = 2m + 1$	axis of order n with a center of symmetry, equivalent to an axis of roto-inversion $\tilde{x}, x = 2n = 4m + 2$
$\bar{n}, n = 4m + 2$	axis of order $\frac{1}{2}n$ and reflection plane, equivalent to a roto-reflection $\tilde{x}, x = \frac{1}{2}n = 2m + 1$
$\bar{n}, n = 4m$	symmetry element that cannot be decomposed (equivalent to a roto-reflection $\tilde{n}$)

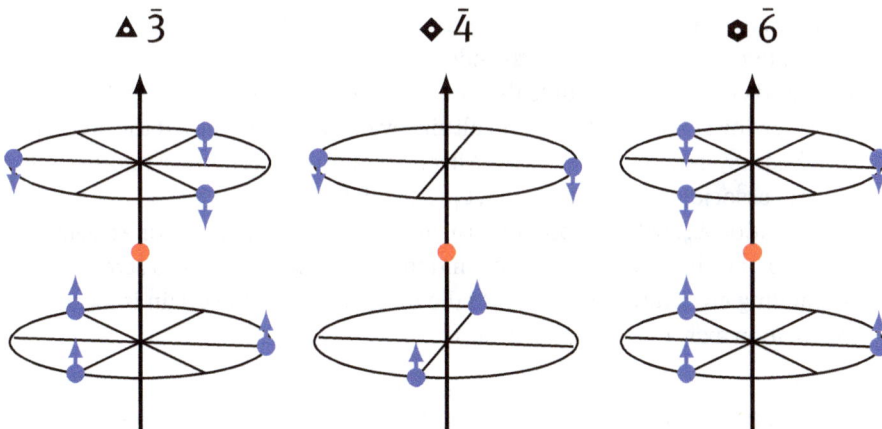

Figure 4.3: Roto-inversion axes $\bar{3}, \bar{4}, \bar{6}$. These axes represent cyclic groups. The inversion center is shown in red.

4.2.2 Reflection planes with glides and helical axes

If a symmetry operation $\mathbb{R}$ is of finite order (n integer), then $\mathbb{R}^n = \mathbb{E}$ is the identity operation. The successive application of n operations of $(\mathbb{R}, \mathbf{t})$ that includes a translation therefore results in a pure translation $\mathbf{T}$: $(\mathbb{R}, \mathbf{t})^n = (\mathbb{E}, \mathbf{T})$. Thus, the operation is expressed by

$$(\mathbb{R}, \mathbf{t})^n = (\mathbb{R}^n, [\mathbb{R}^{n-1} + \mathbb{R}^{n-2} + \cdots + \mathbb{R}^2 + \mathbb{R} + \mathbb{E}]\mathbf{t}) = \left(\mathbb{R}^n, \left[\sum_{x=0}^{n-1} \mathbb{R}^x\right]\mathbf{t}\right) = (\mathbb{E}, \mathbf{t}).$$

By multiplying the sum from the left or the right by $\mathbb{R}$ and by $(\mathbb{R} - \mathbb{E})$,

$$\mathbb{R}\left[\sum_{x=0}^{n-1} \mathbb{R}^x\right] = \left[\sum_{x=0}^{n-1} \mathbb{R}^x\right]\mathbb{R} = \left[\sum_{x=1}^{n} \mathbb{R}^x\right] = \left[\sum_{x=0}^{n-1} \mathbb{R}^x\right]$$

is obtained and

$$(\mathbb{R} - \mathbb{E})\left[\sum_{x=0}^{n-1} \mathbb{R}^x\right] = \left[\sum_{x=0}^{n-1} \mathbb{R}^x\right](\mathbb{R} - \mathbb{E}) = \mathbf{0}$$

with $\mathbf{0}$ the null-matrix, where all matrix terms are zero. Multiplication on the right by the vector $\mathbf{t}$ thus gives $(\mathbb{R} - \mathbb{E})\mathbf{T} = \mathbf{0}$ where $\mathbf{T}$ is an eigenvector of $\mathbb{R}$, and also of $\mathbb{R}^x$, with an eigenvalue of +1. The average of the matrices $\mathbb{R}^x$, Ω, is therefore an *idempotent matrix*[2] with the following properties:

$$\Omega = \frac{1}{n}\left[\sum_{x=0}^{n-1} \mathbb{R}^x\right]; \Omega\Omega = \Omega; \Omega\mathbf{t} = \frac{1}{n}\mathbf{T}$$

The eigenvalue of any idempotent matrix Ω is either 0 or +1; and they correspond to the nonzero and zero eigenvalues of $(\mathbb{R} - \mathbb{E})$.

Four different cases need to be considered:

- The matrix $(\mathbb{R} - \mathbb{E})$ is invertible therefore $\mathbb{R}$ possesses a fixed point and $(\mathbb{R} - \mathbb{E})^{-1}$ exists, giving $\Omega = \mathbf{0}$ and $\mathbf{T} = 0$, and all eigenvalues of Ω are zero. If the origin of a coordinate system coincides with the fixed point, the operation is a roto-inversion without associate translation: $\bar{\mathbb{I}}_1, \bar{\mathbb{I}}_3, \bar{\mathbb{I}}_4, \ldots$, with $\mathbf{t} = 0$.

- For a rotation $\mathbb{A}_n$, with a rotation axis of order n with $n \neq 1$, there is one eigenvalue of $\mathbb{R}$ and of Ω, both with a value of 1 and the other two eigenvalues zero. The corresponding eigenvector is parallel to the rotation axis. If the origin is located on the rotation axis, the vector $\mathbf{t}$ becomes

$$\mathbf{t} = \frac{m}{n}\mathbf{T}$$

with $\mathbf{T}$ a translation parallel to the rotation axis. The symmetry element of the symmetry operation $(\mathbb{R}, \mathbf{t})$ is a helical axis, with the axis defined by the ensemble of

---

2 An idempotent matrix IM is a matrix with the property $IM^2 = IM$.

the fixed points of $\mathbb{A}_n$. There are therefore different helical axes that are described by their numeric values n_m.

− For the operation $\mathbb{I}_2 = \mathbb{S}_1$ (corresponding to a mirror m), two eigenvalues of $\mathbb{R}$ and Ω have the value +1. The corresponding eigenvectors are parallel to the plane of reflection. If the origin is in the plane of reflection, the vector $\mathbf{t}$ reduces to

$$\mathbf{t} = \frac{1}{2}\mathbf{T}_m$$

with $\mathbf{T}_m$ a translation parallel to the plane of reflection. This is therefore a *mirror plane with a glide operation* or a *glide mirror plane* (glide mirror).

− $\mathbb{R}$ is equal to $\mathbb{E}$, $\mathbb{R} = \mathbb{E} = \Omega$, and $\mathbf{t} = \mathbf{T}$ is a translation.

It follows that the only symmetry elements composed of a rotation and a translation, or a roto-inversion and a translation, are either helical axes or reflection planes with glides.

4.2.3 Graphic symbols of symmetry elements

A number of symmetry elements and their respective symbols are listed in Tables 4.4, 4.5 and 4.6, and several depictions of their symmetry operations are shown in Figures 4.4, 4.5, 4.6 and 4.7.

Table 4.4: Symbols of point symmetry elements.

●	binary axis A_2, symbol 2	▲	ternary axis A_3, symbol 3
→	binary axis parallel to projection plane		
◆	quaternary axis A_4, symbol 4	⬣	6-fold axis A_6, symbol 6
none	identity operation	○	symmetry center (inversion), symbol $\bar{1}$
|	reflection plane, symbol m	L	reflection plane parallel to projection plane
◈	4-fold roto-inversion, symbol $\bar{4}$		
◮	3-fold roto-inversion, symbol $\bar{3}$	⬢	6-fold roto-inversion, symbol $\bar{6}$

Table 4.5: Symbols of helical symmetry elements.

⬧	binary screw axis 2_1
→	binary screw axis in the projection plane
▲ ▲	ternary screw axes $3_1, 3_2$
◆ ◆ ◆	4-fold screw axes $4_1, 4_2, 4_3$
⬟ ⬟ ⬟ ⬟ ⬟	6-fold screw axes $6_1, 6_2, 6_3, 6_4, 6_5$

Table 4.6: Symbols of reflection planes with glides.

– – – – – – – –	glide plane perpendicular to projection plane glide **a**/2 parallel to the projection plane
	glide plane parallel to the projection plane
· · · · · · · · · · · · · · ·	glide plane perpendicular to projection plane glide **b**/2 or **c**/2 parallel to the projection plane
	glide plane parallel to the projection plane
– · – · – · – · – ·	glide plane perpendicular to projection plane glide (**a** + **b**)/2, (**b** + **c**)/2 or (**c** + **a**)/2 oblique to the projection plane
	glide plane parallel to the projection plane, arrow gives glide direction

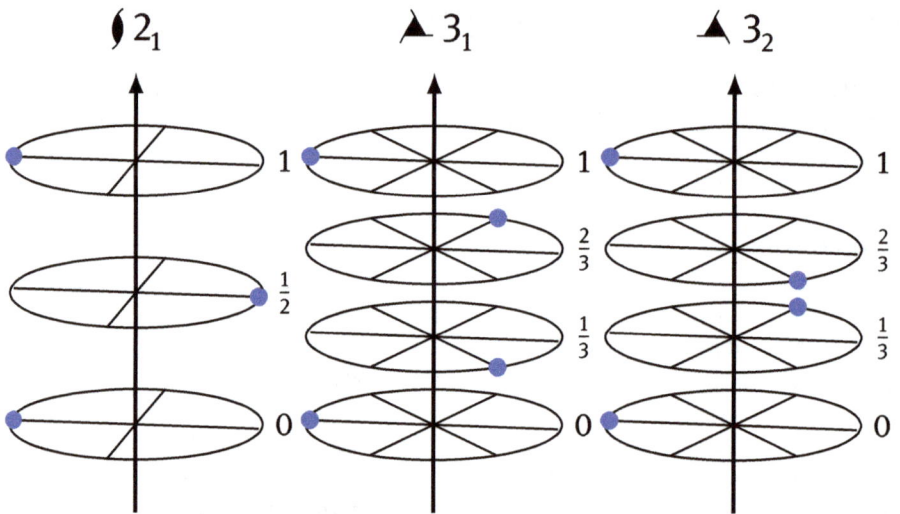

Figure 4.4: Helical axis of order 2 with translation $t = \frac{1}{2}$, and of order 3 with translations $t = \frac{1}{3}, \frac{2}{3}$.

4.3 Symmetry and metric of lattices

4.3.1 Symmetry elements compatible with translations

The number of point groups is infinite, as $\mathbb{A}_n$ and $\mathbb{I}_n$ can have values for $1 \leq n < \infty$. These groups describe the symmetries of macroscopic objects. In contrast, periodic

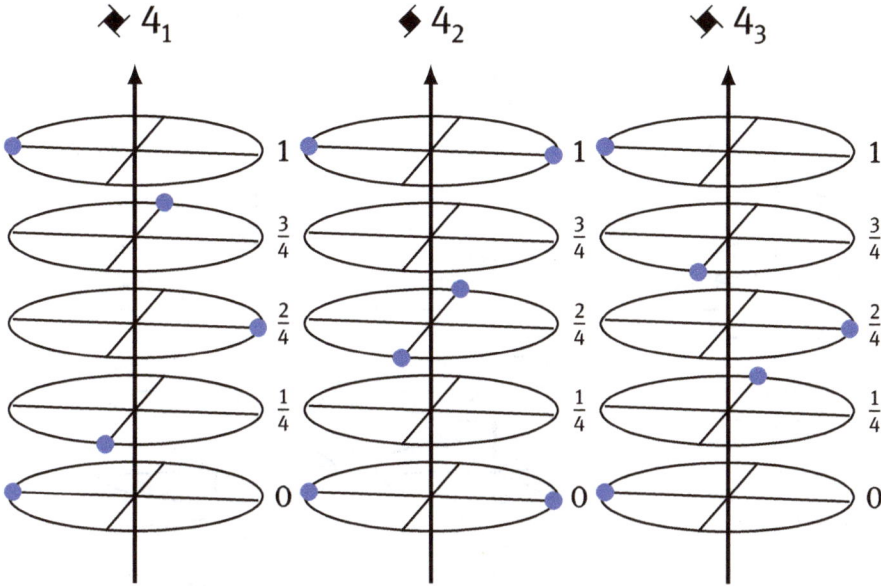

Figure 4.5: Helical axes of order 4, with translations $\mathbf{t} = \frac{1}{4}, \frac{2}{4}$ and $\frac{3}{4}$.

structures do not allow all these symmetry operations; there are only a very restricted number of point symmetries that are invariant in respect to periodic structures. In the previous example of an extended chain (Figure 4.2), one finds a series of symmetry elements, e. g., a series of rotation and roto-inversion axes, as well as mirror planes. The translation symmetry therefore results in a periodic arrangement of symmetry elements (see Figure 4.8) .

The vector $\mathbf{x}$ belongs to the periodic structure with translations $\mathbf{T} = u\mathbf{a} + v\mathbf{b} + w\mathbf{c}$, $-\infty < u, v, w < +\infty$. The endpoints of all the vectors $\mathbf{x} + \mathbf{T}$ are points that are related by translation and are thus equivalent. The symmetry operation $(\mathbb{R}, \mathbf{t})$ transforms the vector $\mathbf{x}$ into $\mathbf{x}'$, with $\mathbf{x}' = \mathbb{R}\mathbf{x} + \mathbf{t}$. This operation therefore transforms $\mathbf{x} + \mathbf{T}$ into $\mathbf{x}' + \mathbb{R}\mathbf{T}$: $\mathbb{R}(\mathbf{x} + \mathbf{T}) + \mathbf{t} = \mathbf{x}' + \mathbb{R}\mathbf{T}$. It thus follows that $\mathbb{R}\mathbf{T} = \mathbf{T}'$ is also a translation of the lattice. The combination of translations with the symmetries discussed previously requires that *the lattice has to be invariant in respect to rotations* $\mathbb{A}_n$ *and roto-inversion* $\mathbb{I}_n$. In a unitary coordinate system, $\mathbb{R}$ is represented by an orthogonal matrix. With translation symmetry in three dimensions, a coordinate system based on the three non-coplanar lattice translations $\mathbf{a}, \mathbf{b}, \mathbf{c}$ is preferred. The coordinates of the lattice nodes u, v, w are therefore integer numbers, and thus all the terms of the corresponding representation of $\mathbb{R}$ are also integer numbers. The orthogonal representation is designated by the matrix $\mathbb{U}$, and its representation in integer numbers is the matrix $\mathbb{N}$. The two matrices $\mathbb{U}$ and $\mathbb{N}$ are therefore *similar* since they represent the same operation: a matrix $\mathbb{X}$

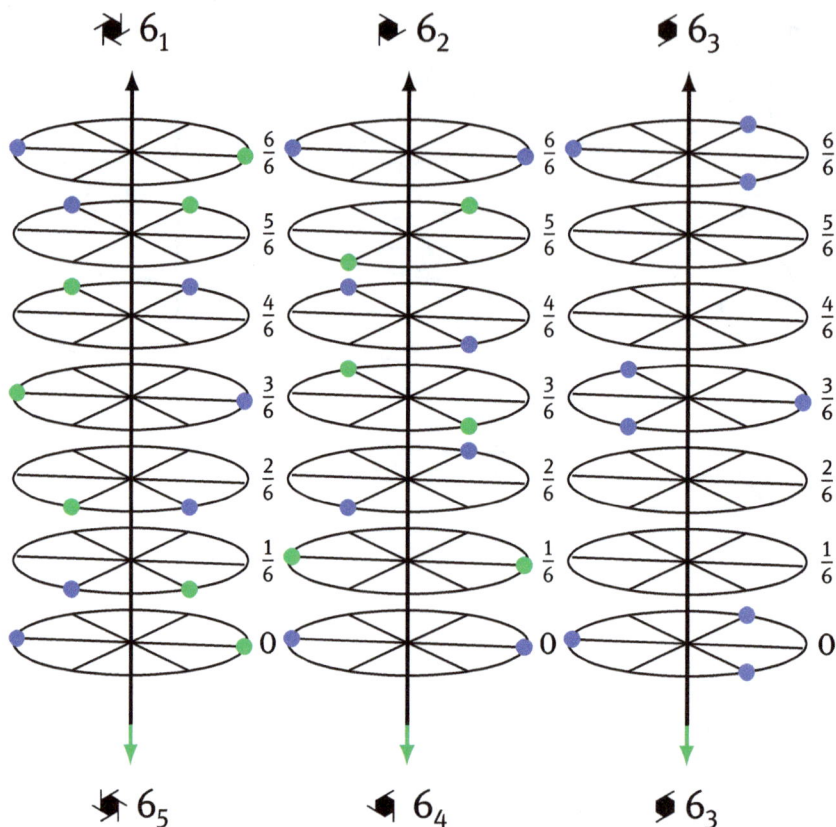

Figure 4.6: Helical axes of order 6, with translations of $t = \frac{1}{6}$ (blue dots), $\frac{2}{6}$ (blue dots), $\frac{3}{6}$ (blue dots), $\frac{4}{6}$ (green dots) and $\frac{5}{6}$ (green dots).

must therefore exist such that $\mathbb{N} = \mathbb{X}^{-1}\mathbb{U}\mathbb{X}$. The matrix $\mathbb{X}$ maps the basis of the lattice onto a unitary basis. Additionally, the matrix $\mathbb{U}$ resembles the matrices given earlier. Since similar matrices possess the same traces, it follows that:

$$\text{trace}(\mathbb{U}) = \pm(2\cos\phi \pm 1) = \text{integer}, \quad \text{and}$$

$$\cos\phi = \cos(2\pi/n) = \frac{1}{2}N, \quad n \text{ and } N \text{ integers} \tag{4.10}$$

Based on these conditions, the *only admissible values for n are n = 1, 2, 3, 4, 6.*

Periodic structures are invariant only in respect to symmetry operations or rotations with 1, 2, 3, 4, and 6-fold rotation, and roto-inversions of $\bar{1}, \bar{2} = m, \bar{3}, \bar{4}, \bar{6}$ (and corresponding roto-reflections).

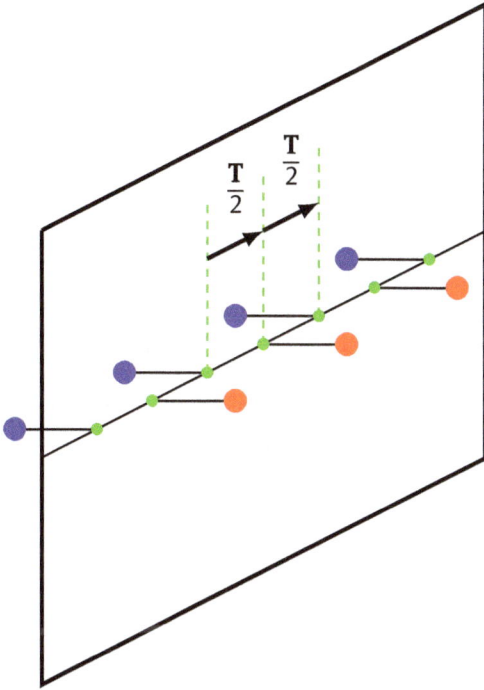

Figure 4.7: Reflection plane with glide, glide plane.

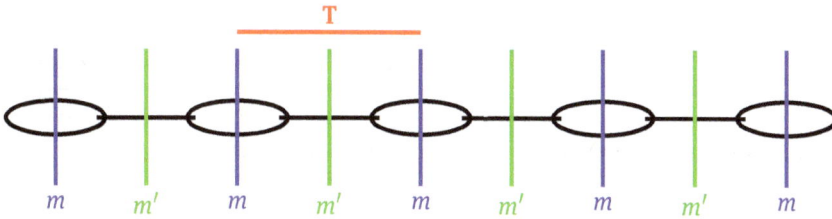

Figure 4.8: Symmetry of a stretched chain: T is the translation. Two types (classes) of mirror symmetries can be distinguished, indicated by m and m' (blue and green). In each class, the mirrors are equivalent via the translation.

4.3.2 Metric imposed by symmetry

The symmetries compatible with a periodic lattice will impose constraints on the metric, the values of the lengths of the vectors **a**, **b**, **c**, and their respective angles α, β, γ of a particular lattice. Binary axes do not impose a particular metric (but fix two angles), in contrast to ternary and 6-fold axes, which impose trigonal and hexagonal symmetry. Similarly, a 4-fold axis imposes a square lattice with angles of 90°. Mirror symmetry imposes constraints on angles; for instance, a single reflection plane has a unique axis defined by the plane normal, thus forcing two angles to be 90° to the two vectors defining the lattice in the plane.

A translation **T** of a lattice that contains a mirror plane $ that is part of the lattice, the translation **T′** produced by the reflection symmetry is also part of the lattice. The translations **T** − **T′** and **T** + **T′** are now perpendicular to, and parallel to the reflection plane, respectively (see Figure 4.10). It follows that *all rotation axes and roto-inversion axes (or roto-reflections) that are symmetry elements of a lattice, are parallel to translations and perpendicular to lattice planes*. With the exception of an inversion center, symmetry elements generate mutually perpendicular translations.

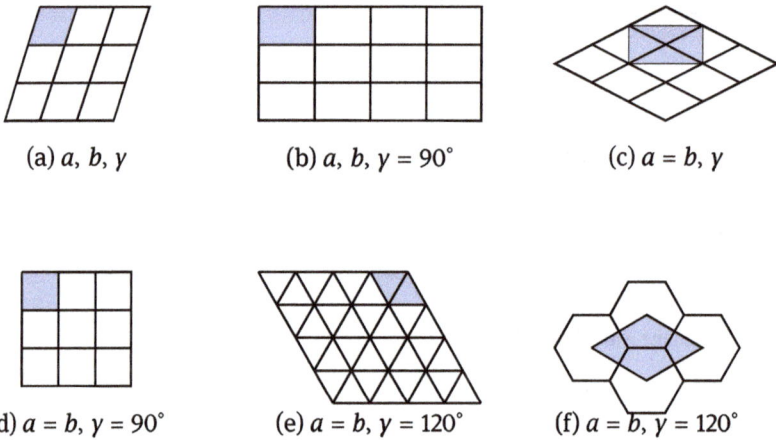

(a) *a, b, γ* (b) *a, b, γ* = 90° (c) *a = b, γ*

(d) *a = b, γ* = 90° (e) *a = b, γ* = 120° (f) *a = b, γ* = 120°

Figure 4.9: Shapes for periodic tiling of the 2-dimensional plane: (a) general lattice with arbitrary axes *a, b* and angle *γ*; binary axis 2; (b) rectangular, reflection lines *m*; (c) rhombus, centered rectangular lattice, reflection lines *m* and glide reflection lines *g*; (d) squares, 4-fold axis 4; (e) triangles, ternary (3-fold) axis 3; (f) hexagonal lattice, 6-fold axis 6, unit cell type equal to (e).

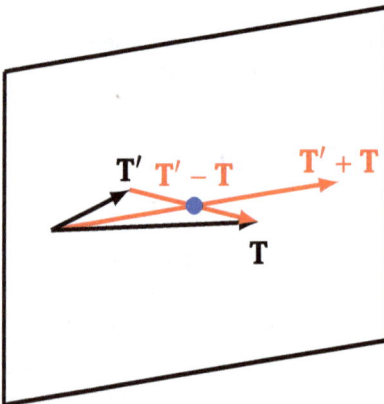

Figure 4.10: Reflection plane and translation.

4.3.3 Point groups and space groups

The groups that are formed by rotations and roto-inversion (roto-reflections) are called *point groups*.

Point groups describe the symmetry of objects with finite dimensions.

The symmetry groups including translations are of infinite order. They can also include rotations and reflections, both also with translations, as well as roto-inversions. In 2-dimensional space, these symmetry groups are called *plane groups*, and in Figure 4.9 the different shapes of periodic tilings are shown, with the repeating motif colored blue. In 3-dimensional space, the combination of the rotations, roto-inversions, mirrors and glide planes with translational symmetry generates the *space groups*. Therefore, *space groups describe the symmetry of periodic structures in 3 dimensions.*

What is the relationship between a space group describing the symmetry of a crystalline structure at the atomic level and may include glide planes and screw axes, and the point group that describes the symmetry of the macroscopic crystal? A crystal is characterized by its properties along different *directions*, for example, elastic constants, electrical conductivity, optical absorption, etc., properties that are usually described as tensors. The macroscopic crystal shapes (crystal habits) do not show the presence of translations, but they do display the presence of rotational and mirror symmetry. In general, a series of rotation or roto-inversion axes manifest themselves as a single rotation axis or roto-inversion axis of the macroscopic crystal, and a series of mirror or glide mirror planes manifest themselves as only one mirror plane. It follows that it is impossible to derive the presence of the translation information from the crystal shape and apparent symmetry. Based on the shapes and symmetry of crystals, a definition of *crystal classes* is possible: they are formed by the operations $\mathbb{A}_1$, $\mathbb{A}_2$, $\mathbb{A}_3$, $\mathbb{A}_4$, $\mathbb{A}_6$, $\mathbb{I}_1$, $\mathbb{I}_2$, $\mathbb{I}_3$, $\mathbb{I}_4$, $\mathbb{I}_6$, and their powers.

4.4 Classes and crystal systems

4.4.1 Classes

Classes are utilized to describe many different ensembles that contain objects that are similar. As an example, a group $\mathbb{G}$ of order m that is produced by the operations $\mathbf{g}_k$, $1 \leq k \leq m$ contains a subgroup $\mathbb{N}$ that regroups the operations $\mathbf{g}_i\mathbb{N}$ into a class. The word **class** is therefore used to describe ensembles that comprise objects of a certain type. The crystal classes group crystals according to their symmetry; and the Laue classes regroup the crystal classes, which may also be groups. Symmetry elements can also be grouped into classes, for instance the two different mirror operations in Figure 4.8.

4.4.2 Group generators

In a space group, some symmetry operations are a result of other symmetry operations. For instance, two mirror planes intersecting at right angles generate a 2-fold
axis. It is therefore possible to describe crystallographic groups with a limited number of symmetry operations or symmetry elements, that are sufficient to generate all
symmetries present in the space group. For example, the cyclic group $\mathbb{A}_n$ is generated
by just one symmetry operation, followed by a repeated application of this symmetry
operation. Similarly, with a 4-fold axis A_4 (rotation by 90°), two consecutive applications of A_4 results in a rotation by 180°, a 2-fold axis A_2, and three consecutive applications give a rotation by −90°. Useful generators are discussed below, and examples
are shown in Figure 4.11.

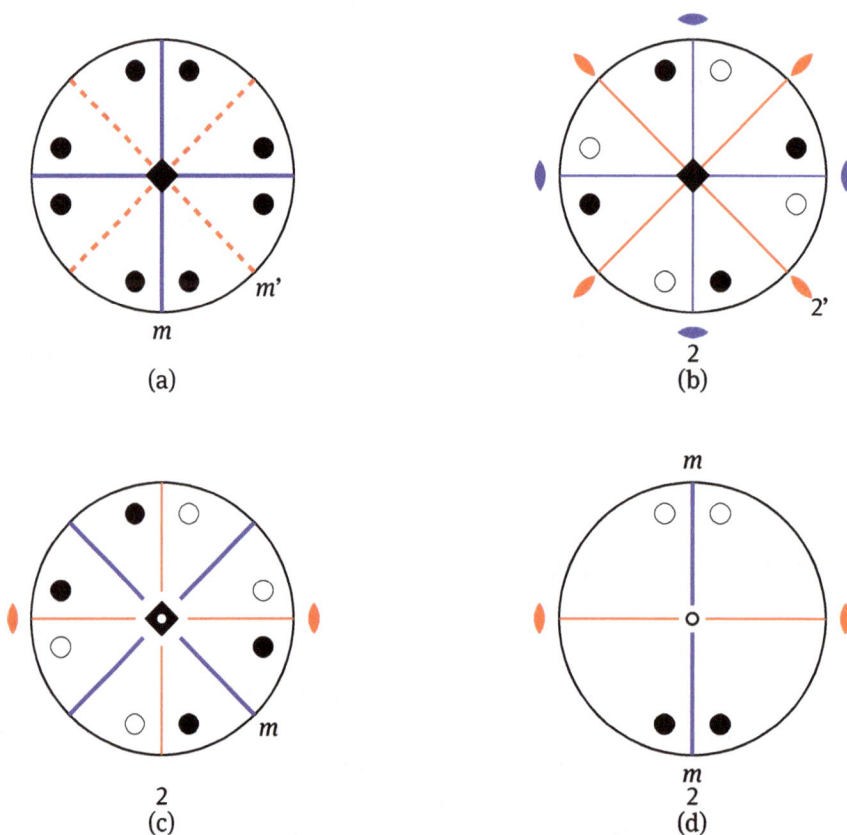

Figure 4.11: Combination of symmetry elements: (a) two mirror planes at 45°, (b) 2 binary axes at 45°,
(c) one binary axis and one mirror plane at 45°, (d) one binary axis perpendicular to a mirror plane.
General positions are indicated by large circles, with the filled circles above the projection plane,
the open circles below the projection plane.

Two reflection planes

Two reflection planes that intersect at an angle ϕ generate a rotation axis with rotation angle 2ϕ. This is shown in Figure 4.11. For example, two reflection planes intersecting at an angle of 45° create a 4-fold rotation axis, with rotation angle 90°. Repeated mirror reflections by the two mirror planes produce not just the 4-fold axis, but also 4 mirror planes that belong to two classes. For each n, two mirror planes intersecting at an angle $\phi = \pi/n$ generate an n-fold axis and a total of n mirror planes that belong to two classes if n is even, and to one class, if n is odd.

Two binary axes

Two binary axes intersecting at an angle ϕ generate a rotation axis 2ϕ. If the angle $\phi = 45°$, then a 4-fold rotation axis is generated. Multiple applications of the binary rotations produce the 4-fold axis plus four binary axes, which belong to two different equivalence classes. For each n, two binary axes at an angle $\phi = \pi/n$ produce a rotation axis $\mathbb{A}_n$ and a total of n binary axes which belong to two classes if n is even, and to one class if n is odd.

One binary axis and one mirror plane

The combination of a reflection plane (mirror plane) and a binary axis that intersect at an angle ϕ generates a roto-reflection axis with rotation angle 2ϕ (or a corresponding roto-inversion axis). For example, a binary axis intersecting a mirror plane at 45° generates a roto-reflection by 90°. Repeat application of the two operations produces a $\bar{4}$-axis, two reflection planes and two binary axes.

One mirror plane and perpendicular even order axis

The combination of a mirror plane and a perpendicular axis of even order (2-, 4-fold, etc.) generate an inversion center at the intersection of the axis with the mirror plane. Two of these three symmetry elements generate the smallest noncyclic (abelian) group of order 4, which comprise the elements $\mathbb{E}$, $\mathbb{A}_2$, $\mathbb{I}_1$ and $\mathbb{I}_2$.

4.4.3 Generation of point groups

The four types of point groups that comprise rotations (operations of the first kind) exclusively and maintain handedness, can be derived. These groups describe the symmetry of *chiral objects* or *enantiomorphs*, as an enantiomorphic object and its mirror image cannot be superimposed, since a left-hand and a right-hand are not superimposable by rotations. Similarly, a right-turning screw cannot be superimposed onto a left-turning screw by rotations only. Two types of rotation groups are defined:

- Cyclic groups characterized by an axis $\mathbb{A}$ of order n
- Groups characterized by an axis $\mathbb{A}$ of order n perpendicular to n binary axes. These binary axes form two equivalence classes if n is even, and a single class if n is odd.

Other rotation groups are derived in the following manner: two rotation axes $\mathbb{P}$ and $\mathbb{Q}$ of order p and q form an angle ϕ. Around $\mathbb{P}$, p axes of order q are generated: $\mathbb{Q}, \mathbb{Q}', \mathbb{Q}'', \ldots, \mathbb{Q}^{(p-1)}$ that belong to the same equivalence class; around $\mathbb{Q}$, q axes of order p are generated, $\mathbb{P}, \mathbb{P}', \mathbb{P}'', \ldots, \mathbb{P}^{(q-1)}$ that belong to an equivalence class. Furthermore, binary axes bisecting the angles formed by the axes from the same class are generated. In Figure 4.12, the combination of a 4-fold axis $p = 4$ and a 3-fold axis $q = 3$ is shown in a stereographic projection. Application of the 4-fold rotation to the 3-fold rotation axis generates four 3-fold axes, and vice versa, the application of the 3-fold rotation to the 4-fold axis produces three 4-fold axes that are mutually perpendicular. In addition, 2-fold axes are generated as well.

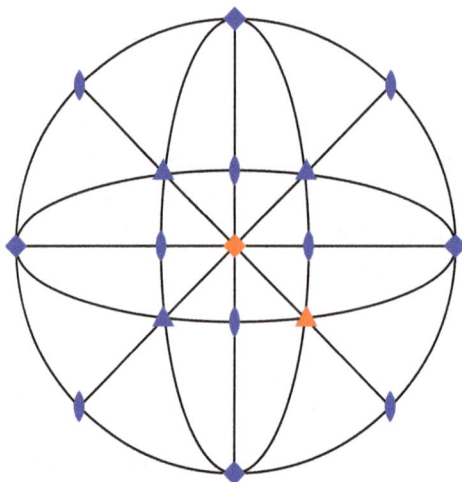

Figure 4.12: Combination of a 4-fold and a 3-fold axis (red) generates four 3-fold axes and three 4-fold axes, as well as binary axes (blue).

Starting with the rotation groups, it is possible to generate other groups by adding mirror planes parallel or perpendicular to the rotation axes in a manner that no new rotation axes are produced. Following this procedure, all 32 point groups are generated.

The literature uses two different nomenclatures for point groups: The first nomenclature, the *Schoenflies* notation, is widely used in chemistry and physics to describe local (point) environments, or the symmetry of finite objects. The *Schoenflies* notation is based the following symmetry elements: *rotation axes* and *roto-reflections*. However, it does not express the close relationships between crystal classes and space groups. The second nomenclature, the *Hermann–Mauguin* notation (or *international notation*) provides a coherent manner to describe the relationship between crystal classes and space groups, and is used in all modern crystallography work and structure determinations. The *Hermann–Mauguin* notation is based on *rotation axes* and *roto-inversions*. Two types of symbols are in use, the full and the abbreviated symbols, with the full symbol explicitly listing the symmetry elements associated with different directions.

4.4.4 The 32-point groups: axial groups

Among the symmetry elements of an axial group, there is only a single axis of order larger than 2. In Table 4.7, these groups are listed.

Table 4.7: The 7 types of axial crystalline classes.

Type	Group generator	Order of the group	International symbol
x	single rotation axis	x	$1, 2, 3, 4, 6$
$\bar{x}$	single roto-inversion axis	x (even) $2x$ (x odd)	$\bar{1}, m, \bar{3}, \bar{4}, \bar{6}$
$\frac{x}{m}$	mirror m *normal* to an axis x	$2x$	$\frac{2}{m}, \frac{4}{m}, \frac{6}{m}$
xm	mirror plane m *parallel* to an axis x	$2x$	$mm2, 3m, 4mm, 6mm$
$x2$	2-fold axis *normal* to an axis x	$2x$	$222, 32.422, 622$
$\bar{x}m$ or $\bar{x}2$	mirror plane m *parallel* to an axis $\bar{x}$ or 2-fold axis *normal* to an axis $\bar{x}$	$2x$ (x even) $4x$ (x odd)	$\bar{3}\frac{2}{m}, \bar{4}2m, \bar{6}m2$
$\frac{x}{m}m$	mirror plane m *parallel* to an axis $\frac{x}{m}$, or 2-fold axis *normal* to an axis $\frac{x}{m}$	$4x$ (x even)	$\frac{2}{m}\frac{2}{m}\frac{2}{m}, \frac{4}{m}\frac{2}{m}\frac{2}{m}, \frac{6}{m}\frac{2}{m}\frac{2}{m}$

In addition:
- The groups of type $\frac{x}{m}$ and $\frac{x}{m}m$ are only defined for $x = $ even. The symmetry element $\bar{x}$ with $x = 4n + 2$ ($x = 2, 6, 10, \ldots$) are equivalent to a rotation axis of order $2n + 1$ normal to a mirror plane m, (e. g., $\frac{3}{m} = \bar{6}$). $\frac{x}{m}$ with x odd represents therefore a cyclic group from the groups $\bar{x}$. Similarly, $\frac{x}{m}m$ with x odd is a member of $\bar{x}m$.
- The groups xm comprise the symmetry operations that correspond to an axis x and x mirror planes m parallel to the axis. These mirror planes are all equivalent if x is odd. For x even, the mirror planes fall into two distinct classes. The international symbols $mm2, 3m, 4mm, 6mm$ label these different groups. Here, the symbol for the rotation axis precedes the symbols for the mirror planes, with the exception for the 2-fold axis in $mm2$.

- The groups $x2$ possess the same structure as the groups of type xm. They comprise the symmetry operation that correspond to an axis x and x 2-fold axes perpendicular to the primary axis. These 2-fold axes are equivalent if x is odd, and they form two distinct classes for x even. These groups do not contain any reflection planes and, therefore, describe enantiomorph objects.

- The groups $\bar{x}m$ and $\bar{x}2$ possess also the same structure as the groups xm. If x is odd, the 2-fold axes are normal to the reflection planes; if x is even, the 2-fold axes are oblique to the reflection planes. $\bar{4}m2$ and $\bar{6}2m$ are alternative symbols to $\bar{4}2m$ and $\bar{6}m2$.

- A reflection plane normal to an axis of even order generates a center of inversion. The groups $\frac{2}{m}, \frac{4}{m}, \frac{6}{m}, \frac{2}{m}\frac{2}{m}\frac{2}{m}, \frac{4}{m}\frac{2}{m}\frac{2}{m}, \frac{6}{m}\frac{2}{m}\frac{2}{m}$ and $\bar{3}\frac{2}{m}$ are therefore centrosymmetric. The *abbreviated symbols* for these groups $2/m, 4/m, 6/m.\ mmm, 4/mmm, 6/mmm$ and $\bar{3}m$, respectively, are often used in text paragraphs due to their compact form.

- The groups of type x, $\bar{x}$ and $\frac{x}{m}$ comprise all groups characterized by a *single* symmetry direction, whereas the other groups possess several symmetry directions.

4.4.5 The 32 crystal classes: tetrahedral and octahedral groups

The *5 regular polyhedra* or *platonic polyhedra* are known since antiquity. The faces of a regular polyhedron are regular polygons; and there is only one type of corner, only one type of edges and only one type of regular polygons. (Table 4.8 lists the regular polyhedra). The Archimedian (or semiregular) polyhedra are composed of regular but inequivalent faces.

Table 4.8: Regular polyhedra.

Polyhedron	Number of faces with (x) edges	Number of corners with (x) edges	Number of edges
Tetrahedron	4(3)	4(3)	6
Octahedron	8(3)	6(4)	12
Cube	6(4)	8(3)	12
Icosahedron	20(3)	12(5)	30
Dodecahedron	12(5)	20(3)	30

The cube and the octahedron are *dual polyhedra* in the sense that if the corners of the octahedron are replaced by faces, the cube results; and starting from the cube, replacing the corners by faces will produce the octahedron. The same holds for the icosahedron and the dodecahedron, which are dual polyhedra. The tetrahedron is *self-dual*, since by replacing the tetrahedron corner by faces, a tetrahedron results. Therefore, the 5 regular polyhedra represent three types of symmetry: tetrahedral, octahedral and icosahedral (see Table 4.8). Since the icosahedral groups comprise 5-fold axes, they are not crystallographic groups.

The symmetry elements of the *tetrahedral* and *octahedral* groups are oriented according to the *characteristic direction of a cube*:

- edges: 3 directions [100], [010], [001]
- body diagonals: 4 directions [111], [$\bar{1}$11], [1$\bar{1}$1], [11$\bar{1}$]
- face diagonals: 6 directions [110], [101], [011], [$\bar{1}$10], [$\bar{1}$01], [0$\bar{1}$1]

The tetrahedron can be inscribed in a cube, since the four body-diagonals in the cube are 3-fold axes and align with the four 3-fold axes of the tetrahedron. Previously, the two enantiomorphic groups, a tetrahedral group, and an octahedral group were derived. Adding mirror planes perpendicular or parallel to the rotation axes, or in bisecting positions, nonenantiomorphic groups are generated. The five groups and the symmetry along different directions are given in Table 4.9.

Table 4.9: The 5 tetrahedral and octahedral groups and symmetry along different directions *(cubic)*.

Symbol	Edges	Body diagonals	Face diagonals	Comments
23	2	3	–	enantiomorph, tetrahedral
$\frac{2}{m}3$	$\frac{2}{m}$	$\bar{3}$	–	23 plus inversion
$\bar{4}3m$	$\bar{4}$	3	m	symmetry of the tetrahedron
432	4	3	2	enantiomorph, octahedral
$\frac{4}{m}\bar{3}\frac{2}{m}$	$\frac{4}{m}$	$\bar{3}$	$\frac{2}{m}$	symmetry of the octahedron

For the two centrosymmetric groups $\frac{2}{m}\bar{3}$ and $\frac{4}{m}\bar{3}\frac{2}{m}$, the *abbreviated symbols m$\bar{3}$ and m$\bar{3}$m* are used. Additionally, the five groups are all characterized by the presence of *3-fold axes along the body diagonals of a cube*.

4.4.6 Noncrystallographic point groups

The labels of noncrystallographic point groups are analogous to the labels for crystalline point groups:

$$x: \quad 5, 7, 8, 9, 10, \ldots, \infty$$
$$\bar{x}: \quad \bar{5}, \bar{7}, \bar{8}, \bar{9}, \bar{10}, \ldots$$
$$x/m: \quad 8/m, 10/m, \ldots, \infty/m \ (x \text{ even})$$
$$xm: \quad 5m, 7m, 8mm, 9m, 10mm, \ldots, \infty m$$

$$x2: \quad 52, 72, 822, 92, 1022, \ldots, \infty 2$$

$$\bar{x}2: \quad 5\frac{2}{m}, 7\frac{2}{m}, \bar{8}2m, 9\frac{2}{m}, \bar{10}m2, \ldots$$

$$\frac{x}{m}\frac{2}{m}: \quad \frac{8}{m}\frac{2}{m}\frac{2}{m}, \frac{10}{m}\frac{2}{m}\frac{2}{m}, \ldots, \frac{\infty}{m}\frac{2}{m} \quad (x \text{ even})$$

$$\text{icosahedral:} \quad 235, \frac{2}{m}\bar{3}\bar{5}$$

$$\text{spherical:} \quad \infty\infty, \frac{\infty}{m}\frac{\infty}{m}$$

The spherical groups are better known by their symbols $\infty\infty = SO_3$, the group of all continuous rotations in the 3-dimensional Euclidean space, represented by all orthogonal matrices with determinants of +1, and $\frac{\infty}{m}\frac{\infty}{m} = O_3$, the group represented by all 3-dimensional orthogonal matrices. The 5 continuous groups possess one unique rotation axis and they can be represented by the following objects:

∞:	rotation cone
∞/m:	cylinder, axial vector (e. g., magnetic field vector)
∞m:	cone, polar vector (e. g., electric polarization)
$\infty 2$:	cylindrical screw of infinite length
$\frac{\infty}{m}\frac{2}{m}$:	cylinder

4.4.7 The 11 Laue classes

The symmetry center (inversion center) plays a particular role among all the symmetry elements. All the physical properties of a centrosymmetric crystal are represented by even functions, which makes them mathematically easy to handle. However, the presence of a center of symmetry may be difficult to establish experimentally, as many anisotropic properties are centrosymmetric even though the crystal does not possess a center of symmetry (for instance, electrical and thermal conductivity, elasticity, etc.). Diffraction of X-rays in general produces a centrosymmetric pattern even in the absence of a center of symmetry (*Friedel law*).

A Laue class comprises all crystalline classes (point groups) that are not distinguishable by methods that do not depend on a symmetry center. The groups belonging to a particular Laue class therefore distinguish themselves by the presence or absence of a symmetry center. Therefore, the Laue classes are identified by the symbols of centrosymmetric groups corresponding to

$$\bar{1}, 2/m, mmm, \bar{3}, \bar{3}m, 4/m, 4/mmm, 6/m, 6/mmm, m\bar{3}, m\bar{3}m$$

Thus, the Laue classes represent a classification of the crystalline classes.

4.4.8 The 7 crystal systems

The *crystal systems* represent another classification of the crystal classes. The relationship between the lattice that is formed by translations and the symmetry elements has been established previously. In particular, the presence of rotation and roto-inversion axes implies a particular metric of the associated lattice, and that the rotation and roto-inversion axes are parallel to translations and perpendicular to lattice planes. For example, if a crystal belongs to the crystal class 2, one can choose the base vectors of the lattice **a**, **b**, **c** with two right angles: with a particular choice of the vector **b** parallel to the 2-fold axis and **a** and **c** parallel to the two translation vectors perpendicular to **b** and belonging to the lattice, two angles are constrained: $\alpha = \gamma = 90°$. The class 2 does not generally allow to choose a unit cell with a higher symmetry; it is therefore a maximal symmetry base. For a crystal of the class *m*, it is possible to choose an analogous base with **b** perpendicular and **a** and **c** parallel to the reflection plane. The maximal symmetry of the unit cell base is easily derived for the other crystalline classes.

> *A crystalline system includes all the crystal classes that have the same maximal symmetry of the lattice.*

Therefore, each crystalline system corresponds to a particular (*canonical*) set of base vectors **a**, **b**, **c**. The unit cell of the structure that results from this choice can either be simple (primitive) or contain multiple units (centered). A centered unit cell is chosen when the primitive unit cell does not reflect the underlying symmetry of the crystalline system.

In the example above, it is the presence of the two-fold axis or the mirror plane that permits the choice of the unit cell with two right angles. Therefore, a crystal with this particular metric does not necessarily possess one of these symmetries: the unit cell metric could be due to an accidental value of the angles of 90°. However, such a unit cell with an accidental metric will only show this particular metric at a given temperature and pressure, whereas a unit cell metric determined by the underlying symmetry will be stable and not depend on the ambient conditions unless a phase transition occurs. It is thus important to *distinguish between the metric imposed by symmetry* and the *accidental metric* that may indicate the presence of a symmetry element that does not exist in the structure.

The crystal systems classify symmetry groups. They are not a classification of the different types of unit cell metrics, as only the underlying symmetry determines the unit cell.

Table 4.10 lists the seven crystal classes. The following remarks pertain to the different symmetries:

- 2/*m* defines *a single direction of 2-fold symmetry*, since the plane normal of a mirror plane (axis $\bar{2}$) coincides with a 2-fold axis 2; 2/*m* admits the same metric type as *m* and 2.

Table 4.10: The 7 crystalline systems.

Name	Definition	Metric constraints
triclinic	classes of 1 and $\bar{1}$	none
monoclinic	one 2-fold axis 2 or $\bar{2}$	$\alpha = \gamma = 90°$ (*b*-axis unique)
orthorhombic	three perpendicular 2-fold axes	$\alpha = \beta = \gamma = 90°$
tetragonal	one 4-fold axis 4 or $\bar{4}$	$a = b$, $\alpha = \beta = \gamma = 90°$
trigonal	one 3-fold axis 3 or $\bar{3}$	$a = b$, $\alpha = \beta = 90°$, $\gamma = 120°$ or ($a = b = c$, $\alpha = \beta = \gamma$)
hexagonal	one 6-fold axis 6 or $\bar{6}$	$a = b$, $\alpha = \beta = 90°$, $\gamma = 120°$
cubic	four 3-fold axes 3 or $\bar{3}$ (body diagonals of the cube)	$a = b = c$, $\alpha = \beta = \gamma = 90°$

– Tradition in mineralogy identifies the unique direction of a 2-fold axis 2 or the plane normal of a mirror plane *m* in the *monoclinic system* with the unit cell vector **b**, giving a monoclinic angle $\beta \neq 90°$ and $\alpha = \gamma = 90°$. In all the other noncubic cases, the *unique axis* is chosen as **c**. It is therefore allowed to use the **c**-axis as the unique direction in the monoclinic system, with $\gamma \neq 90°$. However, the literature follows the traditional nomenclature with monoclinic unique axis **b**. If the non-standard nomenclature is used, it is preferred that the full space group symbols is given, such as $P112/m$.

– The same type of unit cell is used for both the *trigonal* and *hexagonal* systems: $\mathbf{a} = \mathbf{b}, \mathbf{c}$, $\alpha = \beta = 90°$, $\gamma = 120°$; with **c** parallel to the 3-fold or 6-fold axis. It is, however, possible for the *trigonal* system to choose a unit cell with *rhombohedral* axes $a = b = c$, $\alpha = \beta = \gamma$. In this case, the axes **a**, **b**, **c** are equivalent via the 3-fold axis, but neither is parallel nor perpendicular to a symmetry element. For the *hexagonal* system, in contrast, this choice of unit cell is not advantageous. The terms *trigonal* and *rhombohedral* are sometimes used interchangeably. The term *trigonal* designates a crystal system defined by the presence of a 3-fold axis (3 or $\bar{3}$); the term *rhombohedral* designates a choice of the unit cell parameter **a**, **b**, **c** and also a Bravais lattice (see below). The origin of the term *rhombohedral* relates to the rhombohedron that is obtained when a cube is elongated or compressed along one of its body diagonals.

– In the *hexagonal* system, the angle γ is by definition $\gamma = 120°$.

– In the cubic system, the unit cell vectors **a**, **b**, **c** are parallel to the cube edges (axes 2, 4, $\bar{4}$).

4.4.9 International symbols of the point groups

The international symbols for the point groups give a list of inequivalent symmetry elements, i. e., the different classes of symmetry elements, with the symmetry elements given by $1, 2, 3, 4, 6, \bar{1}, m, \bar{3}, \bar{4}, \bar{6}, \frac{2}{m}, \frac{4}{m}, \frac{6}{m}$. The symmetry center is not explicitly mentioned as it is implied by the symbols $\bar{1}, \bar{2}, \frac{2}{m}, \frac{4}{m}$ and $\frac{6}{m}$. The symmetry elements are parallel to the directions spanning the 3-dimensional space, and are defined relative to the canonical coordinates $\mathbf{a}, \mathbf{b}, \mathbf{c}$ of the particular group that is a member of the crystalline system. For a mirror plane, the direction is given by the mirror plane normal. The international symbol of a point group is formed from one to three symmetry element symbols arranged in an order that is particular to each of the crystal systems the group belongs to, and is given in Table 4.11. If needed, the symbols can be completed to include all axes/directions by adding the place holder symbol 1 at the right place. For example, the symbol 121 denotes a monoclinic group with the **b**-axis the unique axis, and no symmetry associated with the **a**- and **c**-axis. Similarly, the symbol 112 also denotes a monoclinic group, but with the **c**-axis the unique axis, and the symbol $m11$ stands for a monoclinic group with a mirror plane perpendicular to the **a**-axis. The symbol $3m1$ indicates that the axes $\mathbf{a}, \mathbf{b}, -(\mathbf{a} + \mathbf{b})$ are normal to mirror planes m; in contrast, the symbol $31m$ indicates that the mirror planes are parallel to these axes. The distinction between 321 and 312 is made in the same way, also for $\bar{3}m1$ and $\bar{3}1m$.

Table 4.11: Order of the symmetry elements in the international point group symbol.

System	1st place	2nd place	3rd place
triclinic	1 or $\bar{1}$		
	edge	edge	edge
monoclinic	a	b	c
orthorhombhic	a	b	c
	unique axis	edges	face diagonals
tetragonal	c	a, b	$a + b, a - b$
trigonal	c	a, b	$2a + b, a + 2b, a + b$
hexagonal	c	a, b	$2a + b, a + 2b, a + b$
	cube edges	4 body diagonals	6 face diagonals
cubic	a, b, c	$a \pm b \pm c$	$a \pm b, b \pm c, a \pm c$

For the centrosymmetric groups $\frac{2}{m}\frac{2}{m}\frac{2}{m}, \frac{4}{m}\frac{2}{m}\frac{2}{m}, \frac{6}{m}\frac{2}{m}\frac{2}{m}, \bar{3}\frac{2}{m}, \frac{2}{m}\bar{3}$ and $\frac{4}{m}\bar{3}\frac{2}{m}$, the abbreviated symbols are used, mmm, $4/mmm$, $6/mmm$, $\bar{3}m$, $m\bar{3}$, and $m\bar{3}m$, respectively. In each of these symbols, the 2-fold axis symbol is omitted, as is the 4-fold axis symbol in the case of $m\bar{3}m$, and only the mirror planes that are perpendicular to these 2-fold axes are indicated. The symmetry elements given in the abbreviated symbol imply the presence of these 2-fold axes.

The international symbol thus provides all the symmetry information and the crystal system of the group can be deduced:

– triclinic: 1 or $\bar{1}$
– monoclinic: one symbol 2 or m, no axis of higher order
– orthorhombic: three symbols 2 or m, no axes of higher order
– trigonal: the symbol starts with 3 or $\bar{3}$
– tetragonal: the symbol starts with 4 or $\bar{4}$
– hexagonal: the symbol starts with 6 or $\bar{6}$
– cubic: the symbol 3 or $\bar{3}$ occupies the second place

Table 4.12 gives the 32 international as well as the Schoenflies symbols for the 32 crystal systems.

4.4.10 The Schoenflies symbols

The Schoenflies symbols use rotations and roto-reflections as the primary symmetry elements. Noncubic enantiomorphic groups based on rotations are labeled by the letters C and D. The label C_x designates a *cyclic group of order x*, and the label D_x represents a *dihedral group* with an axis of order x, and x binary axes perpendicular to it. Other noncubic groups are obtained by adding mirror planes to the enantiomorphic groups. These mirror planes are indicated in the Schoenflies symbol by the letter h (*horizontal*, perpendicular to the primary axis), and the letter v (*vertical*, parallel to the primary axis), or the letter d (*diagonal*, between the 2-fold axes of a dihedral group). If a group comprises both horizontal and vertical mirror planes, then the symbol h is used. The mirror symmetry is designated as C_s (s = *spiegel (mirror)*) instead of S_1 or m, C_{3h} for S_3 or $\bar{6}$, C_i for the inversion (i) instead of S_2 or $\bar{1}$, and C_{3i} for S_6 or $\bar{3}$. By analogy, with $\bar{6} = C_{3h}$, $\bar{6}m2$ becomes D_{3h}. The cubic groups are labeled T (*tetrahedral*) or O (*octahedral*), with the letters h and d added. The Schoenflies symbols are also shown in Table 4.12.

Table 4.12: General, international (Hermann-Mauguin, left) and Schoenflies (right) symbols for the 32 crystal systems. The cubic groups are listed separately at the bottom.

general	Triclinic	Monoclinic	orthorhombic	Tetragonal	Trigonal	Hexagonal
x	$1, C_1$	$2, C_2$		$4, C_4$	$3, C_3$	$6, C_6$
$\bar{x}$	$\bar{1}, C_i$	$m = \bar{2}, C_s$		$\bar{4}, S_4$	$\bar{3}, C_{3i}$	$\bar{6} = 3/m, C_{3h}$
x/m		$2/m, C_{2h}$		$4/m, C_{4h}$		$6/m, C_{6h}$
$x2$			$222, D_2$	$422, D_4$	$32, D_3$	$622, D_6$
xm			$mm2, C_{2v}$	$4mm, C_{4v}$	$3m, C_{3v}$	$6mm, C_{6v}$
$\bar{x}m$				$\bar{4}2m, D_{2d}$	$\bar{3}m, D_{3d}$	$\bar{6}m2, D_{3h}$
x/mmm			mmm, D_{2h}	$4/mmm, D_{4h}$		$6/mmm, D_{6h}$
cubic	$23, T$	$m\bar{3}, T_h$	$\bar{4}3m, T_d$	$432, O$	$m\bar{3}m, O_h$	

4.5 Classification of lattices

4.5.1 The 14 Bravais lattices

The Bravais lattices (or Bravais classes) represent one *classification of lattices accord-ing to their metric* that is imposed by the symmetry (derived by A. Bravais, *Journal de l'Ecole Polytechnique 19, 1–128*, Paris, 1850). The smallest volume (plane) that, when translated, produces the lattice, will contain one lattice point. The choice of such a unit cell does not necessarily reflect the symmetry that imposes a certain metric on the unit cell. An example in two dimensions is given in Figure 4.13.

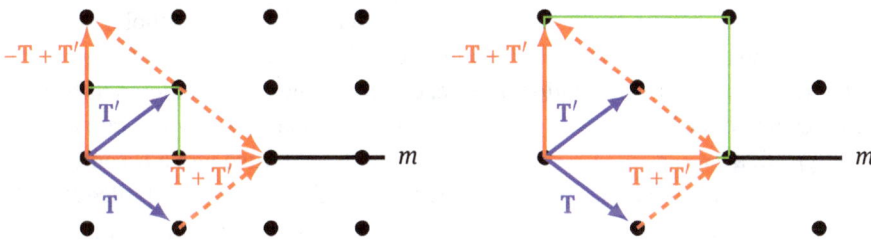

Figure 4.13: Primitive and centered rectangular lattice.

Figure 4.13 shows a 2-dimensional primitive and a centered unit cell, with m represent-ing a reflection line, with $\mathbf{T}$ a primitive translation and $\mathbf{T}'$ representing its mirror image. A translation $\mathbf{T}$ is primitive if the translation $\mathbf{T}/2$ is not a valid translation. Therefore, $\mathbf{T} + \mathbf{T}'$ and $-\mathbf{T} + \mathbf{T}'$ are perpendicular and define a rectangular unit cell. If both $\mathbf{T} + \mathbf{T}'$ and $\mathbf{T}' - \mathbf{T}$ are primitive translations, the unit cell is centered because at the center of the unit cell is lattice point. The primitive unit cell, the rhombus $(\mathbf{T}, \mathbf{T}')$, however, does not show the higher symmetry of the lattice. In contrast, the centered unit cell re-flects the symmetry and the metric constraints imposed by it, in this case right angles. A **Bravais class** is thus characterized by

– the metric of the unit cell imposed by the symmetry
– the type of the unit cell: primitive P, centered A, B, C, F, I, R.

The Bravais class of a lattice is defined by the metric (imposed by the symmetry) and the type of the smallest unit cell obtained by choosing canonical base vectors according to the crystal system that reflect the symmetry of the lattice.

The type of the metric is imposed by the symmetry of the crystal: a monoclinic lattice may accidentally show an orthorhombic unit cell metric (the unique angle = 90° within experimental resolution) at a given pressure and temperature, but the crystal will not possess orthorhombic symmetry.

As shown in Figure 4.13, a rectangular lattice of type c cannot by transformed into a rectangular lattice of type p, whereas an oblique lattice of type c can be transformed into a different oblique lattice of type p. Capital letters are used in three dimensions, P (primitive), A (centering of the (bc)-face: $\frac{b}{2} + \frac{c}{2}$), B (centering of the (ac)-face: $\frac{a}{2} + \frac{c}{2}$), C (centering of the (ab)-face: $\frac{a}{2} + \frac{b}{2}$), F (centering of all faces), I (centering of the unit cell: $\frac{a}{2} + \frac{b}{2} + \frac{c}{2}$), R (centering of the trigonal cell: $\frac{2a}{3} + \frac{b}{3} + \frac{c}{3}$ and $\frac{a}{3} + \frac{2b}{3} + \frac{2c}{3}$, obverse setting) for the 14 3-dimensional lattices. The traditional unit cells of the 14 3-dimensional Bravais lattices are shown in Figure 4.14. Each of these unit cells represent a class of lattices.

The following remarks refer to these Bravais lattices:

– *Monoclinic lattices*: Figure 4.14 shows the traditional orientation with **b** the unique axis (international symbol: $1\,2/m\,1$). In the case of the centered lattice C, the rectangular plane $(hk0)$ is centered. The related primitive cell would therefore be triclinic and will not reflect the higher symmetry. The monoclinic lattices of type A, C, I and F all belong to the monoclinic class denoted by the symbol C since it is possible to transform these lattices into each other without changing the unit cell metric type: the transformation $\mathbf{a}' = -\mathbf{c}$, $\mathbf{c}' = \mathbf{a}$ changes an A-centered lattice into a C centered lattice; $\mathbf{a}' = \mathbf{a} + \mathbf{c}$, $\mathbf{c}' = \mathbf{c}$ changes the a lattice of type I into a lattice of type C; $\mathbf{a}' = \mathbf{a}$, $\mathbf{c} = \frac{1}{2}(\mathbf{a} + \mathbf{c})$ changes an F-centered lattice into a C-centered lattice. Obviously, the B-centered lattice is equivalent to a primitive lattice P. If the alternate orientation with unique axis **c** is used (international symbol $11\,2/m$), it is customary to use the two monoclinic lattice types P and B.

– *Orthorhombic lattices*: The lattices of type A and B are equivalent to a C-centered lattice.

– *Tetragonal lattices*: The C-centered lattice is equivalent to a primitive lattice P, and the F-centered lattice is equivalent to the I-centered lattice. The 4-fold axis forbids the presence of an A- and B-centered lattice.

– *Hexagonal and rhombohedral lattices*: The classification of these lattices does not coincide with the classification of the crystal systems. The *hexagonal lattice P* is compatible with *all trigonal and hexgonal groups*. A hexagonal prism does not represent a unit cell, since a unit cell has to be a parallelepiped, with the angle $\gamma = 120°$. The R Lattice is only allowed for trigonal groups (systems). A projection of the unit cell along its 3-fold axis allows to see the relationship with the hexagonal unit cell, as shown in Figure 4.15. One can choose a centered hexagonal cell, with centering at $(\frac{2}{3}\frac{1}{3}\frac{1}{3})$ and $(\frac{1}{3}\frac{2}{3}\frac{2}{3})$ with the translations $\mathbf{a}_h$, $\mathbf{b}_h$, $\mathbf{c}_h$ with $\mathbf{c}_h$ parallel to the 3-fold axis), or a primitive rhombohedral unit cell with the following symmetrically equivalent translations:

$$\mathbf{a}_r = \frac{1}{3}(2\mathbf{a}_h + \mathbf{b}_h + \mathbf{c}_h), \ \mathbf{b}_r = \frac{1}{3}(-\mathbf{a}_h + \mathbf{b}_h + \mathbf{c}_h), \ \mathbf{c}_r = \frac{1}{3}(-\mathbf{a}_h - 2\mathbf{b}_h + \mathbf{c}_h). \quad (4.11)$$

The hexagonal unit cell is usually chosen since its axes are aligned along high symmetry directions.

The hexagonal primitive lattice P is compatible with two different crystal systems, trigonal and hexagonal, and the two systems should not be confused.

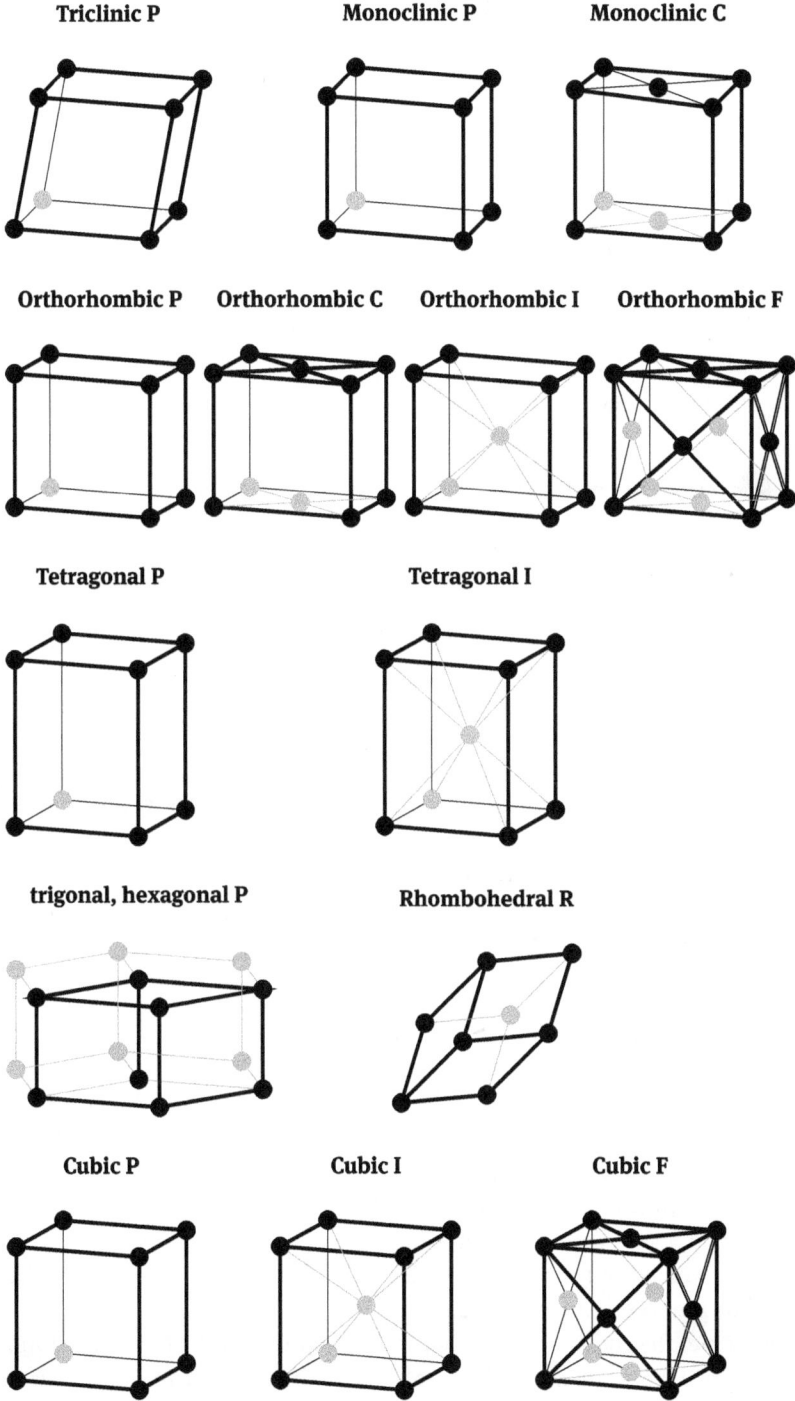

Figure 4.14: The 14 Bravais lattices.

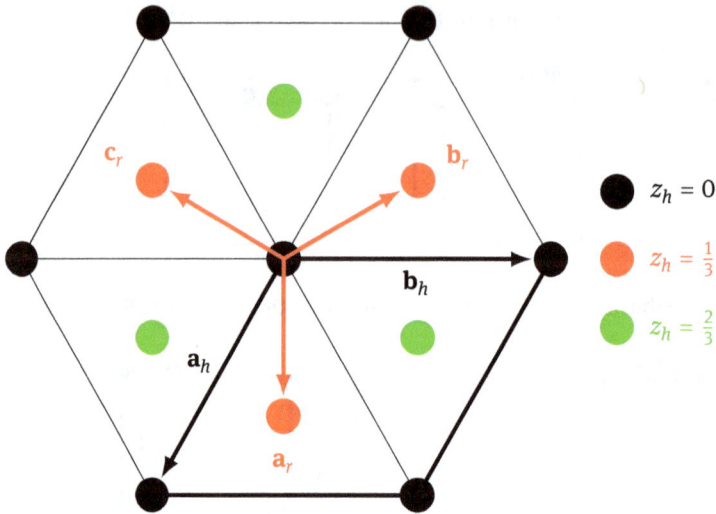

Figure 4.15: R lattice unit cells: Hexagonal and rhombohedral.

The International Union of Crystallography recommends the following:
- The 32 crystal classes are divided into seven *crystal systems*:
  triclinic, monoclinic, orthorhombic, tetragonal, trigonal, hexagonal, cubic.
- The 14 Bravais lattices (classes) are divided into seven *Bravais systems*:
  triclinic, monoclinic, orthorhombic, tetragonal, hexagonal, rhombohedral, cubic.
- It is possible to define six *crystal families* that group the crystal classes and Bravais lattices:
  triclinic, monoclinic, orthorhombic, tetragonal, hexagonal, cubic.
- The term *trigonal* identifies an ensemble of symmetry groups, whereas the term *rhombohedral* identifies a lattice type.

4.5.2 Holohedry and merohedry

The Bravais lattices, as geometrical objects, have a point symmetry, which may be different from the point symmetry of a structure. For instance, a structure with point group symmetry $4/m$ has a tetragonal Bravais lattice that itself has point group symmetry $4/mmm$. This leads to the following definition of holohedry: The point group of a crystal is called *holohedry* if it is identical to the point group of its lattice. This is therefore the maximal symmetry of a Bravais lattice: the holohedry represents the point group of the unit cell containing the lattice points only.

Definition of merohedry: The point group of the unit cell content (the structure) is called *merohedry* if it is a subgroup of the holohedry. The presence of a higher symmetry of the Bravais lattice than the structure itself can lead to merohedral twinning, where the twinned crystal consists of two or more intergrown crystals related to each other by well-defined twin laws: these different orientations are related via symmetry operations that are not part of the crystal class of the untwinned crystal, via rotations around a translation vector [*uvw*] or via a mirror reflection by a lattice plane (*hkl*). An example is the twinning of a crystal with space group $I4/m$, with the twin operation a 180° rotation around [110], resulting in an apparent higher symmetry of $I4/mmm$.

The trigonal groups are simultaneous holohedral and rhombohedral merohedries as well as hexagonal merohedries. They are simultaneous subgroups of $m\bar{3}m$ and $6/mmm$: a trigonal deformation of a cubic structure (elongation or compression along the body-diagonal [111] gives an *R*-lattice); the trigonal deformation of a hexagonal lattice results in a hexagonal *P*-lattice. A merohedral group of order 1/2 of its holohedral group is called *hemihedral*, it is called *tetartohedral* if the order is 1/4, and *ogdohedral* if the order is 1/8. Table 4.13 lists the classification of the holohedry and merohedry for the different crystal systems.

Table 4.13: Classification of crystal classes in holohedries and merohedries.

Bravais system	Holohedry	Hemihedry	Tetartohedry	Ogdohedry
triclinic	$\bar{1}$	1		
monoclinic	$2/m$	2, *m*		
orthorhombic	*mmm*	222, *mm*2		
tetragonal	$4/mmm$	422, 4*mm*, $\bar{4}2m$, $4/m$	4, $\bar{4}$	
rhombohedral	$\bar{3}m$	32, 3*m*, $\bar{3}$	3	
hexagonal	$6/mmm$	622, 6*mm*, $\bar{6}m2$, $6/m$	6, $\bar{6}$	
		$\bar{3}m$	32, 3*m*, $\bar{3}$	3
cubic	$m\bar{3}m$	432, $\bar{4}3m$, $m\bar{3}$	23	

4.6 Symmetry of periodic structures

4.6.1 The 17 planar groups

The planar groups (2-dimensional groups) will be described briefly. In the Euclidean plane, the following symmetry elements are combined:
- rotations around a point in the plane: 1, 2, 3, 4, 6-fold rotations
- reflection line *m*
- translations
- reflection line with translation (glide, parallel to the line), given by *g*

The combination of these symmetry elements give:
- 10 crystal classes: 1, 2, 3, 4, 6, *m*, 2*mm*, 3*m*, 4*mm*, 6*mm* (the international symbol is interpreted similarly to the 3-dimensional groups, with the exception that the first position gives the order of the rotation).
- 4 crystal systems: oblique, rectangular, square, hexagonal

- 5 Bravais lattices: oblique P (2), rectangular P and C (7), square P (3), hexagonal P (5)
- 17 planar groups

The 2-dimensional Bravais lattices are identical to crystal systems. Planar symmetry is present in many everyday objects, such as wallpaper, gift wrapping paper, wall tiling, mosaic designs, building facades, etc. Graphical representations of the planar groups are given in Figure 4.16, with the unit cell indicated, and symmetry equivalent general positions shown as filled circles.

The plane group symbols are the same as discussed before, with the glide-mirror operation denoted by the symbol g. In a periodic structure, the translations also act on the symmetry elements, producing a periodic arrangement of the symmetry elements. In addition, several classes of inequivalent symmetry elements are obtained. The following symmetry classes are found:

- at half-distance between two binary axes, a different binary axis is found. Therefore, the group $p2$ comprises 4 classes of 2-fold rotations.
- in the middle of a triangle defined by three 3-fold rotation axes that are translation equivalent, a different 3-fold axis belonging to a different class is found: the group $p3$ comprises therefore 3 classes of 3-fold rotations.
- in the middle of a square defined by four translation equivalent 4-fold rotations, one finds a 4-fold rotation axis. The group $p4$ therefore comprises two classes of 4-fold rotations; at the midpoint between two translation equivalent four-fold rotations, a 2-fold rotation is found, as a 4-fold rotation is also a 2-fold rotation.
- only one class of 6-fold rotations exist in the group $p6$, but this group also comprises one class of 3-fold rotations and one class of 2-fold rotations.
- at the mid-distance between two reflection lines that are translation equivalent, a reflection line of a different class is found in a rectangular system. The group pm therefore comprises two classes of mirrors.
- at the mid-distance between two reflection lines with glides that are translation equivalent, one finds a reflection line with a glide of a different class in the rectangular system. The group pg comprises therefore two classes of glide mirrors.
- at mid-distance between two reflection lines that are translation equivalent in the rectangular c-centered lattice, one finds a glide mirror, the group cm comprises thus one class of mirrors and one class of glide mirrors.

The planar groups are identified by their *international symbols* by the following:
- *first place*: letter p or c characterizing the type of the unit cell;
- *second place*: international symbol modified by the corresponding crystal class: the letter m is replaced by the letter g if, in the direction given by the symbol, one finds a series of glide mirrors instead of mirrors only.

The symbol of the crystal class is derived form the group symbol in the following way:

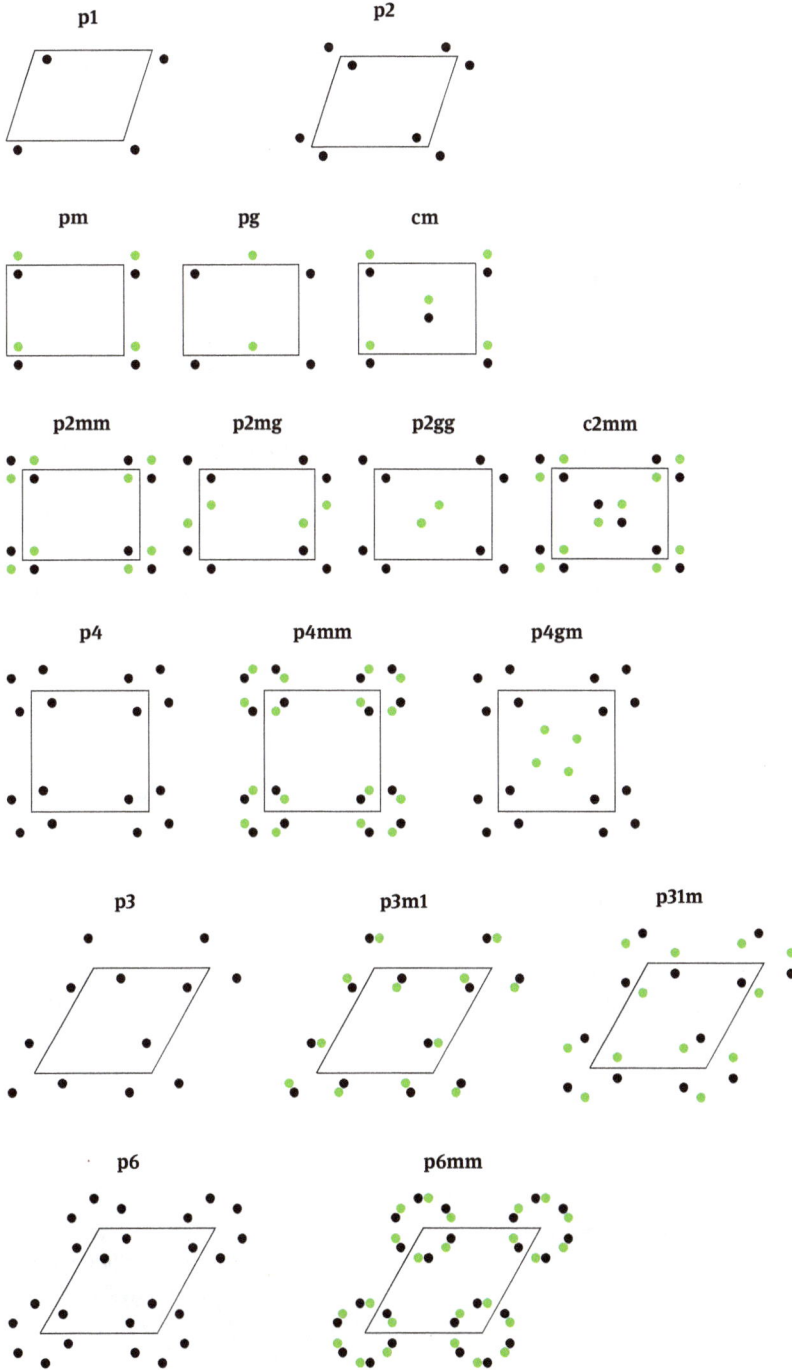

Figure 4.16: The 17 planar groups: Black circles indicate a general position and derived positions via rotations and translations, green circles derive via mirroring and/or translation.

planar group symbol	→	*crystal class symbol*
lattice type symbol (*p* or *c*)	→	remove symbol
1, 2, 3, 4, 6	→	1, 2, 3, 4, 6
m	→	*m*
g	→	*m*

The symbol of a planar group contains all the information necessary to derive the crystal class and the Bravais lattice. To derive the planar groups, the rotations and mirrors in the 10 crystal classes are systematically replaced by a series of rotations, mirrors and glide mirrors, and taking care of the Bravais lattice. From the crystal classes 1, 2, 3, 4 and 6, the groups *p*1, *p*2, *p*3, *p*4 and *p*6 are obtained. The class *m* belongs to the group *pm*, *pg* and *cm*; *cg* is an alternate symbol for *cm*, that is not used, since *cm* contains mirrors and glide mirrors. The groups belonging to 2*mm* are obtained by replacing none, one or both mirrors *m* by a glide mirror *g*; the symbol *p*2*mg* indicates that the **a**-axis is perpendicular to a series of mirrors *m*, and the **b**-axis is perpendicular to a series of glide mirrors *g*; the symbol *p*2*gm* identifies the same group, but with the **a**- and **b**-axes interchanged. The group *c*2*mm* is equivalent to *c*2*mg*, *c*2*gm* and *c*2*gg*; the latter three symbols are not used. The square unit cell is a special parallelogram, and the square groups of the class 4*mm* contain a series of mirrors *m* and glide mirrors *g* parallel to the diagonals of the square (in analogy to the group *c*2*mm*). Therefore, *p*4*mg* and *p*4*gg* are alternate symbols for *p*4*mm* and *p*4*gm*, respectively. The unit cell for the trigonal/hexagonal system is also a special case of a parallelogram, and all mirrors alternate with glide mirrors. It is important to distinguish between *p*3*m*1 and *p*31*m*: in the case of *p*3*m*1, the mirrors are perpendicular to the shortest translation directions of the lattice, whereas in *p*31*m*, the mirrors are perpendicular to the long diagonal of the unit cell. The planar groups are shown in Figure 4.16, with general positions indicated by black and green filled circles. The black circles depict a general position, and corresponding symmetry related positions derived from rotations and translations, while green circles derive from the general position via a mirror or glide mirror symmetry. Symmetry elements are not included.

4.6.2 Equivalent positions

An ensemble of points/positions that are equivalent in respect to a symmetry group is called *orbit* or *crystallographic orbit*. For example, polyhedra are orbits representing point groups. For most groups, there are different types of orbits representing *general positions* and *special positions*. The discussion will use the planar group *p*2*mg* (International Tables for Crystallography, Volume A, p. 98) as an example (Figure 4.17).

General position
An object placed within the unit cell with planar symmetry *p*2*mg* generates an infinite number of symmetry equivalent objects. In the general position (*x*, *y*), one finds four of

International Tables for Crystallography (2006). Vol. A, Plane group 7, p. 98.

$p\,2\,m\,g$ $\qquad 2mm$ $\qquad\qquad$ Rectangular

No. 7 $\qquad\qquad p\,2\,m\,g$ $\qquad\qquad$ Patterson symmetry $p\,2mm$

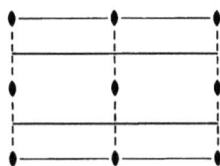

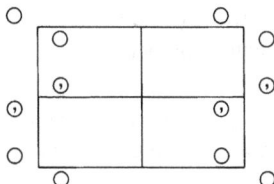

Origin at $21g$

Asymmetric unit $0 \le x \le \frac{1}{4};\ \ 0 \le y \le 1$

Symmetry operations

(1) 1 $\quad$ (2) 2 0,0 $\quad$ (3) m $\frac{1}{4},y$ $\quad$ (4) a $x,0$

Generators selected   (1); $t(1,0)$; $t(0,1)$; (2); (3)

Positions

Multiplicity, Wyckoff letter, Site symmetry		Coordinates			Reflection conditions
4	d	1	(1) x,y $\quad$ (2) $\bar{x},\bar{y}$ $\quad$ (3) $\bar{x}+\frac{1}{2},y$ $\quad$ (4) $x+\frac{1}{2},\bar{y}$		General:
					$h0:\ h=2n$
					Special: as above, plus
2	c	.m.	$\frac{1}{4},y$ $\quad$ $\frac{3}{4},\bar{y}$		no extra conditions
2	b	2..	$0,\frac{1}{2}$ $\quad$ $\frac{1}{2},\frac{1}{2}$		$hk:\ h=2n$
2	a	2..	$0,0$ $\quad$ $\frac{1}{2},0$		$hk:\ h=2n$

Maximal non-isomorphic subgroups

I $\quad$ [2] $p11g\,(pg,4)$ $\quad$ 1; 4
$\qquad$ [2] $p1m1\,(pm,3)$ $\quad$ 1; 3
$\qquad$ [2] $p211\,(p2,2)$ $\quad$ 1; 2

IIa $\quad$ none
IIb $\quad$ [2] $p2gg\,(b'=2b)\,(8)$

Maximal isomorphic subgroups of lowest index

IIc $\quad$ [2] $p2mg\,(b'=2b)\,(7)$; [3] $p2mg\,(a'=3a)\,(7)$

Minimal non-isomorphic supergroups

I $\quad$ none
II $\quad$ [2] $c2mm\,(9)$; [2] $p2mm\,(a'=\frac{1}{2}a)\,(6)$

98

Figure 4.17: Planar Group *p2mg* (reproduced with permission from the IUCR, International Tables for Crystallography, Volume A, p. 98, https://it.iucr.org/A).

these objects at the coordinates (x, y), $(-x, -y)$, $(\frac{1}{2} - x, y)$, $(\frac{1}{2} + x), -y)$; the *multiplicity* of this position is therefore 4. If any of the objects are outside the unit cell, a translation is used to bring them into the unit cell, so that the numerical values of the relative coordinates are between 0 and 1. For example, the position $(-x, -y)$ is translated to its equivalent position $(1 - x, 1 - y)$. Each of these positions indicate the location of a symmetry equivalent object in addition to those that are produced by the translations. Each of these 4 ensembles of translation equivalent points is generated by the symmetry operations of one of the symmetry classes of the normal subgroup of the translations $\mathbb{T}$. *The object itself can be of a general form, for instance, an asymmetric unit with a handedness.*

Special positions

In the example of planar group *p2mg*, an object placed on a location with $x = \frac{1}{4}$, the two coordinates (x, y) and $(\frac{1}{2} - x, y)$ will coincide. Therefore, any point with $x = \frac{1}{4}$ is located on the mirror. The crystallographic orbit comprises only two objects per unit cell that are invariant in respect to the mirror *m*. The multiplicity of this site is therefore 2, with the *site symmetry m*, indicated in the symbol ".*m*.," with the two dots indicating the positions in the international symbol, and the two coordinates $(\frac{1}{4}, y)$ and $(\frac{3}{4}, \bar{y})$. In this case, the first dot is a place holder for the 2-fold axis, the *m* indicates the symmetry associated with the **a**-axis, the perpendicular mirror, and the last dot is a placeholder for the possible symmetry associated with the **b**-axis.

There are two other special positions, both with a multiplicity of 2 and the associated symbol 2.., position $(0, 0)$, and position $(0, \frac{1}{2})$. An object occupying one of these positions is invariant in respect to a 2-fold rotation. In addition, the object that is placed on any of the special positions needs to have at least the symmetry of the particular site; the object may have a higher symmetry, but never lower than the site symmetry. For example, an object that is located on the mirror must itself have mirror symmetry, and an object located on a 2-fold axis must have 2-fold symmetry.

The letters $a, b, c, d, \ldots$ are called *Wyckoff symbols* and are used to label the positions. All the general and special positions of the planar groups are listed, together with a graphical representation and the respective arrangement of the symmetry elements are given in the *International Tables for Crystallography A*, https://it.iucr.org/A/.

4.6.3 The planar group p2mg

The entry for the planar group *p2mg* is shown in Figure 4.17 (reproduced with permission from the IUCR, International Tables for Crystallography, Volume A, p. 98, https://it.iucr.org/A). The left graphical representation has the *x*-axis vertical, the *y*-axis horizontal, and shows the placement of the symmetry elements: 2-fold rotation (●), mirror (——, horizontal) and glide mirror (–––, vertical). On the right, a general position at (x, y) is indicated by a circle, together with all the symmetry generated positions, with operations changing the handedness depicted with a comma inside the circle. With a multiplicity of 4 for the general position, there are 4 circles inside the unit cell, 2 that are related by rotation and 2 related by mirror/glide mirror symmetry operations. The location of the **origin** is at the intersection of the glide mirror line and the 2-fold rotation point: $21g$. The **asymmetric unit** is the smallest area that is needed to generate the full unit cell. In this example, the area is $\frac{1}{4}$ of the unit cell area due to the general multiplicity of 4. The content of this volume needs to be specified, as the other positions are generated by the symmetry operations. The **symmetry operations** are listed, for (1) the identity operation 1, (2) the 2-fold rotation at the origin $(0,0)$, (3) the mirror line m at $(\frac{1}{4}, y)$ and (4) the glide mirror line at $(x, 0)$, with translation by $\frac{a}{2}$, denoted as a. The **positions** are listed, with their multiplicity, Wyckoff letter, site symmetry and coordinates. Subgroups of $p2mg$ are listed further. The **maximal nonisomorphic subgroups I** are obtained by building subgroups using different symmetry operations of $p2mg$. Using symmetry operation (1) and (4) gives $p11g$, operation (1) and (3) give $p1m1$, and operation (1) and (2) gives $p211$, all of order 2. Other relationships are given that include larger or smaller unit cells.

4.7 The 230 space groups

The basics of the notation needed for the 3-dimensional space groups follows the notation for the 2-dimensional planar groups. In view of the large number of space groups, 230, only a few representative examples will be discussed in detail. The 230 space groups are compiled in the *International Tables for Crystallography, Volume A*, https://it/iucr.org/A (previous edition: Volume 1), and web-based tables can be found. (See, for instance, footnote[3].)

The rotation axis of a crystal class corresponds to a series of axes (or helical axes) of a space group, and a mirror plane (reflection plane) of a crystal class corresponds to a series of mirror planes with or without translations:

Crystal class	$\rightarrow$	Space group
1	$\rightarrow$	1
2	$\rightarrow$	$2, 2_1$
3	$\rightarrow$	$3, 3_1, 3_2$
4	$\rightarrow$	$4, 4_1, 4_2, 4_3$
6	$\rightarrow$	$6, 6_1, 6_2, 6_3, 6_4, 6_5$
m	$\rightarrow$	m, a, b, c, n, d

A description of the symmetry elements and their respective symbols has been given previously, and the space group symmetries are set up in the same way as the planar groups. The *international space group symbol* is composed in the following way:

---

3 http://img.chem.ucl.ac.uk/sgp/mainmenu.htm

- one of the letters P, A, B, C, F, I, R, indicating the unit cell type and the centering: single face centering A, B, C, face centering F, body centering I and rhombohedral centering R.
- the international symbol modified for the crystal class

As examples, space groups for the orthorhombic crystal class mmm will be discussed. This class includes space groups $Pmmm$, $Pmma$, $Pbca$, $Pnnm$, $Ccca$, $Fmmm$, $Ibca$, etc. The international symbol for the crystal class mmm denotes the three unique symmetry directions: the first m refers to a mirror plane perpendicular to the **a**-axis, the second m refers to a mirror plane perpendicular to the **b**-axis, while the last m refers to a mirror plane perpendicular to the **c**-axis. In the case of the space group $Pbca$ (# 61), the unit cell is primitive (P); normal to the **a**-axis is a glide mirror plane with translation $\frac{1}{2}\mathbf{b}$; normal to the **b**-axis is a glide mirror with translation $\frac{1}{2}\mathbf{c}$, and normal to the **c**-axis is a glide mirror with translation $\frac{1}{2}\mathbf{a}$. The full space group symbol for $Pbca$ is $P\frac{2_1}{b}\frac{2_1}{c}\frac{2_1}{a}$, with multiplicity of 8. The general positions (Wyckoff 8c) are given in Table 4.14.

Table 4.14: $Pbca$: general positions Wyckoff 8c.

x, y, z	$\bar{x}, \bar{y}, \bar{z}$
$\bar{x} + \frac{1}{2}, \bar{y}, z + \frac{1}{2}$	$x + \frac{1}{2}, y, \bar{z} + \frac{1}{2}$
$\bar{x}, y + \frac{1}{2}, \bar{z} + \frac{1}{2}$	$x, \bar{y} + \frac{1}{2}, z + \frac{1}{2}$
$x + \frac{1}{2}, \bar{y} + \frac{1}{2}, \bar{z}$	$\bar{x} + \frac{1}{2}, y + \frac{1}{2}, z$

It is possible to derive all the space groups belonging to the crystal class mmm by systematically replacing the first mirror m by m, b, c, n, the second mirror m by m, a, c, n, and the third mirror m by m, a, b, n: It is obvious that it is not possible to have a glide mirror perpendicular to the **a**-axis with the translation of $\frac{1}{2}\mathbf{a}$, and analogous, no glide mirror is possible that is perpendicular to the **b**-axis with translation $\frac{1}{2}\mathbf{b}$ and glide mirror perpendicular to **c** with translation $\frac{1}{2}\mathbf{c}$. Glide mirror planes of type d are, in the case of orthorhombic unit cells, only found in space groups of F-centered unit cells. Similarly, the centered space group $Ibca$ is derived from the primitive space group $Pbca$ by adding the centering translation $(\frac{1}{2}, \frac{1}{2}, \frac{1}{2})$. The procedure described produces valid space group symbols, but not necessarily unique space group symbols. For example, the two space groups $Pmma$ and $Pmmb$ are equivalent; with the unit cell transformation $\mathbf{a}' = \mathbf{b}$, $\mathbf{b}' = -\mathbf{a}$ and $\mathbf{c}' = \mathbf{c}$, mapping the first unit cell onto the second one. For this reason, alternate settings for the unit cell are also given in the *International Tables for Crystallography, Volume A*. The alternate setting for $Pbca$ is $Pcab$. Other examples of space group symbols are $P2_1/c, P2_12_12, P2_12_12_1, I\bar{4}c2, I\bar{4}2d, P\bar{3}1c, P6_3/mmc, Pn\bar{3}n$, etc.

From the space group symbol, the crystal class of the system, the Bravais lattice and the crystal system are derived by simply removing all translations: *The crystal*

class is obtained by removing the letter designating the unit cell type, then by replacing all helical axes by their corresponding rotation axes, and substituting any of the letters a, b, c, n, d by the symbol m for the mirror.

For example, the crystal class of the space group $I\bar{4}c2$ is $\bar{4}m2$, the alternate symbol for $\bar{4}2m$; the crystal system is tetragonal, and the Bravais lattice is body-centered tetragonal (I).

In most cases, the international space group symbol contains the necessary information to derive the ensemble of properties of the group, in particular all the crystallographic orbits (general and special positions). The space group $Pnma$ (# 62), the symbol n and a are replaced by the mirror m to give the crystal class mmm: the system is therefore orthorhombic, and the associated unit cell is primitive, giving an orthorhombic primitive Bravais lattice. Placing the glide plane $\hat{n}$ perpendicular to **a** at $x = 0$, the general position (x, y, z) is transformed to $(\bar{x}, y+\frac{1}{2}, z+\frac{1}{2})$ (this is a reflection followed by the translation by $(\frac{1}{2}\mathbf{b} + \frac{1}{2}\mathbf{c})$. The mirror perpendicular to **b** is placed a $y = 0$. This symmetry operation transforms the two positions into $(x, \bar{y}, z)$ and $(\bar{x}, \bar{y} + \frac{1}{2}, z + \frac{1}{2})$. Finally, the glide mirror a perpendicular to **c** is placed at $z = 0$. This symmetry operation transforms the four points previously obtained into $(x+\frac{1}{2}, y, \bar{z})$, $(\bar{x}+\frac{1}{2}, y+\frac{1}{2}, \bar{z}+\frac{1}{2})$, $(x+\frac{1}{2}, \bar{y}, \bar{z})$ and $(\bar{x}+\frac{1}{2}, \bar{y}+\frac{1}{2}, \bar{z}+\frac{1}{2})$. These eight points and their translation equivalent points constitute the general orbit with multiplicity 8. The complete symbol of the space group is $P\frac{2_1}{n}\frac{2_1}{m}\frac{2_1}{a}$; the group therefore includes 2-fold screw axes and inversion symmetry. The positions of the corresponding series of inversion centers, 2-fold axes and 2-fold screw axes are best found sketching the general positions. The origin of the unit cell is moved to one of the eight equivalent inversion centers. This information is found via algebraic operations using the relative coordinates obtained. The symmetry transforming the position (x, y, z) into $(\bar{x} + \frac{1}{2}, \bar{y} + \frac{1}{2}, \bar{z} + \frac{1}{2})$ by inversion requires that the inversion center is located at $(\frac{1}{4}, \frac{1}{4}, \frac{1}{4})$. The full information listed in the International Tables for Crystallography, Volume A (https://it.iucr.org/A) entry for space group $Pnma$ (# 62) is given in Figures 4.18 and 4.19, and an annotated first page of the space group $C2/c$ (# 15), monoclinic, C-centered, centrosymmetric, is given in Figure 4.20 (reproduced with permission from the IUCR (https://it.iucr.org/A)).

4.7.1 Classification of crystals according to their symmetry

Symmetry serves to classify any periodic structure, in this case a crystal, according to its symmetry. Space group, crystal class, Laue class, crystal system and Bravais lattice are all used for this classification. Each crystal therefore possesses a given lattice with corresponding unit cell metric determined by the symmetry at a given temperature and pressure. The 230 space groups are abstractions of the particular metric of the different crystals. The space groups are divided according to the macroscopic symmetry into crystal classes and crystal systems, Bravais lattices and Bravais systems. As the point group symmetry of crystals is amenable to be determined by physical measurements,

International Tables for Crystallography (2006). Vol. A, Space group 62, pp. 298–299.

$Pnma$ D_{2h}^{16} mmm Orthorhombic

No. 62 $P\,2_1/n\,2_1/m\,2_1/a$ Patterson symmetry $Pmmm$

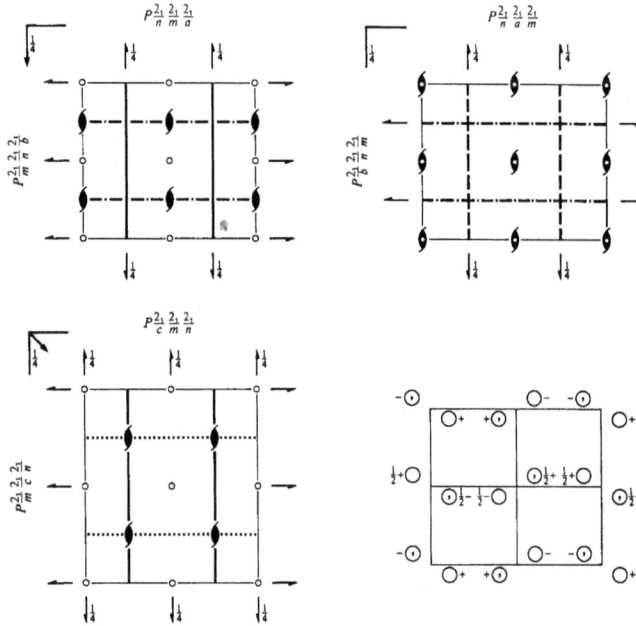

Origin at $\bar{1}$ on $12_1 1$

Asymmetric unit $0 \le x \le \frac{1}{2};\ \ 0 \le y \le \frac{1}{4};\ \ 0 \le z \le 1$

Symmetry operations

(1) 1 (2) $2(0,0,\frac{1}{2})\ \ \frac{1}{4},0,z$ (3) $2(0,\frac{1}{2},0)\ \ 0,y,0$ (4) $2(\frac{1}{2},0,0)\ \ x,\frac{1}{4},\frac{1}{4}$

(5) $\bar{1}\ \ 0,0,0$ (6) $a\ \ x,y,\frac{1}{4}$ (7) $m\ \ x,\frac{1}{4},z$ (8) $n(0,\frac{1}{2},\frac{1}{2})\ \ \frac{1}{4},y,z$

Figure 4.18: Space Group *Pmna*, page 1 (reproduced with permission from the IUCR, International Tables for Crystallography, Volume A, p. 298, https://it.iucr.org/A).

CONTINUED   No. 62   *Pnma*

Generators selected   (1); $t(1,0,0)$; $t(0,1,0)$; $t(0,0,1)$; (2); (3); (5)

Positions

Multiplicity, Wyckoff letter, Site symmetry		Coordinates				Reflection conditions
						General:
8	d	1	(1) x,y,z (2) $\bar{x}+\frac{1}{2},\bar{y},z+\frac{1}{2}$ (3) $\bar{x},y+\frac{1}{2},\bar{z}$ (4) $x+\frac{1}{2},\bar{y}+\frac{1}{2},\bar{z}+\frac{1}{2}$				$0kl$: $k+l=2n$
			(5) $\bar{x},\bar{y},\bar{z}$ (6) $x+\frac{1}{2},y,\bar{z}+\frac{1}{2}$ (7) $x,\bar{y}+\frac{1}{2},z$ (8) $\bar{x}+\frac{1}{2},y+\frac{1}{2},z+\frac{1}{2}$				$hk0$: $h=2n$
						$h00$: $h=2n$
						$0k0$: $k=2n$
						$00l$: $l=2n$
						Special: as above, plus
4	c	$.m.$	$x,\frac{1}{4},z$ $\bar{x}+\frac{1}{2},\frac{3}{4},z+\frac{1}{2}$ $\bar{x},\frac{3}{4},\bar{z}$ $x+\frac{1}{2},\frac{1}{4},\bar{z}+\frac{1}{2}$				no extra conditions
4	b	$\bar{1}$	$0,0,\frac{1}{2}$ $\frac{1}{2},0,0$ $0,\frac{1}{2},\frac{1}{2}$ $\frac{1}{2},\frac{1}{2},0$				hkl : $h+l,k=2n$
4	a	$\bar{1}$	$0,0,0$ $\frac{1}{2},0,\frac{1}{2}$ $0,\frac{1}{2},0$ $\frac{1}{2},\frac{1}{2},\frac{1}{2}$				hkl : $h+l,k=2n$

Symmetry of special projections

Along [001] $p2gm$	Along [100] $c2mm$	Along [010] $p2gg$
$\mathbf{a'}=\frac{1}{2}\mathbf{a}$ $\mathbf{b'}=\mathbf{b}$	$\mathbf{a'}=\mathbf{b}$ $\mathbf{b'}=\mathbf{c}$	$\mathbf{a'}=\mathbf{c}$ $\mathbf{b'}=\mathbf{a}$
Origin at $0,0,z$	Origin at $x,\frac{1}{4},\frac{1}{4}$	Origin at $0,y,0$

Maximal non-isomorphic subgroups

I	[2] $Pn2_1a\,(Pna2_1,33)$	1; 3; 6; 8
	[2] $Pnm2_1\,(Pmn2_1,31)$	1; 2; 7; 8
	[2] $P2_1ma\,(Pmc2_1,26)$	1; 4; 6; 7
	[2] $P2_12_12_1\,(19)$	1; 2; 3; 4
	[2] $P112_1/a\,(P2_1/c,14)$	1; 2; 5; 6
	[2] $P2_1/n11\,(P2_1/c,14)$	1; 4; 5; 8
	[2] $P12_1/m1\,(P2_1/m,11)$	1; 3; 5; 7

IIa   none
IIb   none

Maximal isomorphic subgroups of lowest index
IIc   [3] $Pnma\,(\mathbf{a'}=3\mathbf{a})\,(62)$; [3] $Pnma\,(\mathbf{b'}=3\mathbf{b})\,(62)$; [3] $Pnma\,(\mathbf{c'}=3\mathbf{c})\,(62)$

Minimal non-isomorphic supergroups
I   none
II   [2] $Amma\,(Cmcm,63)$; [2] $Bbmm\,(Cmcm,63)$; [2] $Ccme\,(Cmce,64)$; [2] $Imma\,(74)$; [2] $Pcma\,(\mathbf{b'}=\frac{1}{2}\mathbf{b})\,(Pbam,55)$; [2] $Pbma\,(\mathbf{c'}=\frac{1}{2}\mathbf{c})\,(Pbcm,57)$; [2] $Pnmm\,(\mathbf{a'}=\frac{1}{2}\mathbf{a})\,(Pmmn,59)$

Figure 4.19: ISpace Group *Pnma*, page 2 (reproduced with permission from the IUCR, International Tables for Crystallography, Volume A, p. 299, https://it.iucr.org/A).

International Tables for Crystallography (2006). Vol. A, Space group 15, pp. 192–199.

$C2/c$ C_{2h}^{6} $2/m$ Monoclinic

No. 15 $C12/c1$ Patterson symmetry $C12/m1$ **Comments:**

UNIQUE AXIS b, CELL CHOICE 1

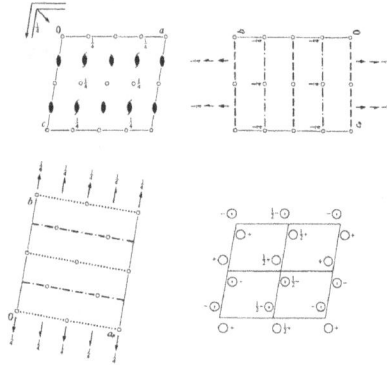

C_{2h}^{6} Schoenflies Symbol

Patterson Symmetry $C12/m1$

b-axis unique

4 projections of the unit cell:
along **b**, **a**, **c**
and crystallographic orbit

position of the origin

asymmetric unit cell:
smallest volume to define

Origin at $\bar{1}$ on glide plane c

Asymmetric unit $0 \leq x \leq \frac{1}{2}$; $0 \leq y \leq \frac{1}{2}$; $0 \leq z \leq \frac{1}{2}$

Symmetry operations

For (0,0,0)+ set
(1) 1   (2) 2 0,y,$\frac{1}{4}$ (3) $\bar{1}$ 0,0,0   (4) c x,0,z

For ($\frac{1}{2}$,$\frac{1}{2}$,0)+ set
(1) $t(\frac{1}{2},\frac{1}{2},0)$ (2) 2(0,$\frac{1}{2}$,0) $\frac{1}{4}$,y,$\frac{1}{4}$ (3) $\bar{1}$ $\frac{1}{4}$,$\frac{1}{4}$,0   (4) $n(\frac{1}{2}$,0,$\frac{1}{2})$ x,$\frac{1}{4}$,z

list of symmetry operations
the cell is centered
two positions $(0,0,0)$ and $(\frac{1}{2},\frac{1}{2},0)$

(1) identity
(2) two-fold axis
(3) inversion
(4) c glide mirror

translation (C-centering)

192

Figure 4.20: *Space Group C2/c* (reproduced with permission from the IUCR, International Tables for Crystallography, Volume A, p. 192, https://it.iucr.org/A).

crystal classification according to symmetry is convenient, and provides a great deal of information.

4.8 Crystal structures

Crystal structures, periodic systems in three dimensions, can now be described in a very condensed "short hand" way. If a symmetry is determined or given, the unit cell metric is known and the necessary unit cell parameters **a**, **b**, **c**, α, β, γ need to be listed. This unit cell needs to be filled with atoms, but instead of listing the position of every atom in the unit cell, only the subset of atoms that are not related by symmetry need to be listed, as all other atom positions are generated by the symmetry. This creates a very concise way of describing any periodic system.

With the unit cell and the stoichiometry of a crystal known, the number of formula units in a unit cell Z can be calculated if the density of the crystal is measured/known.

If M is the molecular mass of the formula unit of the substance and V the unit cell volume as determined by diffraction methods, the density ρ in $[g/cm^3]$ is

$$\rho_{calc} = \frac{ZM}{AV_{unit\ cell}}; \quad Z = \frac{\rho AV_{unit\ cell}}{M}; \quad M = \sum_{formula} M_A;$$

where Z is the number of formula units in the unit cell, $A = 6.022 \times 10^{23} mole^{-1}$ the Avogadro number, M_A the formula unit mass in grams per mole and $V_{unit\ cell}$ the volume of the unit cell in $cm^{-3} = 10^{24} Å$. It is customary to give the following information when describing a crystal structure:

- Symmetry: space group (space group #, sometimes crystal class and crystal system)
- Lattice constants; formula units per unit cell;
- Coordinates of symmetry inequivalent atoms

Two examples will be given: the crystal structure data of Fe_3C and of $PrCl_3$. In both cases, the minimum information is provided, and the unit cell content will be generated. Sketching the structure can then be accomplished and computer based visualization tools may be used to render the structure.

Example 1: Fe_3C
The structure parameters of Fe_3C are given in Table 4.15.
Crystal cystem: orthorhombic, Space Group: *Pnma*
Unit cell: **a** = 5.08 Å, **b** = 6.73 Å, **c** = 4.51 Å, Z = 4

Table 4.15: Fe_3C.

Fe 1 at	4c	x = 0.040;		z = 0.667;
Fe 2 at	8d	x = 0.183;	y = 0.065;	z = 0.167;
C 1 at	4c	x = 0.36;		z = 0.47;

The space group symbol *Pnma* (Figures 4.18 and 4.19) indicates that the crystal class is *mmm* and that the crystal system is orthorhombic. Therefore, the angles α, β, γ are all 90° and do not need to be listed. The unit cell contains 4 formula units, resulting in a total of 12 Fe atoms and 4 C atoms. The *International Tables for Crystallography, Volume A* show that the multiplicity of the general position is 8 (Wyckoff symbol 8d), and that there are 3 positions with multiplicity of 4 (Wyckoff positions 4a, 4b, 4c). The 12 Fe atoms can be distributed in the following way: The Fe atoms must occupy either the general position (multiplicity 8) and one special position (multiplicity 4), or all 3 of the special positions with multiplicity of 4. The carbon atoms (C) will be found in one of the positions with a multiplicity of 4. The coordinates given in Table 4.15 show that the Fe atoms occupy the position 4c (Fe 1) and 8d (Fe2), and the C atoms occupy the 4c position, but with different coordinates than the Fe 1 atom. With this information, the unit cell content can be drawn to visualize the structure, and is shown in Figure 4.21.

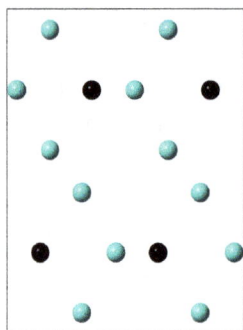

View of Fe_3C, projection along the **c**-axis

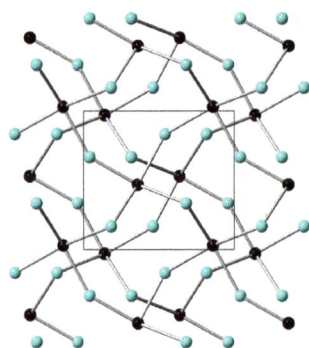

View of Fe_3C, projection along the **b**-axis

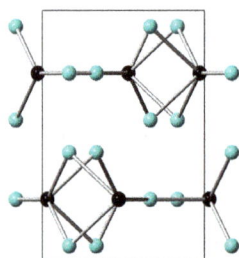

View of Fe_3C, projection along the **a**-axis

Figure 4.21: Three views of Fe_3C (Fe atoms in blue, C atoms in black), projection along the **c**-axis, no bonds (top), projection along the **b**-axis with bonds (middle), and projection along **a**-axis with bonds (bottom). The figures were generated using Crystalmaker®.

Example 2: $PrCl_3$

The crystal structure of the rare earth salt $PrCl_3$ was initially determined by W. H. Zachariasan [5]. The compound $PrCl_3$ crystallizes in the space group $P6/m$ (# 176), $Z = 2$, and lattice parameters were given as $a = 7.41\text{Å}$ and $c = 4.25\text{Å}$. A more recent structure determination using single crystals of $PrCl_3$ gave lattice parameter values of $a = 7.4208(2)\text{Å}$ and $c = 4.2787(1)\text{Å}$, values within the expected error range. The praseodymium atom is located in the 2d Wyckoff position at $(\frac{2}{3}, \frac{1}{3}, \frac{1}{4})$, and the Cl atom resides in the 6h Wyckoff position $(x, y, \frac{1}{4})$, with $x = 0.3879(1)$ and $y = 0.3015(1)$. The space group as listed in the *International Tables for Crystallography A, pages 564–565* is shown in Figures 4.22 and 4.23, and the structure is shown using a projection approximately along the 6-fold axis (Figure 4.24). The atomic positions generated by the symmetry are listed in Table 4.16.

Table 4.16: $PrCl_3$: Atomic position in the unit cell.

Atom	Wyckoff	x	y	z
Pr 1	2d	2/3	1/3	1/4
Pr 2		1/3	2/3	1/4
Cl 1	6h	0.3879	0.3015	1/4
Cl 2		0.6985	0.0864	1/4
Cl 3		0.9136	0.6121	1/4
Cl 4		0.6121	0.6985	3/4
Cl 5		0.3015	0.9136	3/4
Cl 6		0.0864	0.3879	3/4

The praseodymium atom coordination by chlorine atoms is trigonal prismatic, with the prism sides capped, giving a coordination number of 9. There are six Pr–Cl distances of 2.903Å forming the trigonal prism, and three Pr–Cl distances of 2.933Å for the capping chlorine atoms. The capping chlorine atoms are part of the trigonal prism of the adjacent praseodymium atoms. The tri-capped trigonal prisms share faces and form staggered columns along the c-axis, and the columns arrange in a honey-comb lattice forming channels along the c-axis. In Figure 4.24, a ball-and-stick model with bonds indicated is shown on top, and the unit cell drawn in the center. The figure below shows the praseodymium coordination polyhedra and their connections. Depending on the structural features that need to be emphasized, different structure representations may be chosen. The polyhedral representation emphasizes the coordination polyhedron as the structural building block, whereas the ball-and-stick model emphasizes the bonding. Structure drawing were made using the program CrystalMaker® [6].

International Tables for Crystallography (2006). Vol. A, Space group 176, pp. 564–565.

$P6_3/m$ $\qquad$ C_{6h}^2 $\qquad\qquad$ $6/m$ $\qquad\qquad$ Hexagonal

No. 176 $\qquad\qquad$ $P6_3/m$ $\qquad\qquad\qquad$ Patterson symmetry $P6/m$

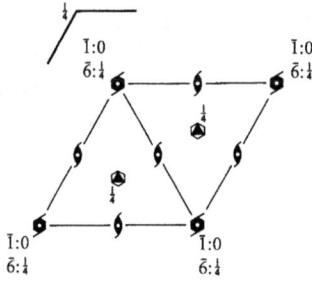

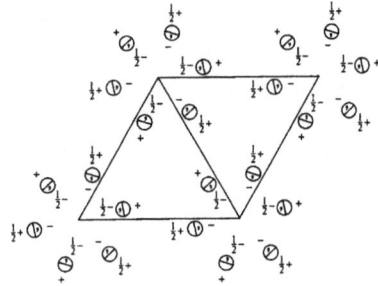

Origin at centre ($\bar{3}$) on 6_3

Asymmetric unit $\quad 0 \le x \le \frac{2}{3}; \quad 0 \le y \le \frac{2}{3}; \quad 0 \le z \le \frac{1}{4}; \quad x \le (1+y)/2; \quad y \le \min(1-x,(1+x)/2)$

Vertices $\quad 0,0,0 \quad \frac{1}{2},0,0 \quad \frac{2}{3},\frac{1}{3},0 \quad \frac{1}{3},\frac{2}{3},0 \quad 0,\frac{1}{2},0$

$\qquad\qquad 0,0,\frac{1}{4} \quad \frac{1}{2},0,\frac{1}{4} \quad \frac{2}{3},\frac{1}{3},\frac{1}{4} \quad \frac{1}{3},\frac{2}{3},\frac{1}{4} \quad 0,\frac{1}{2},\frac{1}{4}$

Symmetry operations

(1) 1 $\qquad\qquad$ (2) 3^+ $0,0,z$ $\qquad\qquad$ (3) 3^- $0,0,z$
(4) $2(0,0,\frac{1}{2})$ $\;0,0,z$ $\qquad$ (5) $6^-(0,0,\frac{1}{2})$ $\;0,0,z$ $\qquad$ (6) $6^+(0,0,\frac{1}{2})$ $\;0,0,z$
(7) $\bar{1}$ $\;0,0,0$ $\qquad\qquad$ (8) $\bar{3}^+$ $0,0,z;\; 0,0,0$ $\qquad$ (9) $\bar{3}^-$ $0,0,z;\; 0,0,0$
(10) m $\;x,y,\frac{1}{4}$ $\qquad\quad$ (11) $\bar{6}^-$ $0,0,z;\; 0,0,\frac{1}{4}$ $\qquad$ (12) $\bar{6}^+$ $0,0,z;\; 0,0,\frac{1}{4}$

Figure 4.22: Space Group $P6_3/m$, page 1 (reproduced with permission from the IUCR, International Tables for Crystallography, Volume A, p. 564, https://it.iucr.org/A).

CONTINUED   No. 176 $P6_3/m$

Generators selected   (1); $t(1,0,0)$; $t(0,1,0)$; $t(0,0,1)$; (2); (4); (7)

Positions

Multiplicity, Wyckoff letter, Site symmetry			Coordinates				Reflection conditions
							General:
12	i	1	(1) x,y,z (2) $\bar{y},x-y,z$ (3) $\bar{x}+y,\bar{x},z$				$000l$: $l=2n$
			(4) $\bar{x},\bar{y},z+\frac{1}{2}$ (5) $y,\bar{x}+y,z+\frac{1}{2}$ (6) $x-y,x,z+\frac{1}{2}$				
			(7) $\bar{x},\bar{y},\bar{z}$ (8) $y,\bar{x}+y,\bar{z}$ (9) $x-y,x,\bar{z}$				
			(10) $x,y,\bar{z}+\frac{1}{2}$ (11) $\bar{y},x-y,\bar{z}+\frac{1}{2}$ (12) $\bar{x}+y,\bar{x},\bar{z}+\frac{1}{2}$				

								Special: as above, plus
6	h	$m..$	$x,y,\frac{1}{4}$ $\bar{y},x-y,\frac{1}{4}$ $\bar{x}+y,\bar{x},\frac{1}{4}$ $\bar{x},\bar{y},\frac{3}{4}$ $y,\bar{x}+y,\frac{3}{4}$ $x-y,x,\frac{3}{4}$					no extra conditions
6	g	$\bar{1}$	$\frac{1}{2},0,0$ $0,\frac{1}{2},0$ $\frac{1}{2},\frac{1}{2},0$ $\frac{1}{2},0,\frac{1}{2}$ $0,\frac{1}{2},\frac{1}{2}$ $\frac{1}{2},\frac{1}{2},\frac{1}{2}$					$hkil$: $l=2n$
4	f	$3..$	$\frac{1}{3},\frac{2}{3},z$ $\frac{2}{3},\frac{1}{3},z+\frac{1}{2}$ $\frac{2}{3},\frac{1}{3},\bar{z}$ $\frac{1}{3},\frac{2}{3},\bar{z}+\frac{1}{2}$					$hkil$: $l=2n$ or $h-k=3n+1$ or $h-k=3n+2$
4	e	$3..$	$0,0,z$ $0,0,z+\frac{1}{2}$ $0,0,\bar{z}$ $0,0,\bar{z}+\frac{1}{2}$					$hkil$: $l=2n$
2	d	$\bar{6}..$	$\frac{2}{3},\frac{1}{3},\frac{1}{4}$ $\frac{1}{3},\frac{2}{3},\frac{3}{4}$					$hkil$: $l=2n$ or $h-k=3n+1$ or $h-k=3n+2$
2	c	$\bar{6}..$	$\frac{1}{3},\frac{2}{3},\frac{1}{4}$ $\frac{2}{3},\frac{1}{3},\frac{3}{4}$					$hkil$: $l=2n$ or $h-k=3n+1$ or $h-k=3n+2$
2	b	$3..$	$0,0,0$ $0,0,\frac{1}{2}$					$hkil$: $l=2n$
2	a	$\bar{6}..$	$0,0,\frac{1}{4}$ $0,0,\frac{3}{4}$					$hkil$: $l=2n$

Symmetry of special projections

Along [001] $p6$
$\mathbf{a}'=\mathbf{a}$ $\mathbf{b}'=\mathbf{b}$
Origin at $0,0,z$

Along [100] $p2gm$
$\mathbf{a}'=\frac{1}{2}(\mathbf{a}+2\mathbf{b})$ $\mathbf{b}'=\mathbf{c}$
Origin at $x,0,0$

Along [210] $p2gm$
$\mathbf{a}'=\frac{1}{2}\mathbf{b}$ $\mathbf{b}'=\mathbf{c}$
Origin at $x,\frac{1}{2}x,0$

Maximal non-isomorphic subgroups

I	[2] $P\bar{6}$ (174)	1; 2; 3; 10; 11; 12
	[2] $P6_3$ (173)	1; 2; 3; 4; 5; 6
	[2] $P\bar{3}$ (147)	1; 2; 3; 7; 8; 9
	[3] $P2_1/m$ (11)	1; 4; 7; 10
IIa	none	
IIb	none	

Maximal isomorphic subgroups of lowest index
IIc   [3] $P6_3/m$ ($\mathbf{c}'=3\mathbf{c}$) (176); [3] $H6_3/m$ ($\mathbf{a}'=3\mathbf{a},\mathbf{b}'=3\mathbf{b}$) ($P6_3/m$, 176)

Minimal non-isomorphic supergroups
I   [2] $P6_3/mcm$ (193); [2] $P6_3/mmc$ (194)
II   [2] $P6/m$ ($\mathbf{c}'=\frac{1}{2}\mathbf{c}$) (175)

Figure 4.23: Space Group $P6_3/m$, page 2 (reproduced with permission from the IUCR, International Tables for Crystallography, Volume A, p. 298, https://it.iucr.org/A).

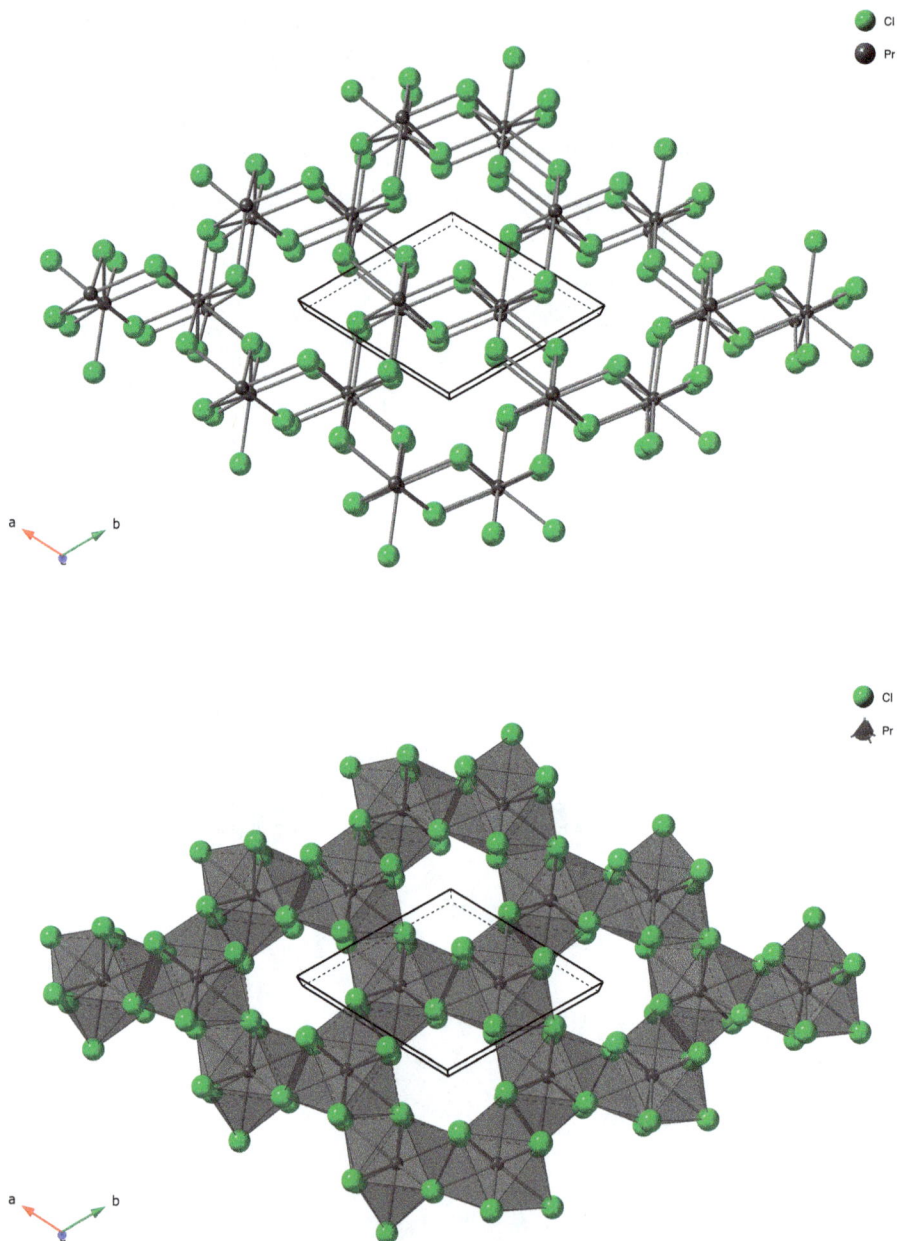

Figure 4.24: Projection of the PrCl₃ structure along the 6-fold axis. Praseodymium atoms are shown in grey, chlorine atoms in green. Upper panel: ball and stick model. Lower panel: polyhedral representation showing the interconnection of the tri-capped trigonal prisms around praseodymium. Structure figures were generated using CrystalMaker®.

4.8.1 Miller–Bravais indices for the hexagonal lattice

The indices of equivalent faces are obtained via the symmetry operations of its crystal class.

crystal class	face indices
222	: $(hkl), (h\bar{k}\bar{l}), (\bar{h}, k, \bar{l}), (\bar{h}k\bar{l})$
4	: $(hkl), (\bar{k}hl), (\bar{h}\bar{k}l), (k\bar{h}l)$
23	: $(hkl), (h\bar{k}\bar{l}), (\bar{h}k\bar{l}), (\bar{h}\bar{k}l)$
	$(klh), (k\bar{l}\bar{h}), (\bar{k}l\bar{h}), (\bar{k}\bar{l}h)$
	$(lhk), (l\bar{h}\bar{k}), (\bar{l}h\bar{k}), (\bar{l}\bar{h}k)$
	(the 3-fold axis parallel to [111] permutes the indices)

The hexagonal unit cell for a 3-fold axis presents a difficulty, with the $\mathbf{a}_1$, $\mathbf{a}_2$ and the $(-\mathbf{a}_1 - \mathbf{a}_2) = \mathbf{a}_3$ axes all being symmetry equivalent and, therefore, simple permutations and sign changes will not generate the indices for symmetry equivalent faces. For this reason, a four index notation is used, with the index i inserted at the third place to give $(h\,k\,i\,l)$. The three indices $(h\,k\,i)$ refer to the axes $\mathbf{a}_1$, $\mathbf{a}_2$ and $\mathbf{a}_3$ perpendicular to the 3-fold axis, while the index l refers to $\mathbf{c}$-axis parallel to the 3-fold (or 6-fold) axis. For $h, k > 0$, the intercept for the axis $\mathbf{a}_3$ is negative with the value:

$$i = -(h + k); \quad h + k + i = 0 \tag{4.12}$$

Since the 3-fold axis permutes the vectors $\mathbf{a}_1$, $\mathbf{a}_2$ and $\mathbf{a}_3$, the following algorithm is obtained to generate the indices of symmetry equivalent faces:

- insert the fourth index $i = -(h + k) = \overline{(h + k)}$ to the normal three-letter index (hkl)
- the indices of the symmetry equivalent faces are obtained by permutation: $(hkil)$, $(kihl)$, $(ihkl)$
- since the fourth index in the third position is redundant, it is omitted in calculations
- the three-letter indices of symmetry equivalent faces are $(h\,k\,l)$, $(k\,\overline{(h+k)}\,l)$, $(\overline{(h + k)}\,h\,l)$.

Unfortunately, this algorithm cannot be applied to the directions indices $[uvw]$ (for zones and edges). A method to define a four-index notation has been published [7], but is not often utilized. It is usually more convenient to use the three-index notation, with symmetry equivalent directions to $[uvw]$ given as $[(v - u)\,\bar{u}\,w]$ and $[\bar{v}\,(u - v)\,w]$.

Bibliography

[4] D. Schwarzenbach. Cristallographie, 1st edition. Presses polytechniques et universitaires romandes, 1996. ISBN: 2-88074-246-3.
[5] W. H. Zachariasan. J. Chem. Phys., 16:254, 1948. https://doi.org/10.1063/1.1746856.
[6] D. Palmer CrystalMakerCrystalMaker®, CrystalMaker Software Ltd., www.crystalmaker.com.
[7] L. Weber. Z. Kristallogr., 53:200–203, 1922.

5 Scattering and diffraction

5.1 Scattering of electromagnetic radiation by an electron

5.1.1 Classic scattering from an electron (Thompson scattering)

The interaction of an electromagnetic wave with an electron in matter produces two effects:

- absorption of the wave energy by changing the energy level of the electron
- scattering of the wave in different directions, one part without changing the wavelength (energy, elastic scattering, Thompson scattering), and a part where the wavelength changes to a longer wavelength (Compton scattering)

Scattering from a periodic structure (crystal, grating, etc.) is called *diffraction*.

The theory of coherent scattering of electromagnetic radiation by a free electron (Thompson scattering) needs to be treated quantum mechanically, but the classic theory is sufficient to describe the process. The free electron, treated as a classic particle with charge $-e$ and mass m is accelerated by the oscillating electric field of the electromagnetic wave. The accelerated charge will emit electromagnetic radiation, acting in fact like an antenna. The radiation pattern is the same as that from a dipole.

Figure 5.1 shows the scattering conditions where an incident wave with propagation vector $\mathbf{s_0}$ interacts with an electron. At a point P, the wave scattered by the free electron is polarized in the plane defined by the incident wave with $\mathbf{E_0}$ and $\mathbf{s}$. The amplitude is proportional to the perpendicular projection of the electron acceleration on $\mathbf{s}$:

$$\|\mathbf{E}(\phi)\| = \|\mathbf{E_0}\|\frac{1}{r} \cdot \frac{1}{4\pi\varepsilon_0}\frac{e^2}{mc^2}\sin\phi \tag{5.1}$$

with c the speed of light and ε_0 the dielectric constant of the vacuum (in SI units). Furthermore, $\|\mathbf{E}(\phi)\|$ is independent of the wavelength. The quantity $e^2/4\pi\varepsilon_0 mc^2 = d_e = 2.818 \times 10^{-15}$m is called the *electron scattering length* or the *classical electron radius*. It is assumed that the electron is a conducting sphere of diameter d_e with charge $-e$ uniformly distributed over the surface. Its electrostatic energy $e^2/4\pi\varepsilon_0$ is set equal to the energy mc^2. The amplitude $\|\mathbf{E}(\phi)\|$ is maximal in all directions $\mathbf{s}$ that are perpendicular to $\mathbf{E_0}$ and, as a consequence, $\|\mathbf{E}(\phi)\| = 0$ along the direction of $\mathbf{E_0}$.

5.1.2 Polarization factor

The discussion in the previous paragraph showed that the interaction of an incident wave with an electron has an angle dependence (equation (5.1)). This angular dependence produces a partial polarization of the total scattered intensity if unpolarized

https://doi.org/10.1515/9783110610833-005

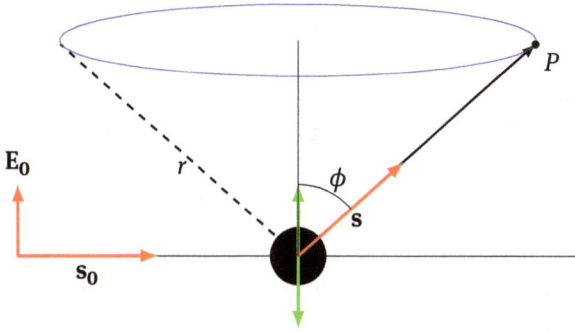

Figure 5.1: Electromagnetic wave interaction with an electron (classical).

radiation interacts with matter. To calculate the angular dependence of the partial polarization, a coordinate system has to be set up to describe the different directions. As before, the incident wave with wave vector $\mathbf{s_0}$ and amplitude $\mathbf{E_0}$ are perpendicular. The scattering plane is defined by the noncollinear vectors $\mathbf{s_0}$ and $\mathbf{s}$, the incident and scattered wave vectors. The scattered wave with wave vector $\mathbf{s}$ has its amplitude $\mathbf{E}$ perpendicular. Therefore, the coordinate system shown in Figure 5.2 will be used.

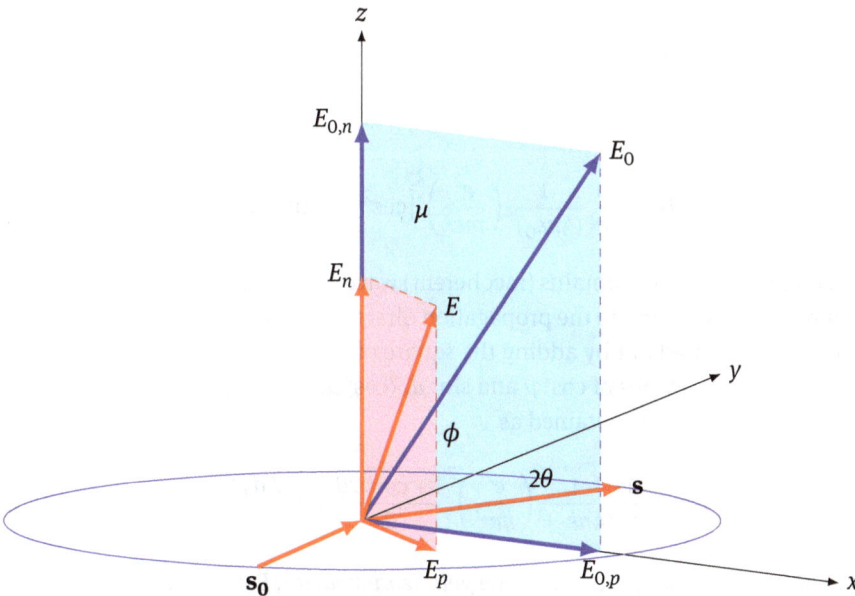

Figure 5.2: Polarization of radiation scattered by an electron.

The vector $\mathbf{E}$ is perpendicular to $\mathbf{s}$, the angle between $\mathbf{s_0}$ and $\mathbf{s}$ is 2θ, the angle between $\mathbf{E_0}$ and $\mathbf{s}$ is ϕ, the angle between $\mathbf{E_0}$ and z is μ (the polarization angle). The amplitude $\mathbf{E}$ of the scattered wave in direction $\mathbf{s}$ is given by the angle ϕ, and $\mathbf{E}$ and $\mathbf{E_0}$ are coplanar.

The amplitude of $\mathbf{E_0}$ that is perpendicular to the scattering plane ($\mathbf{s_0}$ $\mathbf{s}$) is $\mathbf{E_{0,n}}$ (parallel to z). Therefore, $\|\mathbf{E_{0,n}}\| = \|\mathbf{E_0}\| \cos \mu$, and the scattered amplitude $\|\mathbf{E_n}\|$ is

$$\|\mathbf{E_n}\| = \|\mathbf{E_{0,n}}\| \frac{1}{r} \frac{1}{4\pi\varepsilon_0} \frac{e^2}{mc^2} = \|\mathbf{E_0}\| \frac{1}{r} \frac{1}{4\pi\varepsilon_0} \frac{e^2}{mc^2} \cos \mu \tag{5.2}$$

The in-plane amplitude $\|\mathbf{E_{0,p}}\| = \|\mathbf{E_0}\| \sin \phi$ is at an angle of $\pi/2 - 2\theta$, the angle between $\mathbf{s_0}$ and $\mathbf{s}$. Thus, the amplitude $\|\mathbf{E_p}\|$ becomes

$$\|\mathbf{E_p}\| = \|\mathbf{E_{0,p}}\| \frac{1}{r} \frac{1}{4\pi\varepsilon_0} \frac{e^2}{mc^2} \cos 2\theta = \|\mathbf{E_0}\| \frac{1}{r} \frac{1}{4\pi\varepsilon_0} \frac{e^2}{mc^2} \sin \mu \cos 2\theta$$

The total scattered amplitude is the sum of the two vectors $\|\mathbf{E_n} + \mathbf{E_p}\|$, and is therefore

$$\|\mathbf{E_\mu}\| = \|\mathbf{E_n} + \mathbf{E_p}\| = \|\mathbf{E_0}\| \frac{1}{r} \frac{1}{4\pi\varepsilon_0} \frac{e^2}{mc^2} [\cos^2 \mu + \sin^2 \mu \cos^2 2\theta]^{1/2}$$

The intensity of the scattered wave is proportional to the absolute square of the amplitude, or

$$I(\phi) = I_0 \frac{1}{r^2} \frac{1}{(4\pi\varepsilon_0)^2} \left(\frac{e^2}{mc^2} \right)^2 \sin^2 \phi$$

and

$$I(\mu, 2\theta) = I_0 \frac{1}{r^2} \frac{1}{(4\pi\varepsilon_0)^2} \left(\frac{e^2}{mc^2} \right)^2 [\cos^2 \mu + \sin^2 \mu \cos^2 2\theta]$$

An unpolarized wave contains (incoherent) wave trains distributed over all possible angles μ perpendicular to the propagation direction $\mathbf{s}$. Summing up all these incoherent waves is carried out by adding the square of their respective amplitudes. Furthermore, the mean values of $\cos^2 \mu$ and $\sin^2 \mu$, $\langle \cos^2 \mu \rangle = \langle \sin^2 \mu \rangle = \frac{1}{2}$, the scattering of an unpolarized beam is obtained as

$$I_e = I_0 \frac{1}{r^2} \frac{1}{(4\pi\varepsilon_0)^2} \left(\frac{e^2}{mc^2} \right)^2 \frac{1 + \cos^2 2\theta}{2} = I_0 \left(\frac{d_e}{r} \right)^2 P \tag{5.3}$$

The factor $P = \frac{1}{2}(1 + \cos^2 2\theta)$ is called the *polarization factor*. For the case of a partially polarized primary beam, the above given derivation needs to be adjusted to account for the deviation of the averages of $\langle \cos^2 \mu \rangle \neq \langle \sin^2 \mu \rangle \neq \frac{1}{2}$. It is interesting to note that at a scattering angle of $2\theta = 90°$, the scattered wave is fully polarized. This corresponds to the *Brewster law* stating that a mirror-reflected wave is fully polarized if the incoming and reflected beams are perpendicular. Furthermore, the law of refraction states that the half-angle between incident and reflected wave is given by $\tan\theta = 1/n$, with n the index of refraction. For an electromagnetic wave traveling in air with an index of refraction of $n = 1$, it follows that $2\theta = 90°$. In addition, the reflection of light from any nonconducting surface is partially polarized, such as reflection of light off a water surface.

In the derivation of equation (5.3), the assumption was made that the electrons in an atom behave like free electrons that oscillate with the frequency of the incoming electromagnetic wave. This is a simplification, but in most cases, it is not unreasonable. If the incident wave energy is close to the energy of a resonance such as an electronic transition in the atom, then the phase of the scattered wave is also affected and the scattering power of an atom containing many electrons becomes a complex number. This energy (wavelength, frequency) dependent scattering power is called *dispersion*. For visible light, this effect is used in a glass prism to separate the different wavelengths. However, in the case of X-rays, the dispersion is often so small that it can be neglected to first order, simplifying calculations.

5.2 Optical lattice

A plane wave is assumed to propagate along the x-axis and impinges on an absorbing screen with two infinitely small holes. According to the *Huyghens principle*, the holes become sources of spherical waves that interfere with each other. At a distance that is large versus the wavelength λ of the wave and the distance d between the holes, the spherical waves can be treated as plane waves in the observation direction $\mathbf{s}$ (*Fraunhofer* approximation). The phase difference Δ of the two waves ξ_1 and ξ_2 in the direction of $\mathbf{s}$ needs to determined to obtain the amplitude. The phase difference is the projection of d onto $\mathbf{s}$:

$$\Delta = d\sin\theta = \lambda\mathbf{d}\cdot\mathbf{s}$$

The vector product $\mathbf{d}\cdot\mathbf{s}$ is given as $|d|\times|s|\times\cos(\frac{\pi}{2}-\theta)$. In the direction of the incoming wave $\mathbf{s_0}$, the calculated phase difference is zero, consistent with the fact that the vector product $\mathbf{s}\cdot\mathbf{d} = 0$. Figure 5.3 depicts the schematic of the two-hole experiment. As only the phase difference is relevant, the phase of the first spherical wave can be set to $\phi = 0$. Therefore, the resultant wave ξ will be

$$\xi = \xi_1 + \xi_2 = A(e^{2\pi i 0} + e^{2\pi i \Delta/\lambda}) = A(1 + e^{2\pi i \mathbf{d}\cdot\mathbf{s}})$$

The intensity I is proportional to the square of the amplitude and, therefore,

$$I = |\xi|^2 = 4A^2\cos^2(\pi\mathbf{d}\cdot\mathbf{s}) = 4A^2\cos^2\left(\pi\frac{d}{\lambda}\sin\theta\right)$$

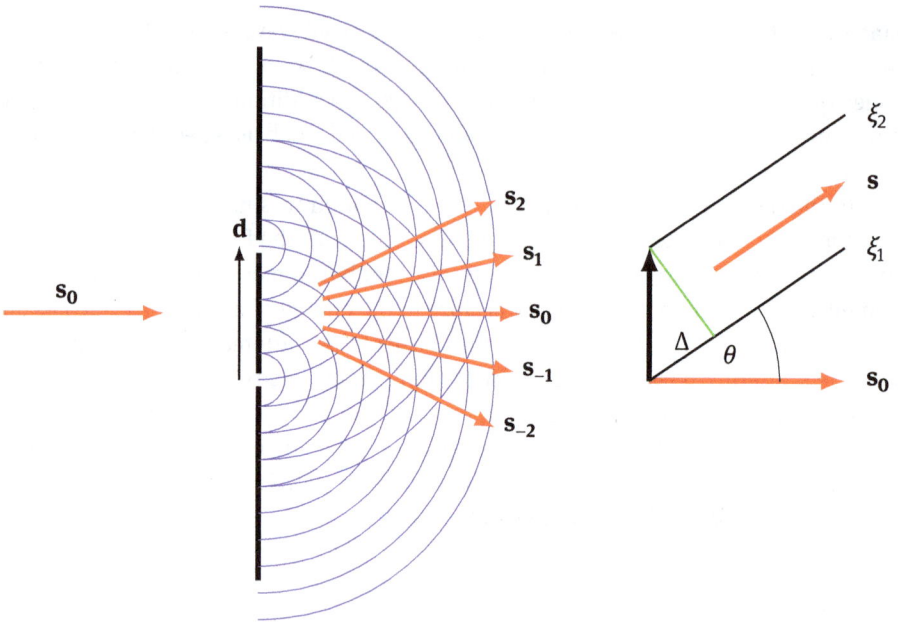

Figure 5.3: Wave propagation through two holes.

The intensity is constant for every direction **s** that is situated on a cone with axis **d**. If a screen is placed perpendicular to **s**, then diffuse bands of intensity will be observed.

This setup of two holes separated by the distance d is generalized to a series of N equidistant holes, located at points $0, \mathbf{d}, 2\mathbf{d}, 3\mathbf{d}, \ldots (N-1)\mathbf{d}$. The resulting wave is calculated in the same way, except that the sum is now over all waves, and the summation is over all the phase differences, including $\mathbf{sd}, 2\mathbf{sd}, \ldots, (N-1)\mathbf{sd}$. This is a geometric series, and the total amplitude can be expressed by

$$\xi = \sum_{n=1}^{N} \xi_n = A \sum_{n=0}^{N-1} e^{2\pi i \mathbf{sd}} = A\frac{1 - e^{2\pi i N\mathbf{sd}}}{1 - e^{2\pi i \mathbf{sd}}}$$

Consequently, the total intensity is expressed by

$$I = |\xi|^2 = \xi\xi^* = A^2\frac{1 - \cos 2\pi N\mathbf{sd}}{1 - \cos 2\pi \mathbf{sd}} = A^2\frac{\sin^2 \pi N\mathbf{sd}}{\sin^2 \mathbf{sd}} = A^2 J_N^2(\mathbf{sd})$$

with

$$J_N^2(\mathbf{sd}) = \frac{\sin^2 \pi N\mathbf{sd}}{\sin^2 \pi \mathbf{sd}}, \quad \mathbf{sd} = \frac{d}{\lambda}\sin\theta \tag{5.4}$$

The function J_N^2 is the *interference function* of a unidimensional crystal with N unit cells of periodicity **d**. This function is periodic, such that

$$J_N^2(\mathbf{sd}) = J_N^2(\mathbf{sd} + n), \quad n = \text{integer}$$

and, therefore,

$$J_N^2(\sin\theta) = J_N^2(\sin\theta'), \quad \sin\theta' = \sin\theta + n\frac{\lambda}{d} \leq 1$$

Using the rule of *Hopital*, it is shown that the function has principal maxima of height

$$\lim_{x\to 0}\frac{\sin^2 Nx}{\sin^2 x} = N\lim_{x\to 0}\frac{\sin 2Nx}{\sin 2x} = N^2\lim_{x\to 0}\frac{\cos 2Nx}{\cos 2x} = N^2$$

$$J_N^2 = N^2 \quad \text{for } \mathbf{sd} = n, \ d\sin\theta = n\lambda, \ n = \text{integer}$$

The minima are found if $J_N^2 = 0$,

$$J_N^2 = 0 \quad \text{for } \mathbf{sd} = \frac{m}{N}, \ m \neq nN, \ m, n = \text{integer}$$

Therefore, the interference function has $N-1$ minima between two principal maxima, and a total of $N-2$ secondary maxima.

The principal maxima height grows with N^2, while the number of minima increases linearly. As a consequence, the secondary maxima are reduced. In Figure 5.4, the interference function J_N^2 for $N = 11$ is plotted. The maxima have a height of $N^2 = 121$, and the secondary maxima in between the primary maxima are reduced in height. In the limit for $N \to \infty$, J_N^2 will approach a series of δ functions, and the secondary maxima vanish. Thus, the number of repeating units determines the sharpness of the principal maxima. A generalization to three dimensions is straightforward and produces a 3-dimensional array of principal maxima. For large N_x, N_y, N_z, this array will approach a 3-dimensional array of δ functions.

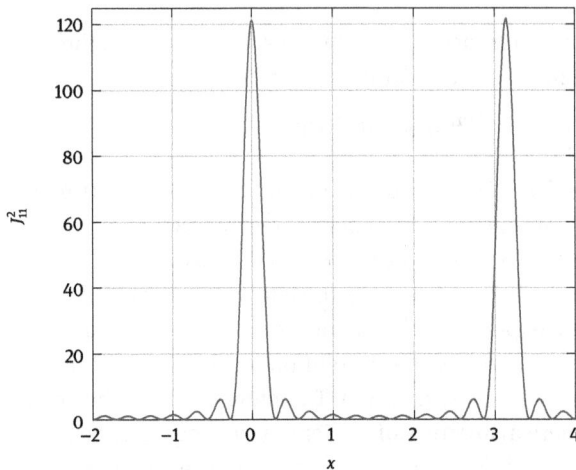

Figure 5.4: Function J_{11}^2.

5.3 Scattering of light by matter

The distribution of electrons in crystalline, amorphous, liquid or gaseous states of matter is described by an electronic density distribution function $\rho(\mathbf{r})$, with values given in units of electrons per volume, such as e/nm^3. The number of electrons in a volume element d^3r is thus $\rho(\mathbf{r})d^3r$. This function possesses maxima at the location of atoms (core electrons), and is small in between atoms (valence electrons). Additionally, this function also represents the scattering power of the volume d^3r, with the amplitude of the scattered waves proportional to the number of electrons within a volume element. To calculate the total distribution of scattered radiation, summation (integration) over all possible volume elements under consideration of the phase differences is required. This summation (integration) is straightforward, with boundaries given by the shape of the body of matter. For an incident wave with wave vector $\mathbf{s_0}$ and the direction of the scattered wave $\mathbf{s}$, two volume elements separated by the distance $\mathbf{r}$ have a relative phase difference that is given by

$$\Delta = \lambda \mathbf{r} \cdot (\mathbf{s} - \mathbf{s_0})$$

where Δ is the projection of $\mathbf{r}$ onto $\mathbf{s}$ (Figure 5.5). The wave emanating from the volume element at $\mathbf{r}$ with wave vector $\mathbf{s}$ is thus written as

$$\xi = \rho(\mathbf{r})e^{2\pi i \mathbf{r} \cdot (\mathbf{s} - \mathbf{s_0})}$$

The total diffracted wave in direction of $\mathbf{s}$ is the integral (sum) over all the waves emanating from the volume elements in the scattering body, and using $\mathbf{S} = \mathbf{s} - \mathbf{s_0}$ ($\|\mathbf{S}\| = 2\sin\theta/\lambda$, Figure 5.6) is given by

$$G(\mathbf{S}) = \int \rho(\mathbf{r})e^{2\pi i \mathbf{r} \cdot \mathbf{S}} \, d^3r = \Phi[\rho(\mathbf{r})] \qquad (5.5)$$

$G(\mathbf{S})$ (equation (5.5)) is the *Fourier transformation* Φ of $\rho(\mathbf{r})$. It follows that the electron distribution function is the inverse of the Fourier transform

$$\rho(\mathbf{r}) = \int G(\mathbf{S})e^{-2\pi i \mathbf{r} \mathbf{S}} \, d^3\mathbf{S} = \Phi^{-1}[G(\mathbf{S})] \qquad (5.6)$$

The scattered waves described by $G(\mathbf{S})$ (equation (5.5)) represent the density $\rho(\mathbf{r})$ in a mathematically rigorous way. Consequently, for a perfect reconstruction of the density $\rho(\mathbf{r})$ (equation (5.6)), all values of $\mathbf{S}$ need to be known. In the case of a microscope (optical or otherwise) relying on far-field optics where the integration is carried out by a lens system, some vectors $\mathbf{S}$ are experimentally not accessible since the maximum value of $\|\mathbf{S}\| = 2/\lambda$. This results in a reduction in resolution for the density function $\rho(\mathbf{r})$, and $\rho(\mathbf{r})$ can only be known to an approximation. Therefore, the finer the details to be imaged, the larger the numerical aperture of the lens system needs to be to collect the scattered light at high scattering angles. In general, $G(\mathbf{S})$ is a complex entity, and can be written as

$$G(\mathbf{S}) = |G(\mathbf{S})|e^{2\pi i \phi(\mathbf{S})}$$

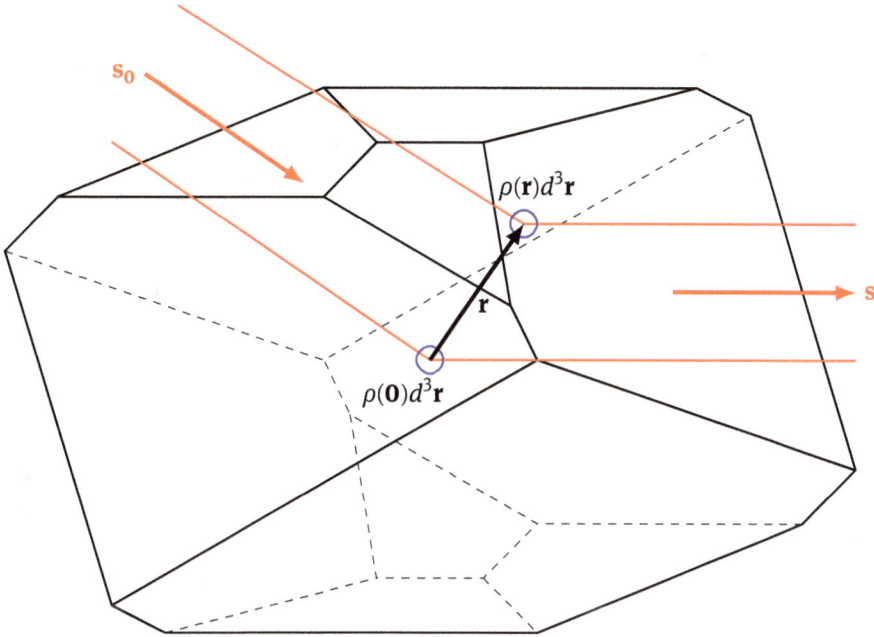

Figure 5.5: Scattering in a finite body of matter.

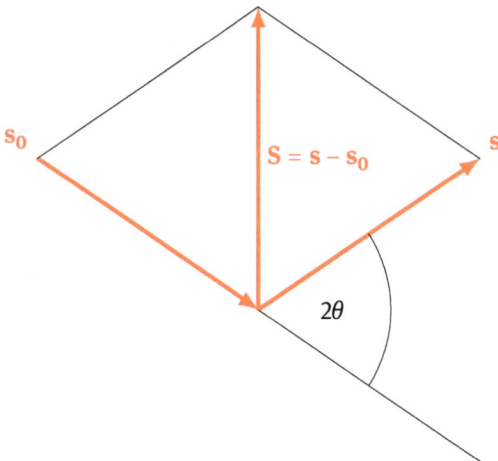

Figure 5.6: Definition of the vector $\mathbf{S} = \mathbf{s} - \mathbf{s_0}$, with $\|\mathbf{s}\| = \|\mathbf{s_0}\| = 1/\lambda$.

with $|G(\mathbf{S})|$ the amplitude and $\phi(\mathbf{S})$ the phase of the wave. Furthermore, if $\rho(\mathbf{r})$ is a real function, $G(-\mathbf{S})$ is the complex conjugate of $G(\mathbf{S})$, $G(-\mathbf{S}) = G^*(\mathbf{S})$. The diffracted intensity in direction of $\mathbf{s} = \mathbf{S} + \mathbf{s_0}$ is proportional to the amplitude squared:

$$I(\mathbf{S}) = |G(\mathbf{S})|^2$$

This relationship shows the origin of the *phase problem*, since any detector is measuring an intensity and, therefore, does not register the associated phase value. The phase information is therefore needed to determine unambiguously the scattering density $\rho(\mathbf{r})$ using the inverse Fourier transform. Unfortunately, an infinite number of scattering densities $\rho(\mathbf{r})$ do exist that give identical intensity functions $I(\mathbf{S})$. While it is always possible to calculate $|G(\mathbf{S})|$ for a known $\rho(\mathbf{r})$, the inverse is not possible without constructing models, using methods to reconstruct phase information, and using chemical knowledge about the scattering body.

Approximations

A number of approximations were implicitly made in the derivation of the mathematical description of scattering processes.

- As electrons are part of atoms, their scattering power is only approximately a function of the electron density, as the electronic energy levels are not considered. By replacing $\rho(\mathbf{r})$ with a complex function, dispersion effects can be included.
- Absorption has been neglected. Both the incident as well as the scattered waves suffer partial absorption, which can be accounted for if the exact shape of the scattering body is determined.
- The theory developed is called the *kinematical theory* and is fully elastic, as the scattered wave has the same wavelength as the incident wave (energy is redistributed, not converted). The scattering of a 3-dimensional body made up of different atoms is more complex than indicated. For instance, the incident beam is attenuated since energy is removed from this wave and redistributed into the scattered waves, and the scattered waves undergo scattering as well. Therefore, the *kinematical theory* does not obey the conservation of energy. These effects reduce the scattered intensity compared to the calculated intensity obtained using the kinematical theory. The attenuation phenomena of the scattered radiation is termed *extinction*, and is distinguished into *primary* and *secondary* extinction. The *primary extinction* includes the effects of coherent interference between waves, while *secondary extinction* describes the incoherent effects. The scattered intensities are small compared to the incident wave due to the small value of the classical electron diameter. For small scattering volumes with a small path length through the solid compared to the absorption length, the kinematical theory is an excellent approximation. This is at the heart of the success of the scattering methods in analyzing the electron distribution in solids. The exact theory, holding for perfect crystals without any defects, is called *dynamical theory*. Observation have shown that most diffraction and scattering effects of crystalline systems can be treated successfully using the kinematical theory. However, for high quality crystals such as Czochralsky grown silicon and thin film layers formed on perfect substrate crystals, the dynamical theory needs to be applied.

Taking into account the above mentioned effects, the scattered (diffracted) intensity is expressed by

$$I(\mathbf{S}) = Kg(\theta)Ay|G(\mathbf{S})|^2 \tag{5.7}$$

where K is a constant that includes the factor $(e^2/4\pi\varepsilon_0 mc^2)^2$ and the volume of the scattering body, $g(\theta)$ is a function independent of the details of the material and includes the effects of the polarization and the experimental conditions. The absorption factor $A \le 1$ and the extinction factor $y \le 1$ are material dependent and can be determined experimentally.

5.4 Atomic model

Matter is composed of individual atoms that are held together by interatomic forces. The electrons in a material are therefore located at the atoms, and the electron distribution in the atom needs to be known for a model of $\rho(\mathbf{x})$. The electronic distribution in a free (nonbonded) atom can be calculated using quantum mechanical methods. To first order, it is found that a spherical electron distribution is a good approximation for atoms containing more than 11 electrons, and a spherical model is acceptable for atoms with less than 10 electrons as nonspherical atoms, for instance atoms with partially filled shells, do not differ greatly from spherical symmetry. Therefore, the electron density in a material can be described as a sum of individual atomic electron densities (see Figure 5.7):

$$\rho(\mathbf{r}) = \sum_m^{\text{atoms}} \rho_m(\mathbf{r} - \mathbf{r_m})$$

In this approximation, the total electron density in the material is equal to the superposition of atomic electron distributions $\rho(\mathbf{R})$ centered at points $\mathbf{r}_m$. For a spherical atom, $\rho(\mathbf{R}) = \rho(R)$ and can be expressed as a sum of Gaussian distributions as

$$\rho_m(R) \approx \sum_g K_{m,g} e^{-\alpha_{m,g} R^2}$$

Neglecting the effects of the chemical bonding on the electron distribution greatly simplifies the calculation, and it can be shown that the results are within a few percent of experimental values. The Fourier transformation of an assembly of atoms is therefore

$$G(\mathbf{S}) = \int \sum_m \rho_m(\mathbf{r} - \mathbf{r_m}) e^{2\pi i \mathbf{r} \cdot \mathbf{S}} \, d^3\mathbf{r} = \sum_m f_m e^{2\pi i \mathbf{r} \cdot \mathbf{S}}$$

with $f_m(\mathbf{S})$ defined as

$$f_m(\mathbf{S}) = \int_{\text{atom } m} \rho_m(\mathbf{R}) e^{2\pi i \mathbf{R} \cdot \mathbf{S}} \, d^3\mathbf{R} = \Phi[\rho_m] \tag{5.8}$$

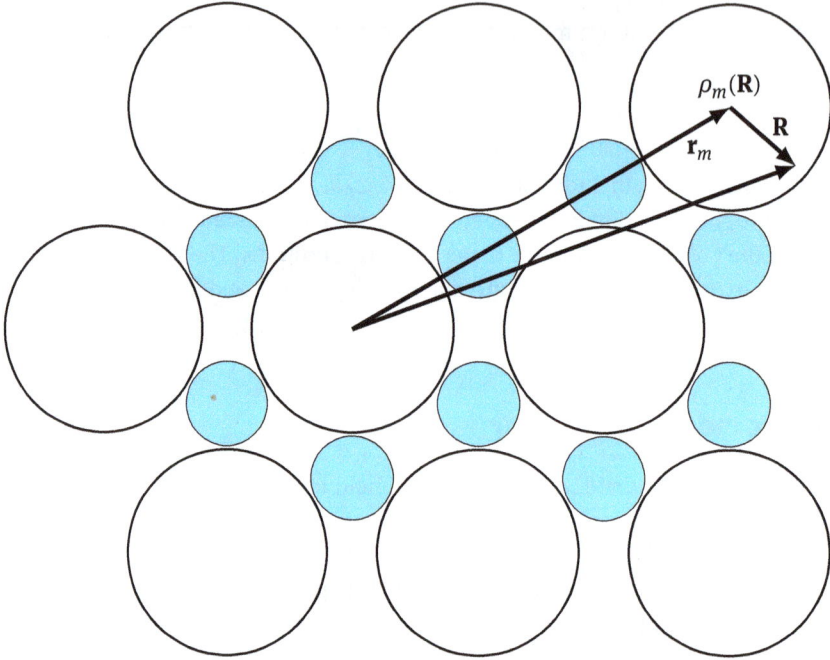

Figure 5.7: Schematic structure composed of spherical atoms. Each circle represents a spherical electron distribution $\rho_m(\mathbf{R})$ at position $\mathbf{r}_m$ of the mth atom.

For a spherical atom, using spherical coordinates and integrating over the angles, the scattering power of the atom becomes

$$f_m(\|\mathbf{S}\|) = f_m(S) = \int_0^\infty 4\pi R^2 \rho_m(R)\frac{\sin 2\pi RS}{2\pi RS}\,dR \tag{5.9}$$

where $S = \|\mathbf{S}\| = 2\sin\theta/\lambda$ and $f_m(\mathbf{S})$ the *atomic form factor* derived as the Fourier transform of the atomic charge distribution (equations (5.8) and (5.9)). The scattering power of an atom depends on the scattering angle and the wavelength and the drop in scattering power with scattering angle is due to the interference effects of scattered waves from different parts of the electron density around the nucleus. The form factor can be extended to include dispersion effects by adding wavelength dependent terms $f_{total} = f + \Delta f' + i\Delta f''$. The atomic form factors of atoms and ions are tabulated in the *International Tables of Crystallography C, pages 477–503*, as well as the dispersion corrections for commonly used wavelengths (*pp. 219–222*).

The atomic form factor has the following properties:
- it is a function of $\sin\theta/\lambda$.
- its values are given in units of classical electrons.
- $f_m(0)$ is equal to the total number of electrons of the atom or ion on position m.
- $f_m(\sin\theta/\lambda)$ decreases faster if the electron distribution is diffuse.

- the form factor of neutral atoms and their respective ions differ only at small values of $\sin\theta/\lambda$.

The Fourier transform of a diffuse electron distribution falls off rapidly with scattering angle, a consequence of the spatially extended electron density. Thus, the bonding electrons, which are in a diffuse distribution, do not strongly affect the overall scattering. The contribution of the valence electrons is, however, most pronounced for the light atoms, such as boron, carbon, nitrogen, oxygen and fluorine. The overall contribution of the valence electrons to the scattering/diffraction effects are therefore minor, but can be observed in careful measurements. The results of quantum mechanical calculations to solve the Schrödinger equation for many electrons, including relativistic effects for heavy atoms, has been parametrized, using the following nine-term expansions:

$$f(\sin\theta/\lambda) = \sum_{i=1}^{4} a_i e^{-b_i(\sin\theta/\lambda)^2} + c \qquad (5.10)$$

This approximation is excellent in the range of $0 \leq \sin\theta/\lambda \leq 2\,\text{Å}^{-1}$ and the range is sufficient for most applications. Care has to be taken when high angle data at short wavelengths need to be measured, where $\sin\theta/\lambda$ is large. The values for the expansion coefficients a_i, b_i and c are tabulated (*International Tables for Crystallography, Volume C, pp. 500–502*). For example, the coefficients for the neutral and ionic fluorine are given in Table 5.1, and Figure 5.8 shows the resulting form factors as a function of $\sin\theta/\lambda$ (Table 5.1 and equation (5.10)).For instance, with molybdenum radiation ($\lambda = 0.70932\,\text{Å}$), at a value of $\sin\theta/\lambda = 0.5$ ($2\theta = 41.55°$), the contribution of fluorine to the overall scattering is equivalent to approximately 3 electrons.

Table 5.1: Form factor table entries for neutral and ionic fluorine.

El	a_1	b_1	a_2	b_2	a_3	b_3	a_4	b_4	c
F	3.53920	10.2825	2.64120	4.29440	1.51700	0.261500	1.02430	26.1476	0.277600
F⁻	3.63220	5.27756	3.51057	14.7353	1.26064	0.442258	0.940706	47.3437	0.653396

5.5 Atomic model: thermal vibrations in a material

The atomic form factor is the Fourier transform of the electron distribution of an atom at rest. However, thermal effects cause the atoms to oscillate around their position in a material, with oscillation frequencies of the order of 10^{12} to 10^{14} Hz. Even at a temperature of 0 K, the quantum mechanical zero point motion ensures that an atom is not at rest. The instantaneous structure, therefore, is constantly changing, and only the temperature averaged structure can be observed. Due to these vibrations/displace-

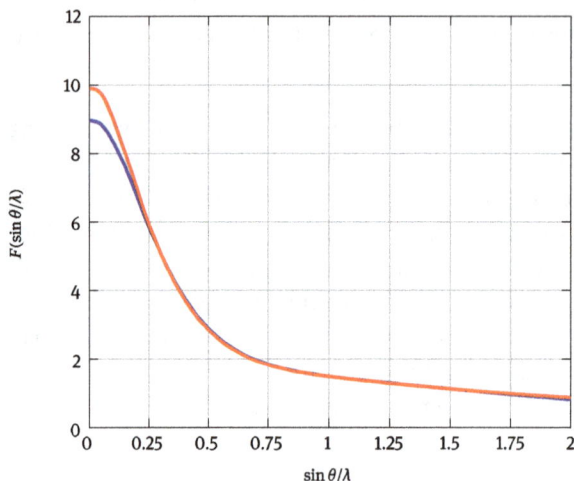

Figure 5.8: Form factors of neutral fluorine (F, in blue) and ionic fluorine (F⁻, in red).

ments, the time averaged structure of a material $\langle \rho \rangle_t$ gives rise to the scattering pattern/diffraction pattern that can be observed. The effect of these thermal vibrations will be considered in the following. It is assumed that during the oscillations, the atom remains spherical, and deviations from the spherical shape are not considered. The displacement of an atom m at the average position X (displacement Δ) to the position X' is described by the probability $P(\Delta)d^3\Delta$ for a given moment in time (Figure 5.9). The electron density therefore depends on the displacement Δ, and the contribution of the displaced atom is given by

$$P(\Delta)\rho_m(\mathbf{R} - \Delta)d^3\Delta$$

with $\rho_m(\mathbf{R})$ the electron density of the atom at the average position. The total mean electron density of the vibrating atom is calculated by integrating over all possible displacement vectors Δ,

$$\langle \rho_m(\mathbf{R}) \rangle_t = \int P(\Delta)\rho_m(\mathbf{R} - \Delta) \, d^3\Delta = P * \rho_m(\mathbf{R})$$

The form of this integral is the *convolution product* of the functions $P(\Delta)$ and $\rho_m(\mathbf{R})$. The convolution product is commutative, $P * \rho_m = \rho_m * P$. The average form factor $\langle f_m \rangle$ is the Fourier transform of the $\langle \rho_m \rangle_t$. Since the Fourier transform of a convolution product is the product of the Fourier transforms of each individual function, the average form factor becomes

$$\langle f_m \rangle_t = \Phi[\rho_m * P] = \Phi[\rho_m]\Phi[P] = f_m(\mathbf{S})T(\mathbf{S})$$

and

$$T(\mathbf{S}) = \int P(\Delta)e^{2\pi i \Delta \cdot \mathbf{S}} \, d^3\Delta \qquad (5.11)$$

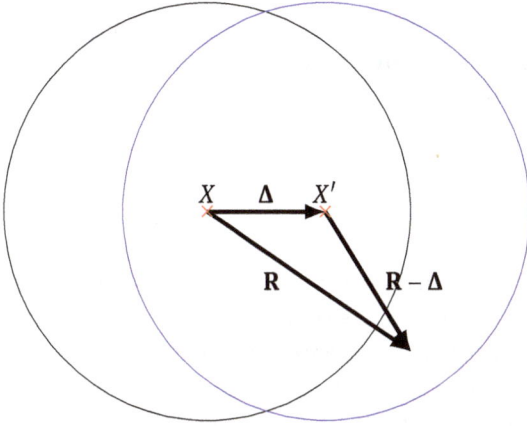

Figure 5.9: A spherical atom in averaged position X is displaced to position X' with probability $P(\Delta)d^3\Delta$.

With $T(\mathbf{S})$ the Fourier transform of the probability function. The probability function $P(\Delta)$ may not only describe dynamic displacements such as vibrations, but also static displacements. In the latter case, atomic disorder, where different positions are statically occupied, is described in the same way. For this reason, the quantity $T(\mathbf{S})$ is called the *displacement factor* rather than the more traditional *temperature factor*.

The probability function $P(\Delta)$ can be quite general under condition that a Fourier transform exists, but it can be shown that if vibrations are harmonic (true for small vibration amplitudes), the probability distribution is Gaussian:

$$P(\Delta) = (2\pi)^{-3/2}|\mathbf{V}|^{-1/2}e^{-(\Delta^T\mathbf{V}^{-1}\Delta)/2}$$

where the matrix $\mathbf{V}$ is a variance-covariance matrix with the averaged terms $V_{ij} = \langle\Delta_i\Delta_j\rangle_t$. It is possible to define a coordinate system that is based on the eigenvectors V_i of this matrix, so that the expression simplifies to

$$P(\Delta) = (2\pi)^{-2/3}(V_1V_2V_3)^{-1/2}e^{-(\Delta_1^2/V_1+\Delta_2^2/V_2+\Delta_3^2/V_3)/2}$$

With a Gaussian probability distribution, the expression for $T(\mathbf{S})$ becomes

$$T(\mathbf{S}) = e^{-2\pi\mathbf{S}^T\mathbf{VS}} = \exp(-2\pi(V_{11}s_1^2 + V_{22}s_2^2 + V_{33}s_3^2 + 2V_{12}s_1s_2 + 2V_{13}s_1s_3 + 2V_{23}s_2s_3)) \quad (5.12)$$

The term $T(\mathbf{S})$ in equation (5.12) is called the *Debye–Waller factor*. It is therefore possible to define the mean square displacement U_s parallel to $\mathbf{S}$ as

$$U_\mathbf{s} = \frac{\langle(\mathbf{S}\cdot\Delta)^2\rangle_t}{\|\mathbf{S}\|^2} = \frac{\mathbf{S}^T\mathbf{VS}}{\|\mathbf{S}\|^2}$$

Including the definition for $\|\mathbf{S}\| = 2\sin\theta/\lambda$, the Debye–Waller factor is given by

$$T(\mathbf{S}) = e^{-8\pi^2 U_\mathbf{s}(\sin\theta/\lambda)^2} \quad (5.13)$$

If the probability function is isotropic, meaning that the displacement amplitude is independent of the direction Δ, the matrix $\mathbf{V}$ greatly simplifies, since $V_{11} = V_{22} = V_{33} = U_{iso}$ and $V_{12} = V_{13} = V_{23} = 0$, and the Debye–Waller factor from equation (5.13) becomes

$$T(\sin\theta/\lambda) = e^{-8\pi^2 U_{iso}(\sin\theta/\lambda)^2} \tag{5.14}$$

Typically, the mean square displacement values U_{iso} in equation (5.14) at room temperature are of the order of $0.01\,\text{Å}^2$ to $0.1\,\text{Å}^2$. In diamond, the hardest material known, U_{iso} is of the order $0.002\,\text{Å}^2$. In NaCl, one finds $U_{iso} \approx 0.02\,\text{Å}^2$. The Debye–Waller factor reduces the diffracted intensities since the interference function is perturbed, and the phase relationships are no longer exact. Since $T(\sin\theta/\lambda)$ depends on the scattering angle θ, intensities at higher scattering angle are more strongly affected. This is most apparent in van der Waals bonded molecular crystals, where molecular motion strongly attenuates the reflection at higher angles.

The tensor $\mathbf{V}$ describes an ellipsoid with principle axes V_{11}, V_{22} and V_{33}. Thus, the displacement tensor can be represented as an ellipsoid at, for instance, the 30 or 50 % probability level.

$$\mathbf{X}^T\mathbf{V}\mathbf{X} = \text{const.}$$

The determination of the displacement parameters requires the measurement of intensities over an extended range of $\sin\theta/\lambda$ so that the square exponential drop-off in intensities is determined with sufficient precision. Therefore, using data with a large angular range is expected to yield more precise information about the dynamic and static atomic displacements.

5.6 Diffraction by a periodic structure

Laue equations
In a crystal, the time averaged electron density $\langle\rho(\mathbf{r})\rangle_t$ is periodic in three dimensions

$$\langle\rho(\mathbf{r})\rangle_t = \rho(\mathbf{r} + u\mathbf{a} + v\mathbf{b} + w\mathbf{c})$$

with u, v, w integers and

$$\mathbf{r} = x\mathbf{a} + y\mathbf{b} + z\mathbf{c}$$

with $0 \leq x, y, z \leq 1$ the relative coordinates. The vectors $\mathbf{a}$, $\mathbf{b}$, $\mathbf{c}$ form the basis of a lattice. Taking the Fourier transform of the charge density gives

$$G(\mathbf{S}) = \sum_{cells}\left[\int_{1\,cell}\langle\rho(\mathbf{r})\rangle_t e^{2\pi i\mathbf{r}\mathbf{S}}\right]e^{2\pi i(u\mathbf{a}+v\mathbf{b}+w\mathbf{c})\cdot\mathbf{S}}$$

$$= F(\mathbf{S})\sum_u e^{2\pi i u\mathbf{a}\mathbf{S}}\sum_v e^{2\pi i v\mathbf{b}\mathbf{S}}\sum_w e^{2\pi i w\mathbf{c}\mathbf{S}}.$$

If the crystal form is simplified to be a parallelepiped with M unit cells in the direction of **a**, N unit cells in the direction of **b** and P unit cells in the direction of **c**, then the sums can be evaluated since they each form finite geometric series:

$$\sum_{u=0}^{M-1} e^{2\pi i u \mathbf{aS}} = \frac{\sin \pi M \mathbf{aS}}{\sin \pi \mathbf{aS}} e^{\pi i (M-1)\mathbf{aS}} = J_M e^{\pi i (M-1)\mathbf{aS}}$$

The phase term $1/2(M-1)\mathbf{aS}$ depends on the choice of the origin. Placing the origin at the center of the crystal, one finds for odd values of M,

$$\sum_{-(M-1)/2}^{+(M-1)/2} e^{2\pi i u \mathbf{aS}} = \sum_{-(M-1)/2}^{+(M-1)/2} \cos 2\pi u \mathbf{aS} = \frac{\sin M\pi \mathbf{aS}}{\sin \pi \mathbf{aS}} = J_M(\mathbf{aS})$$

The intensity of the diffracted waves is proportional to $|G(\mathbf{S})|^2$; therefore, the total intensity from a periodic three-dimensional periodic object is proportional

$$I(\mathbf{S}) \propto |G(\mathbf{S})|^2 = |F(\mathbf{S})|^2 J_M^2(\mathbf{aS}) J_N^2(\mathbf{bS}) J_P^2(\mathbf{cS})$$

and

$$F(\mathbf{S}) = \int_{1\,\text{cell}} \langle \rho(\mathbf{r})\rangle_t e^{2\pi i \mathbf{rS}} \, d^3\mathbf{r} = \sum_{\text{atoms}}^{1\,\text{cell}} [f_m]_t e^{2\pi i \mathbf{r}_m \cdot \mathbf{S}}$$

The form factor $[f_m]_t$ includes vibration and displacement terms. The new quantity $F(\mathbf{S})$ is called the *structure factor*, and the functions J^2 are characteristic of the periodicity. This equation shows that *the diffraction pattern of a periodic structure is the product of the diffraction pattern of a single unit cell and the interference functions J^2 describing the periodicity in three dimensions.* Since the number of unit cells MNP is large, the intensity I is generally zero, except for the case meeting the following conditions:

$$\mathbf{a} \cdot \mathbf{S} = h$$
$$\mathbf{b} \cdot \mathbf{S} = k; \quad h, k, l \text{ integers} \qquad (5.15)$$
$$\mathbf{c} \cdot \mathbf{S} = l$$

These three equations are called *Laue equations* and give the condition for positive interference. Recalling the definition of the reciprocal lattice

$$\mathbf{r}^* = h\mathbf{a}^* + k\mathbf{b}^* + l\mathbf{c}^*; \quad h, k, l \text{ integers}$$
$$\mathbf{a} \cdot \mathbf{a}^* = \mathbf{b} \cdot \mathbf{b}^* = \mathbf{c} \cdot \mathbf{c}^* = 1$$
$$\mathbf{a} \cdot \mathbf{b}^* = \mathbf{a} \cdot \mathbf{c}^* = \mathbf{b} \cdot \mathbf{a}^* = \mathbf{b} \cdot \mathbf{c}^* = \mathbf{c} \cdot \mathbf{a}^* = \mathbf{c} \cdot \mathbf{b}^* = 0$$

the condition for **S** is

$$\mathbf{S} = \mathbf{s} - \mathbf{s}_0 = \mathbf{r}^*$$

meaning that the scattering vector **S** has to be a vector of the reciprocal lattice. *The Fourier transform of a periodic function is zero, except at the nodes of the reciprocal lattice, where it is proportional to the value obtained for a single unit cell.* The scattering of radiation from a periodic structure is called diffraction, for example, diffraction from a crystal or from an optical lattice.

5.7 Structure factor

Modeling of the diffraction pattern of a periodic crystal is reduced to determining the reciprocal lattice for a given unit cell based on the geometry, and calculating the structure factor $F(\mathbf{S})$ of a single unit cell based on the unit cell content. Unit cell and atomic coordinate data of the structure are required, and symmetry information allows a "shorthand" notation. The structure factor $F(\mathbf{S})$ is the vector sum of all the waves interfering within the unit cell that originate from all the atoms at their respective positions. This requires that the incident wave has a coherence length that exceeds the unit cell dimensions. The vector **S** is a reciprocal lattice vector, and is expressed as $\mathbf{S} = h\mathbf{a}^* + k\mathbf{b}^* + l\mathbf{c}^*$, while the coordinates in the unit cell are given by $\mathbf{r} = x\mathbf{a} + y\mathbf{b} + z\mathbf{c}$. This allows to replace $F(\mathbf{S})$ by $F(hkl)$, since **S** is a reciprocal lattice vector of the reciprocal coordinate system that is spanned by the reciprocal axes $\mathbf{a}^*$, $\mathbf{b}^*$ and $\mathbf{c}^*$. Therefore, $F(\mathbf{S}) = F(hkl)$ and can be expressed as

$$F(hkl) = \sum_{m=1}^{N} f_m e^{2\pi i (hx_m + ky_m + lz_m)}. \tag{5.16}$$

The structure factor $F(hkl)$ is complex and it expresses both amplitude and phase of the diffracted wave in direction **S**. The complex notation is easily expanded into the respective cosine and sine components $F = a + ib$, such that

$$a = \sum_m f_m \cos 2\pi (hx_m + ky_m + lz_m)$$
$$b = \sum_m f_m \sin 2\pi (hx_m + ky_m + lz_m)$$

5.7.1 Structure factors for a primitive lattice with one atom

In the case of a unit cell that contains exactly one atom, it can be placed at the origin without any loss of generality. The structure factor (equation (5.16)) becomes

$$F = f e^{2\pi i 0} = f$$

$|F(hkl)|^2 = |f|^2$ is thus independent of (hkl), and it is the same for all reflections. Since f depends on $\sin\theta/\lambda$, the angle dependence of the scattered intensities observed is due to the atomic form factor and the displacement parameters.

5.7.2 Structure factors for a primitive lattice with several atoms

Extending from the example of a primitive lattice with only one atom, the structure factor is the sum over all atoms present, and is expressed as

$$F(hkl) = \sum_{atoms} f_m e^{2\pi i (hx_m + ky_m + lz_m)}$$

This expression is simplified if a center of inversion is present at the origin of the unit cell: for each atom at a position (x, y, z), there is a symmetry related atom at $(-x, -y, -z)$. By combining the sum of these two atoms, the structure factor becomes real, facilitating calculations.

5.7.3 Structure factors for a C-centered lattice

A C-centered lattice with one atom at the origin is considered. The unit cell contains two atoms, one at $(0, 0, 0)$ and one at $(\frac{1}{2}, \frac{1}{2}, 0)$. The structure factor (equation (5.16)) is the sum of the two waves originating from the two atoms giving

$$F(hkl) = f e^{2\pi i 0} + f e^{2\pi i (h/2 + k/2)} = f \left[1 + e^{\pi i (h+k)} \right]$$

This expression imposes constraints for the sum $(h+k)$ = integer. If $(h+k) = 2N$ (even), the structure factor becomes

$$F(hkl) = 2f \quad \text{and} \quad F^2(hkl) = 4f^2$$

Similarly, if the sum $(h + k) = 2N + 1$ (odd), then $e^{\pi i (h+k)} = -1$, and

$$F(hkl) = 0 \quad \text{and} \quad F^2(hkl) = 0$$

The value of l does not have any effect on the structure factor. Therefore, the reflections (111), (112), (201), (202), etc. are allowed. Consequently, the reflections (100), (101), (010), (012), (013), etc. are not observed $F(hkl) = 0$. Due to the base centering (C-Centering), the diffraction pattern shows systematic absences for general hkl with condition $h + k = 2N + 1$, defining a *systematic absence*.

5.7.4 Structure factor for a unit cell with two different atoms

A primitive unit cell containing two different atoms is considered. As an example, the CsCl structure (cubic, $Pm\bar{3}m$ consists of a Cs ion at $(0, 0, 0)$ and a Cl ion at $(\frac{1}{2}, \frac{1}{2}, \frac{1}{2})$). The structure factor is thus

$$F(hkl) = f_{Cs} e^{2\pi i 0} + f_{Cl} e^{2\pi i (h/2 + k/2 + l/2)} = f_{Cs} + f_{Cl} e^{\pi i (h+k+l)}$$

In this case, there are two choices for the sum $(h+k+l)$, even or odd. For $(h+k+l) = 2N$, the structure factor (equation (5.16)) becomes

$$F(hkl) = f_{Cs} + f_{Cl}e^{\pi i(h+k+l)} = f_{Cs} + f_{Cl}$$

the two atoms interfere constructively. Consequently, for $(h+k+l) = 2N+1$, the structure factor becomes

$$F(hkl) = f_{Cs} + f_{Cl}e^{\pi i(h+k+l)} = f_{Cs} - f_{Cl}$$

the two atoms interfere destructively. Therefore, the condition of $(h + k + l) = 2N$ results in stronger reflections, where the waves emanating from the two atoms are in phase, whereas the condition $(h + k + l) = 2N + 1$ results in weaker reflections where the waves are out-of-phase (offset by π or 180°). In the case of a body centered unit cell (I-centered), where the two atoms at the corner and the center are identical, the condition $(h+k+l) = 2N+1$ leads to a *systematic extinction*, characteristic of the space group centering symbol I.

5.7.5 The NaCl structure

The rock salt or NaCl structure consists of 2 fcc lattices that are offset by $(\frac{1}{2}, 0, 0)$. The unit cell contains 4 Na atoms and 4 Cl atoms with fractional coordinates:

$$\text{Na} \quad 000 \quad \tfrac{1}{2}\tfrac{1}{2}0 \quad \tfrac{1}{2}0\tfrac{1}{2} \quad 0\tfrac{1}{2}\tfrac{1}{2}$$

$$\text{Cl} \quad \tfrac{1}{2}00 \quad 0\tfrac{1}{2}0 \quad 00\tfrac{1}{2} \quad \tfrac{1}{2}\tfrac{1}{2}\tfrac{1}{2}$$

The sum over all eight atoms, with the proper phase relationships is

$$F(hkl) = f_{Na}e^{2\pi i 0} + f_{Na}e^{2\pi i(h/2+k/2)} + f_{Na}e^{2\pi i(h/2+l/2)} + f_{Na}e^{2\pi i(k/2+l/2)}$$
$$+ f_{Cl}e^{2\pi i(h/2)} + f_{Cl}e^{2\pi i(k/2)} + f_{Cl}e^{2\pi i(l/2)} + f_{Cl}e^{2\pi i(h/2+k/2+l/2)}$$
$$= f_{Na}[1 + e^{\pi i(h+k)} + e^{\pi i(h+l)} + e^{\pi i(k+l)}] + f_{Cl}[e^{\pi i h} + e^{\pi i k} + e^{\pi i l} + e^{\pi i(h+k+l)}]$$

The common translation, in this case, the F-centering, allows rewriting the above equation:

$$F(hkl) = f_{Na}[1 + e^{\pi i(h+k)} + e^{\pi i(h+l)} + e^{\pi i(k+l)}]$$
$$+ f_{Cl}e^{\pi i(h+k+l)}[1 + e^{\pi i(-h-k)} + e^{\pi i(-h-l)} + e^{\pi i(-k-l)}]$$

The signs in the expression for the chlorine atoms can be changed since $e^{n\pi i} = e^{-n\pi i}$ for all integer numbers. The expression for $F(hkl)$ simplifies to

$$F(hkl) = [1 + e^{\pi i(h+k)} + e^{\pi i(h+l)} + e^{\pi i(k+l)}][f_{Na} + f_{Cl}e^{\pi i(h+k+l)}]$$

The face-centering terms (F) are separated from the content of the unit cell (two atoms), and is consistent with the structural information that needs to be given for the NaCl structure: the F-centering plus the positions for Na at $(0,0,0)$ and Cl at $(\frac{1}{2}, \frac{1}{2}, \frac{1}{2})$. The first term is nonzero only of the following conditions for the indices hkl are met:

$$h + k = 2N \text{ even}$$
$$h + l = 2N \text{ even}$$
$$k + l = 2N \text{ even}$$

In this case, the three indices need to be either all even or all odd. This is called *mixed* and *unmixed* indices, and describes the systematic extinction conditions for the F-centered lattice. As a result, the NaCl structure has three different values for the structure factor $F(hkl)$:

$$F(hkl) = 0 \qquad \text{for mixed indices}$$
$$F(hkl) = 4[f_{Na} + f_{Cl}] \quad \text{for unmixed indices} \quad h + k + l = 2N$$
$$F(hkl) = 4[f_{Na} - f_{Cl}] \quad \text{for unmixed indices} \quad h + k + l = 2N + 1$$

The calculations are best automated using programs that eliminate the need for the ever more complex algebra if the number of atoms becomes large. Small unit cells, often found for simple metals that crystallize in the body-centered cubic (bcc) or face-centered cubic (fcc) structure, can easily be handled, but simple molecular systems with low symmetry can contain of the order of 20 atoms or more, making calculations exceedingly tedious. For instance, the molecule naphthalene $(C_{10}H_8)$ has a total of 18 atoms and crystallizes in a monoclinic unit cell with space group $P2_1/a$ and $Z = 2$, requiring extensive calculations.

5.7.6 Example: cubic perovskite SrTiO$_3$

The perovskite structure of $SrTiO_3$ crystallizes in the cubic space group $Pm\bar{3}m$, $Z = 1$ (one formula unit per unit cell), and lattice parameter $a = 3.905$ Å. The atomic positions are given in Table 5.2.

Table 5.2: SrTiO$_3$ atomic positions.

Atom	x	y	z
Sr	0	0	0
Ti	1/2	1/2	1/2
O	1/2	1/2	0

The *structure factor* F(hkl) calculation sums over all the different atoms in the unit cell. Using the symmetry information, the list of all five atoms in the unit cell is given in Table 5.3.

Table 5.3: SrTiO$_3$ all atomic positions in the unit cell.

Atom	x	y	z
Sr	0	0	0
Ti	1/2	1/2	1/2
O1	1/2	1/2	0
O2	1/2	0	1/2
O3	0	1/2	1/2

The structure factor becomes

$$F(hkl) = f(Sr) + f(Ti)e^{2\pi i(\frac{h}{2}+\frac{k}{2}+\frac{l}{2})} + f(O1)e^{2\pi i(\frac{h}{2}+\frac{k}{2})} + f(O2)e^{2\pi i(\frac{h}{2}+\frac{l}{2})} + f(O3)e^{2\pi i(\frac{k}{2}+\frac{l}{2})}$$

The phases of the individual waves emanating from the atoms in the unit cell are real, with values ±1. Calculations of the structure factor F(hkl) for the first few (hkl)s are shown in Table 5.4.

Table 5.4: Phases for the calculation of F(hkl) for SrTiO$_3$

h	k	l	f(Sr)	f(Ti)	f(O1)	f(O2)	f(O3)
1	0	0	+	−	−	−	+
1	1	0	+	+	+	−	−
1	1	1	+	−	+	+	+
2	0	0	+	+	+	+	+
2	1	0	+	−	−	+	−
2	1	1	+	+	−	−	+
2	2	0	+	+	+	+	+
2	2	1	+	−	+	−	−
3	0	0	+	−	−	−	+
2	2	2	+	+	+	+	+

For instance, the F(100) structure factor is small, given by F(100) = f(Sr)−f(Ti)−f(O1)− f(O2)+f(O3), with Sr and Ti interfering destructively, and only one of the three oxygen atoms in phase with the strontium atom. In contrast, the structure factor for F(200), F(220) and F(222) are large, with all atoms interfering constructively, their individual form factors adding up. For other indices (hkl), the structure factors F(hkl) values fall in between. If, to first order, the oxygen form factor is neglected, then this perovskite system will have larger structure factors for (hkl) with the condition h + k + l = 2N, where the strontium and titanium atoms are in phase, with their respective large form factors adding up.

5.8 Diffraction condition

There are two ways to describe the diffraction geometry of elastically scattered waves off a periodic structure: in direct space, or in reciprocal space. Both descriptions are equally valid, and depending of the task, each description has advantages and disadvantages. In the following, the *Bragg law* is derived, as well as the *Ewald construction*. The first is based on direct space, whereas the latter is a reciprocal space construction.

5.8.1 Bragg law

The *Bragg law* or *Bragg equation* is contained in the *Laue equations* (equations (5.16)). Recall that the plane normal given by $\mathbf{r}^* = h\mathbf{a}^* + k\mathbf{b}^* + l\mathbf{c}^*$, with $(hkl) = n(HKL)$, is perpendicular to a series of planes (HKL) of the crystal lattice. The values HKL are the smallest integers possible, and n is the largest common factor of $(hkl) = n(HKL)$. Therefore, the length of the reciprocal vector is $\|\mathbf{r}^*\| = n/d_{HKL}$, with d_{HKL} the distance between the planes. For different orders n, the d-spacing becomes $d_{hkl} = d_{HKL}/n$. The scattering vector $\mathbf{S} = \mathbf{r}^*_{hkl} = \mathbf{r}^*_{nH,nK,nL}$ is bisecting the angle between $\mathbf{s}_0$ and $\mathbf{s}$, with $\|S\| = 2\sin\theta/\lambda$. The *Bragg equation* becomes

$$2d_{hkl}\sin\theta = \lambda \quad \text{or} \quad \sin\theta = \frac{\lambda}{2d_{hkl}} \tag{5.17}$$

It is important to realize that the *Bragg equation* references lattice planes, not planes formed by atoms. It is derived from the Laue equations, and is a consequence of the periodicity of the lattice. The formulas for calculating d-spacings d_{hkl} as functions of the indices (hkl) for different crystal systems with unit cell parameters a, b, c, α, β and γ are given in Table 5.5.

5.8.2 Ewald construction

The *Ewald construction* defines the direction of the diffracted wave by the intersection of 2 geometric points. The absolute value of $\|\mathbf{s}\| = \|\mathbf{s}_0\| = 1/\lambda$ defines a sphere of radius $1/\lambda$ in reciprocal space. The sphere intersects the origin (000) of the reciprocal lattice, and if properly oriented, another reciprocal lattice point. Therefore, the vector between $\mathbf{s}$ and $\mathbf{s}_0$ is, by virtue of these conditions, a member of the reciprocal lattice, and thus satisfies the Laue equations (equation (5.16)).

Table 5.5: Calculation of d_{hkl}.

cubic	$\frac{1}{d^2} = \frac{h^2+k^2+l^2}{a^2}$
tetragonal	$\frac{1}{d^2} = \frac{h^2+k^2}{a^2} + \frac{l^2}{c^2}$
hexagonal	$\frac{1}{d^2} = \frac{4}{3}\left(\frac{h^2+hk+k^2}{a^2}\right) + \frac{l^2}{c^2}$
rhombohedral	$\frac{1}{d^2} = \frac{(h^2+k^2+l^2)\sin^2\alpha + 2(hk+kl+hl)(\cos^2\alpha - \cos\alpha)}{a^2(1-3\cos^2\alpha + 2\cos^3\alpha)}$
orthorhombic	$\frac{1}{d^2} = \frac{h^2}{a^2} + \frac{k^2}{b^2} + \frac{l^2}{c^2}$
monoclinic	$\frac{1}{d^2} = \frac{1}{\sin^2\beta}\left(\frac{h^2}{a^2} + \frac{k^2\sin^2\beta}{b^2} + \frac{l^2}{c^2} - \frac{2hl\cos\beta}{ac}\right)$
triclinic	$\frac{1}{d^2} = \frac{1}{V^2}(S_{11}h^2 + S_{22}k^2 + S_{33}l^2 + 2S_{12}hk + 2S_{13}hl + 2S_{23}kl)$

$V = abc(1 - \cos^2\alpha - \cos^2\beta - \cos^2\gamma + 2\cos\alpha\cos\beta\cos\gamma)$

$S_{11} = b^2c^2\sin^2\alpha$

$S_{22} = a^2c^2\sin^2\beta$

$S_{33} = a^2b^2\sin^2\gamma$

$S_{12} = abc^2(\cos\alpha\cos\beta - \cos\gamma)$

$S_{13} = ab^2c(\cos\gamma\cos\alpha - \cos\beta)$

$S_{23} = a^2bc(\cos\beta\cos\gamma - \cos\alpha)$

In Figure 5.10, a crystal in a random orientation will in general *not* satisfy the *Ewald construction*, and thus the Laue equations, and no diffracted intensity will be observed. However, moving the Ewald sphere by changing $\mathbf{s}_0$, or rotating the crystal, moving the reciprocal lattice, both allow to satisfy the Laue equations by fulfilling the conditions of the Ewald construction. There remains a further option, which is changing the wavelength λ of the radiation. In reciprocal space, this is equivalent to changing the radius of the Ewald sphere to a wavelength value that satisfies the Ewald construction. It is also clear that if the smallest reciprocal lattice point is further away from the origin than the radius of the Ewald sphere $1/\lambda$, no diffraction is possible.

The Ewald construction is useful to understand the instrumentation for measuring $I(hkl)$s. For a fixed wavelength, the crystal (reciprocal lattice) needs to be oriented so that the Ewald construction is satisfied for a given reciprocal lattice point (hkl). For this, two axes are needed, which are often based on the Euler angles, and the corresponding orientation device is called an *Euler cradle*. It is thus possible to orient any direct lattice plane (reciprocal vector) against an incident wave $\mathbf{s}_0$. A detector needs to be positioned to record the $I(hkl)$, which is achieved using one or two more axes. Therefore, three or four axes are needed, and such a system is called a *diffractometer*.

The diffraction pattern contains the information about the atom positions within a unit cell in the diffracted intensities, whereas the positions of the intensities (angu-

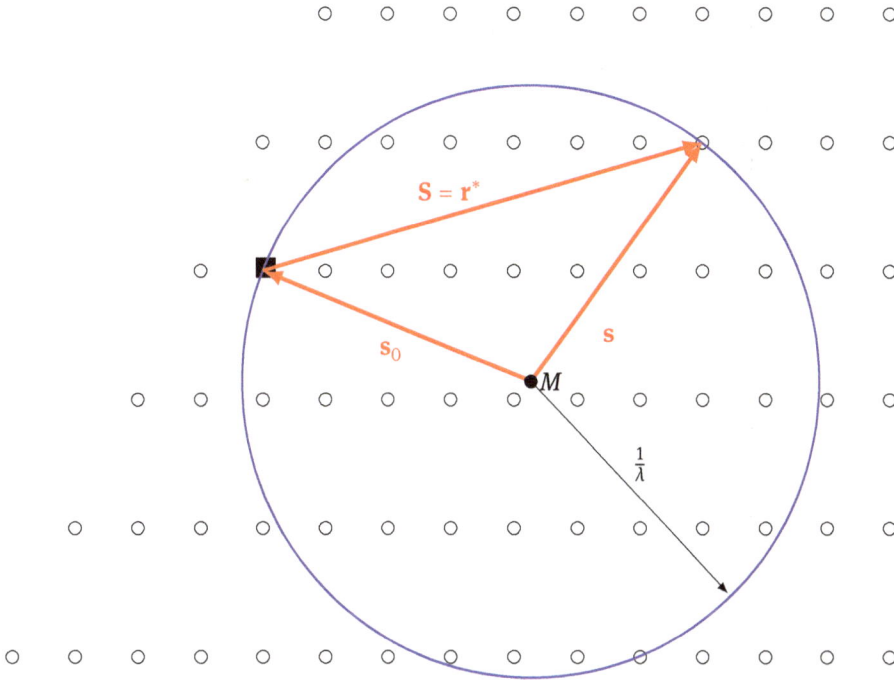

Figure 5.10: Ewald construction: the Ewald sphere is shown as a circle in a plane of reciprocal lattice points. The Ewald sphere always includes the origin (000), and intersects a reciprocal lattice point. If this condition is not met, then the reciprocal lattice may be rotated to satisfy the condition.

lar positions) define by the unit cell dimensions. By measuring $I(hkl)$s, the structural motif of the unit cell is sampled at all the reciprocal lattice positions, where diffracted intensities are allowed by the Laue equations. The periodicity of the structure allows measuring the same structural motif many times, resulting in well-defined diffraction intensities.

5.8.3 Calculation of a diffraction pattern

A procedure for the quantitative calculation of a diffraction pattern can be set up:

- A list of (hkl) indices is generated, taking into account the symmetry, unit cell, and maximum 2θ.
- For each (hkl), the Bragg angle θ is determined, as well as the values of $\sin\theta/\lambda$.
- The form factor for each atom is determined using the nine coefficients for the form factor calculation for each (hkl), to account for the $\sin\theta/\lambda$ dependence.
- Using the symmetry information and the shorthand notation for the atom positions in the unit cell, all atomic positions within the unit cell are generated.

- The structure factor is obtained by summing the form factors of all the atoms in the unit cell with their respective phase and displacement parameter.
- The values may be ordered according to their (hkl)s, or θs.
- Experimental correction can be applied, such as the polarization factor, geometrical details of the diffraction system, etc.

Such procedures are implemented in a number of open source programs, but are easily carried out by hand for simple structures.

6 X-ray physics

6.1 Production of X-rays

Electrons

An accelerated electron emits electromagnetic radiation, in this case dipole radiation (see Figures 6.1 and 6.2). Therefore, if the electron is accelerated in an electric field to high enough energies, it eventually will emit radiation with a wavelength λ that is in the range of the size of an atom: $\lambda = 0.2\,\text{Å}$ to $2.5\,\text{Å}$.

6.1.1 Laboratory X-ray sources

A simple device to produce X-rays is a vacuum tube (diode tube), where electrons are emitted from a heated tungsten filament via thermionic emission. The filament is at a negative potential so that the electrons are accelerated in a high electric field toward the grounded anode. The electrons travel at high speed to the anode target and are stopped (decelerated) by the electrons in the material. Since this impact will stop the electron in a very short time over a short distance, the deceleration (negative acceleration) is large, producing radiation up to the kinetic energy of

$$E = eV = \frac{1}{2}mv^2 = hv = h\frac{c}{\lambda}$$

For the wavelength to be in the desired region, the acceleration voltage is between 5 kV and 60 kV. This requires a high voltage source with the capability to deliver a high current. For example, medical X-ray systems may work at voltages up to 150 keV to produce radiation sufficiently energetic to penetrate tissue and bone. For these voltages, the electron will reach a speed that is about 1/3 to 1/2 of the speed of light and, therefore, can be treated as nonrelativistic. Unfortunately, the deceleration of the electrons is not a very favorable emission process, only about 1 % of the kinetic energy of the electrons is converted to radiation, while 99 % is converted to heat. This requires efficient cooling to prevent the X-ray tube target from melting. For a traditional X-ray tube, the power load is of the order of 1–3 kW for a stationary target, with typical voltages up to $V = 50\,\text{kV}$ and currents up to $I = 40\,\text{mA}$, for a tube power of about 2 kW. If the target is moving/rotating at a high speed (40 to 50 Hz), the heat deposited by the electron beam is removed efficiently, and power levels up to about 18 kW are routinely achieved (60 kV, 300 mA). Heat transfer is the ultimate limit, as the material stopping the electrons should have either high melting point or a high thermal conductivity, or both. A smaller focal spot where the electrons impinge on the material makes the heat removal easier. For stationary and rotating targets, copper is an excellent material due to its high thermal conductivity allowing a high power load. Traditional stationary

https://doi.org/10.1515/9783110610833-006

copper anode tubes can have power loads up to 3 kW depending on the focal spot geometry. Similarly, molybdenum and tungsten, due to their high melting temperatures, can also tolerate high power loads.

X-ray tube spectrum and power loads

The radiation spectrum that is produced by the deceleration of the electrons will be continuous, with the short wavelength limit (SWL) due to the maximum accelerating voltage. The short wavelength limit is given by

$$\lambda_{SWL}[\text{Å}] = \frac{hc}{eV} = \frac{12398}{V}$$

with V given in Volts, h the Planck constant, c the velocity of light and e the electron charge. Typically, an X-ray tube is constructed using a tungsten filament and a Wehnelt cylinder to focus the electrons onto a rectangular spot (focal spot) of the order of $10 \times 1\,\text{mm}^2$ (normal focus), giving a specific power load at 3 kW total power of the order of $300\,\text{W/mm}^2$ or $3 \times 10^8\,\text{W/m}^2$. This rectangular shape is traditionally observed under an angle of $6°$, given a foreshortening factor of 10. Different X-ray source geometries are therefore possible: if the rectangle is viewed along the short or the long dimension, either a $1\times 1\,\text{mm}^2$ spot or a $10 \times 0.1\,\text{mm}^2$ line is observed. A *fine focus tube* has a rectangular spot size of 8 mm $\times$ 0.4 mm, and at 1000 W total power, it has a specific power load of $312\,\text{W/mm}^2$. The *long fine focus tube* has an extended rectangle: $12 \times 0.4\,\text{mm}^2$, and can handle a 50 % higher total power than a fine focus tube. Rotating targets spinning at high speed can increase the specific power load by almost an order of magnitude. For instance, a 18 kW rotating anode X-ray generator with a normal focal spot size may have a power load of $1800\,\text{W/mm}^2$.

The X-ray tube is under vacuum, and the X-ray ports are X-ray transparent windows using thin beryllium metal. The traditional tube design has standardized dimensions, and four X-ray windows, two for line and two for spot focus. Depending on the application, either spot or line focus is chosen. Due to the presence of metallic beryllium, disposal of an X-ray tube has to follow local regulations.

Microfocus tubes

Microfocus tubes are more widely adopted for use in single crystal diffraction applications, where only a small sample volume is irradiated with a brilliant X-ray beam. The small spot diameter facilitates heat removal and enables higher specific power loads. This has resulted in shorter data acquisition times and enables to routinely study small samples of less than 100 μm size. The smaller size and reduced electrical power requirements has resulted in very compact X-ray source designs with minimal cooling requirements. Rotating anode microfocus sources have pushed the specific power load to high levels; for instance, a 7 kW power rotating anode tube with a 100 μm diameter focal spot can have a power load of the order of 0.2×10^6 W/mm², significantly higher than a traditional stationary X-ray tube.

A novel technique to remove the heat from a small focal spot is implemented in the *MetalJet* design, where a liquid alloy of gallium-indium is formed into a fast moving jet that forms the anode material in the X-ray tube. The liquid jet is very efficient in continuously removing the heat from a focal spot with a size of the order of 20 µm diameter. With an electron beam power of 250 W a specific power load of the order of 0.2×10^6 W/mm² can be achieved. Since the anode material contains gallium and indium, two different emission wavelengths are available.

Continuous X-ray spectrum

The X-ray spectrum of a tube is continuous, with the high energy cutoff (SWL) as discussed above. The total X-ray energy emitted per unit time depends on the atomic number Z of the target material, since elements with a large number of electrons are more efficient in stopping the incoming electrons. The total intensity is given by

$$I_{tot} = AZIV^m$$

where I is the tube current and V the tube voltage, and the exponent m is of the order of $1.8 - 2$. Therefore, the emitted intensity has a linear dependence on the tube current, but has almost a quadratic dependence on the tube voltage. The optimal tube for a continuous spectrum will therefore use a heavy element as the target material, such as tungsten. For a given power, the voltage should be maximized, with the voltage limit for a tube at 60 kV. Therefore, for a tube power of 1 kW, a setting of 50 kV/20 mA is preferable over a setting of 20 kV and 50 mA. Voltages in excess of 50 kV do have a higher propensity for arcing, and a good compromise uses the settings 40 kV/25 mA. The continuous spectral distribution of wavelengths is called *white radiation* or *polychromatic radiation*, even though not every wavelength is present with an equal intensity. In particular, the radiation produced in an X-ray tube, where electrons are impinging on a target material, is called *Bremsstrahlung*, German for "stopping radiation." Since *Bremsstrahlung* is a dipole radiation, it has the familiar anisotropic intensity distribution and it is partially polarized (see Figures 6.1 and 6.2).

The time averaged Poynting vector in the far field, far away from the emitting dipole, has the form

$$\langle \mathbf{S} \rangle = \frac{\mu_0 p_0^2 \omega^4}{32\pi^2 c} \times \frac{\sin^2 \theta}{r^2} \mathbf{r}$$

with μ_0 the vacuum magnetic permeability, p_0 the dipole moment, ω the frequency, c the velocity of light, θ the angle between the dipole moment and the direction of the vector $\mathbf{r}$. The emission pattern is such that in the direction of the acceleration/deceleration, there is no Bremsstrahlung intensity. The continuous spectrum that is produced by the deceleration of the electron beam on a target that is positioned perpendicular

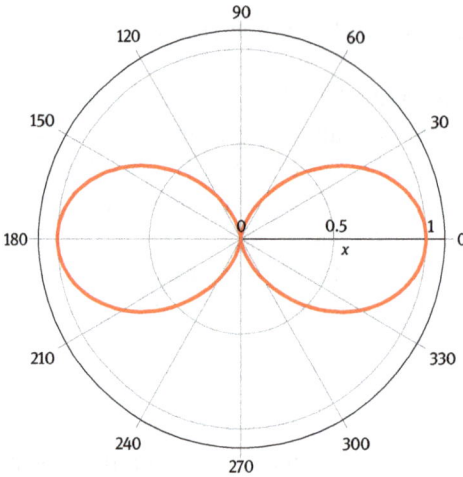

Figure 6.1: Dipole radiation of an accelerated electron. The acceleration is along the y-axis.

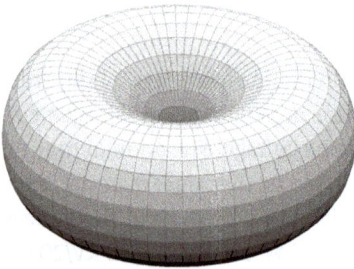

Figure 6.2: 3-dimensional representation of the dipole radiation field.

to the electron beam axis therefore has the maximum intensity parallel to the target surface. The take-off angle has thus to be small if the continuous spectrum is to be used, with take-off angles between 3 to 12°.

White radiation techniques for diffraction can be understood if the range of usable wavelengths is represented by a variable Ewald sphere in the Ewald construction. It immediately follows that for a single crystal, a wavelength can be found that satisfies the Ewald construction and therefore diffraction will occur without the need to move the crystal into a diffraction position. This is the *Laue* method and it is often used when a sample needs to be stationary. A *Laue pattern* using the white radiation from a tube is utilized for the precise alignment of a single crystal. The pattern will exhibit the symmetry of a given direction. In the case of a hexagonal crystal, the 6-fold c-axis will be easily recognized, and alignment of the axis in respect to the incoming beam can be achieved within a fraction of a degree. Additionally, the quality of a crystal can be assessed; well-defined reflections in the Laue pattern indicate a well-crystallized sample, whereas excessive streaking indicates a poorly crystallized sample.

6.1.2 Synchrotron sources

A magnetic device that bends the electrons into a circular orbit provides a continuous acceleration perpendicular to the direction of travel. This is achieved in a circular storage ring fed by a particle accelerator, where charged particles travel in the ring. Since the trajectory of the charged particles is bent, they emit radiation, and thus lose energy. The energy loss is offset by an acceleration section to maintain the kinetic energy of the particles. In particular, in an electron or positron accelerator, magnetic fields are used to bend the particle beam (bending magnets), using magnets distributed along the circumference of the particle track. The storage ring therefore consists of straight sections and bending sections. At a bending magnet, the electron (positron) experiences the acceleration perpendicular to its trajectory and will emit *synchrotron radiation*, which is continuous over a large wavelength range. Due to the high particle energy of the order of 1–8 GeV, the electrons/positrons travel at a velocity that is close to the speed of light and, therefore, need to be treated relativistically. In the local reference frame of the charged particle, the radiation emitted is the dipole radiation as shown in Figures 6.1 and 6.2. In the reference frame of the outside observer, this produces a strong emission at high frequency due to the Doppler effect, which is concentrated in the forward direction. This results in a much higher energy density and brilliance (energy per area and solid angle) of X-ray radiation, several orders of magnitude larger than a laboratory source. The large range of wavelengths with high brilliance is coupled to a monochromator system, producing a spectrally pure X-ray beam with a wavelength spread of the order of $\Delta\lambda/\lambda \approx 10^{-3}$, and allowing tuning of the wavelength over a large range. The emitted beam is highly polarized in the plane of the circular motion and, therefore, diffraction and scattering experiments are oriented perpendicular to this plane. New types of experiments are possible, such as resonant scattering, anomalous diffraction, magnetic X-ray diffraction and magnetic circular dichroism to mention just a few.

The success of the beam lines at the first generation synchrotrons led to second generation synchrotrons, where the storage ring was optimized to generate synchrotron radiation and straight sections were included in the electron/positron path for *insertion devices* to increase the flux of the radiation.

Two types of insertion devices are used: *wigglers* and *undulators*. Both combine magnetic fields in a way to produce oscillatory lateral deflections of the electrons/positrons along the straight path. A *wiggler* induces an oscillatory motion (wiggles) the electrons/positrons. This electron motion occurs at a higher frequency than in a bending magnet, and leads to an increase of the emitted radiation, generally at a shorter wavelength. In the wiggler, the emitted radiation adds incoherently, but has a large wavelength range. An *undulator* is operating in a similar fashion, but is set up so that the oscillating electrons/positrons and the emitted light are in-phase, producing coherent addition of the emitted electromagnetic waves. This leads to a higher brilliance and higher intensity, since out-of-phase radiation is reabsorbed. The undulator is in effect a special wiggler that is tuned to produce coherent radiation in a high brilliance beam. While the emitted radiation from a wiggler is "white," the radiation from an undulator is at a specific wavelength and its harmonics.

With such a high intensity/high brilliance beam, the power absorbed by the sample can lead to sample degradation. This is pronounced for organic and biological samples, and can be can be overcome by cooling the sample to low temperatures and by reducing the time for the experiment. In addition, the high brilliance allows to study smaller samples.

Synchrotrons are large machines: the Advanced Photon Source (APS) at Argonne National Laboratory has a ring circumference of 1104 m with more than 1000 electromagnets and associated equipment to store charged particles on a circular path (http://www.aps.anl.gov). Currently, there are several large synchrotrons in operation in the world. Among them, ESRF (Grenoble, France; www.esrf.fr), APS (Argonne, US; www.aps.gov), Spring-8 (Sayo, Japan; www.spring8.or.jp/en/), NSLS-II (Brookhaven, US; www.bnl.gov/ps/nsls2/about-NSLS-II.asp), Diamond (Didcot, UK, www.diamond.ac.uk), Petra III at DESY (Hamburg, Germany, www.desy.de) are large operations with several hundreds of scientists developing new diffraction and scattering techniques to study materials. More information can be found at http://www.lightsources.org.

6.1.3 Free electron laser

A further development is the free electron laser, where an undulator of the order of 50 m in length is fed a single train of high energy electrons from a linear accelerator. To achieve coherent radiation in the X-ray region, a single pass through the undulator is required, since no high reflectivity X-ray mirrors are available. As the electrons travel along the undulator, the electrons start to interact with the radiation they emit. This interaction feeds back on the electrons, accelerating some, and slowing down others. As a result, the electrons "bunch" up and start to constructively add their emitted radiation. With increasing coherence, the interaction becomes stronger, producing further bunching. The wavelength of the emitted radiation is given by

$$\lambda = \frac{\lambda_u}{2\gamma^2}\left(1 + \frac{K^2}{2}\right)$$

where λ is the wavelength of the emitted radiation, λ_u the periodicity of the magnetic field of the undulator, γ the relativistic Lorentz factor, and K a parameter describing the strength of the undulator as

$$K = \frac{eB\lambda_u}{2\pi m_e c}$$

with B the magnetic field strength of the undulator, e the elementary charge, m_e the electron mass and c the speed of light. For λ_u of the order of centimeter and B of the order of 1 Tesla, the factor K is of the order of unity. Therefore, a λ_u of the order of one centimeter requires a y value of the order of 10^4 to reach a wavelength of the emitted radiation of the order of 1 Å. This requires that the electrons travel close to the speed of light, and a large linear accelerator is required. Furthermore, due to the finite length of the electron train, the radiation is pulsed, with pulse lengths of the order of 10^{-15}s, allowing to study physical behavior at very short time scales.

There are now several XFEL sources in operation, such as the electron linac at SLAC that has been converted into a free electron laser *Linac Coherent Light Source (LCLS)* (lcls.slac.stanford.edu), *FLASH* at DESY in Hamburg (flash.desy.de), the European XFEL in Hamburg, Germany (www.xfel.eu), SACLA at RIKEN, Harima (xfel.riken.jp), the PAL-XFEL (Pohang Accelerator Laboratory X-ray Free-Electron Laser) in Korea (pal.postech.ac.kr), and the SwissFEL in Switzerland (www.psi.ch/en/swissfel).

The brilliance of a free electron laser can exceed the brilliance of a synchrotron by several orders of magnitude (www.lcls.slac.stanford.edu), with femtosecond pulse length. Free electron lasers have not yet moved into the hard X-ray wavelength range, but are expected to do so in the near future. Thus, time resolved experiments with coherent radiation will become possible, expanding our understanding of the dynamics of complex systems.

6.2 Electron beam interactions with matter

If an accelerated electron beam impinges on a solid, the electrons are decelerated and scattered. Impinging electrons interact with the electrons that are bound in the atoms making up the target. If the impinging electron can transfer enough energy to free a bound electron, then the hole in the electron orbital of the atoms will be filled by another electron from a higher lying orbital of the atom. It is therefore necessary to recall the electronic level structure of an atom.

- The principal quantum numbers $n = 1, 2, 3, \ldots$ denote the electronic shells $K, L, M, \ldots$
- the quantum number l of the angular moment takes values of $l = 0, 1, n \ldots$ corresponding to the types of orbitals $s, p, d, \ldots$, respectively
- the orbital quantum number m takes the values $-l < m < +l$, thus, there is/are one s orbital, three p orbitals, five d orbitals, etc.
- the spin s of the electron has values $-\frac{1}{2}$ and $+\frac{1}{2}$
- the total angular momentum j of an electron is given by $j = l \pm s$; $j = 1/2$ for an s state, $j = 1/2$ or $3/2$ for a p state, $j = 3/2$ or $5/2$ for a d state, etc.

For the heavier atoms, the innermost electron shells are all filled, and the electrons in these shells are all paired. If an electron is ejected from the s state of an inner shell, the atom is then in a $S_{1/2}$ state due to the unpaired electron. In particular, for the K shell, the state is a $1S_{1/2}$ state, and for the L shell, a $2S_{1/2}$ state, respectively. If a p electron is ejected, the possible states are either $P_{1/2}$ or $P_{3/2}$, with the angular moment $l = 1$, and total angular moment of either $j = 1/2$ or $3/2$. Ionization, the creation of a vacancy in an atomic shell, will lead to a rearrangement of the electrons, where electrons from a higher level will transition into the lower level. In particular, a vacancy in the K shell will be filled from electrons from the L or M shell. These transitions are labeled $K\alpha$ for $L \rightarrow K$ and $K\beta$ for $M \rightarrow K$. Furthermore, atomic physics shows that for an optical transition (absorption or emission of a photon associated with an electronic transition), the quantum numbers are constrained to

$$\Delta l = \pm 1; \quad \Delta j = 0; \text{ or } \pm 1 \tag{6.1}$$

To fill a vacancy in the K shell ($l = 0$), a vacancy in the L shell with $l = 1$ is generated. This means that the states are either $P_{\frac{1}{2}}$ or $P_{\frac{3}{2}}$. Therefore, the characteristic emission that is due to a $L \rightarrow K$ transition is a doublet from $P_{\frac{1}{2}} \rightarrow S_{\frac{1}{2}}$ or $P_{\frac{3}{2}} \rightarrow S_{\frac{1}{2}}$. The new vacancies that are now created in the L shell are filled from the higher shells, producing a cascade of transitions and therefore characteristic emission lines due to $P_{\frac{1}{2}} \rightarrow S_{\frac{1}{2}}, D_{\frac{3}{2}} \rightarrow P_{\frac{1}{2}}, P_{\frac{3}{2}} \rightarrow S_{\frac{1}{2}}, D_{\frac{1}{2}} \rightarrow P_{\frac{3}{2}}, D_{\frac{5}{2}} \rightarrow P_{\frac{3}{2}}$, etc. As can be seen, such a cascade will never produce a $2S_{\frac{1}{2}}$ state in the L shell. It can only be created by ionization via absorption of a photon, or by electron impact. In Figure 6.3, an energy level map is presented, with the possible transitions indicated by vertical arrows. It has to be kept in mind that radiation with sufficient energy can also ionize an atom.

When bombarded with electrons or irradiated with hard X-rays, each element emits characteristic radiation at different energies, that allow to identify the element. The different characteristic emissions can therefore be used for chemical analysis, semiquantitative in the case of energy dispersive X-ray emission (EDX) spectroscopy or quantitative in the case of an electron microprobe where the analysis of the characteristic emission is wavelength dispersive. EDX is easily implemented in a scanning electron microscope, with a solid state energy dispersive detector, such as a PIN diode or a silicon drift detector. Emission of $K\alpha$ radiation from elements lighter than sodium are hard to detect due to their low energies, and used to required window-less detectors. Recent advances in detector technology managed to push the cut-off for light atoms to carbon and even lower. For this, detector surfaces are directly mounted to thin X-ray transparent window, or used in a window-less setup to reduce the absorption of the characteristic radiation.

To activate the characteristic emission line, the energy of an impinging electron beam or incident radiation needs to be able to eject an electron from the respective shell, for instance the K or the L shell. As an example, the energy of the copper (Cu) K shell energy is 8979 eV, the incoming electron beam or incoming radiation needs to

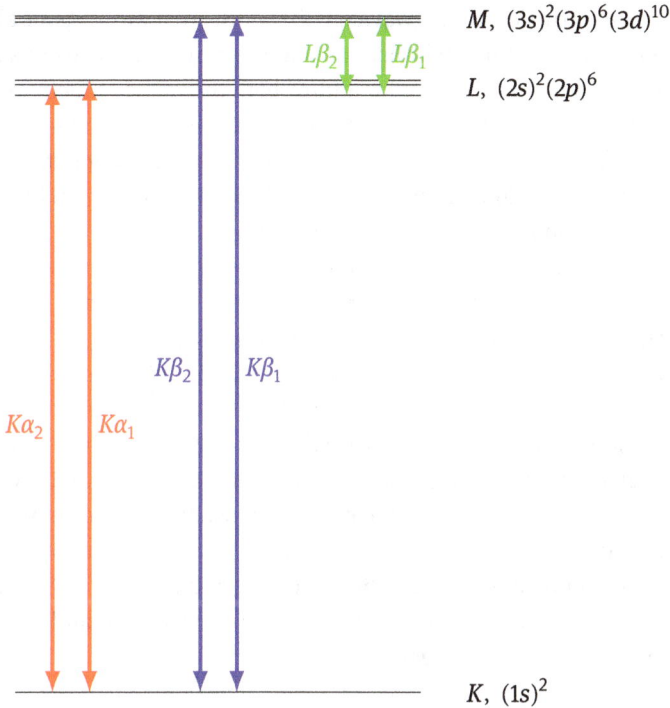

Figure 6.3: Energy levels and allowed transitions between electron shells.

have an energy of at least 8979 eV as this energy is needed to release an electron from the K shell and remove it from the atom. For an electron beam case, only the top layers of atoms can be ionized, since any energy loss experienced by the impinging beam will inhibit further ionization. Similarly, radiation will be strongly absorbed (absorption edge) in the top layers, limiting the penetration of the radiation into the bulk. The intensity of the characteristic emission lines satisfy the following expression for an electron beam accelerated by a voltage V impinging on a target that has K shell energy of eV_0:

$$I(K\alpha) = k(V - V_0)^n; \quad n \approx 1.8, \quad \frac{V}{V_0} < 4 \qquad (6.2)$$

with k a proportionality factor and $1.5 < n < 2$ for most elements. This relationship shows that, for a given power level, it is advantageous to increase the voltage instead of the current to increase the overall intensity.

The characteristic emission in the X-ray region has a small line width in energy (wavelength), and can be used as a source for X-ray radiation. This characteristic emission is strong, and exceeds the white radiation by orders of magnitude, and can be treated in most cases as a single wavelength source. This holds for the $K\alpha$ radiation pair $K\alpha_1$ and $K\alpha_2$, that have intrinsic energy (wavelength) widths that are in general much smaller than the instrumental line widths. The materials that can be used as suitable targets need to have a high melting point and/or high thermal conductivity, properties not often simultaneously met. Therefore, the most common target materials with high thermal conductivity are copper and silver, and with high melting temperatures molybdenum and tungsten. Copper gives a $K\alpha_1$ = 1.54059421 Å emission line, which is well suited to many applications [8]. Copper radiation is primarily used for laboratory powder diffraction and single crystal diffraction of organic crystals. Molybdenum, with a high melting point, is often used for single crystal diffraction of inorganic and small molecule crystals, with a characteristic radiation $K\alpha_1$ = 0.70932 Å. Similar, silver radiation with a wavelength for $K\alpha_1$ of 0.5594075 Å is also used for single crystal diffraction application. Compared to molybdenum radiation, silver radiation has a higher penetration, and is preferentially used for strongly absorbing materials. Tungsten, a very high melting point material, has an excitation energy for its $K\alpha$ lines in excess of 60 kV, exceeding the upper voltage limit of most high voltage power supplies used for diffraction. Furthermore, tungsten has a high Z, and is therefore an excellent choice for the production of white radiation. A tungsten target is therefore a good choice for the Laue method, where a continuous spectrum is required and only the weaker L-lines are present for tube voltages below 60 kV.

Figure 6.4 shows a spectrum obtained from an aged copper tube source operated at 20 kV, without any filter and the detector tuned to accept a range of wavelengths around copper radiation. The characteristic lines of copper, $K\alpha$ = 1.54059 Å and $K\beta$ = 1.392 Å, are superimposed on the Bremsstrahlung background. Since the plot is logarithmic in intensity, the weaker lines are emphasized.

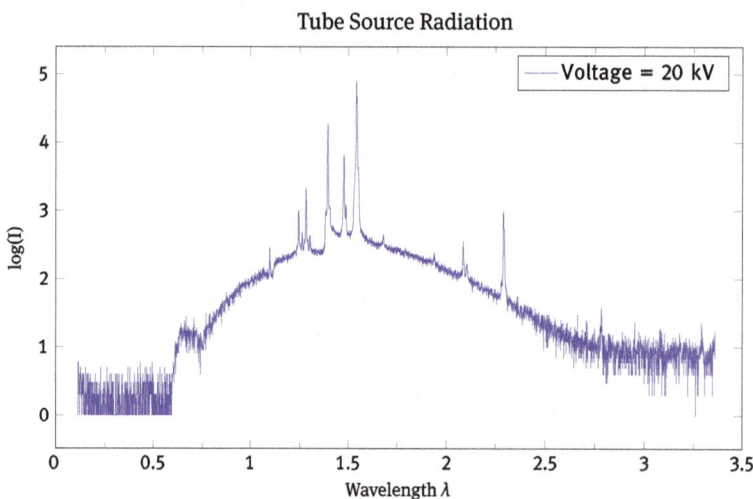

Figure 6.4: Radiation from an aged tube at 20 kV analyzed by a Ge (111) crystal. The SWL is at 0.62 Å.

Since several transitions between different shells are allowed, a laboratory source always emits multiple wavelength besides Bremsstrahlung. In suitable target materials, three wavelengths are most prominent, the $K\alpha 1$ and $K\alpha 2$ wavelengths that are close together, and the $K\beta$ wavelength. Of these, the $K\alpha 1$ line is most intense, and the $K\alpha 2$ line is half of the $K\alpha 1$ line intensity. The $K\beta$ line is about one-tenth of the intensity of the two $K\alpha$ lines (Figure 6.4). X-ray tubes that have been in service for an extended period often show additional lines that stem from contamination of the focal spot with materials present in the tube itself. In particular, tungsten contamination, with tungsten originating from the hot filament, produces a series of wavelengths due to electron transitions between the tungsten M and L shells. Contamination of the focal spot by iron is often observed as well, with iron originating from the Wehnelt cylinder and other tube surfaces constructed from steel. In addition, the more volatile elements in steel, chromium ($K\alpha$ = 2.290 Å) and manganese ($K\alpha$ = 2.102 Å) also give rise to spectral lines in the incident beam. It is therefore often necessary to filter the X-ray beam, or to use a monochromator system to obtain a "clean" incident spectrum. The effect of a filter is shown in Figure 6.5, where the X-ray beam emanating from the focal spot of an aged copper tube is analyzed with and without nickel filter, with the tube operated at 40 kV.

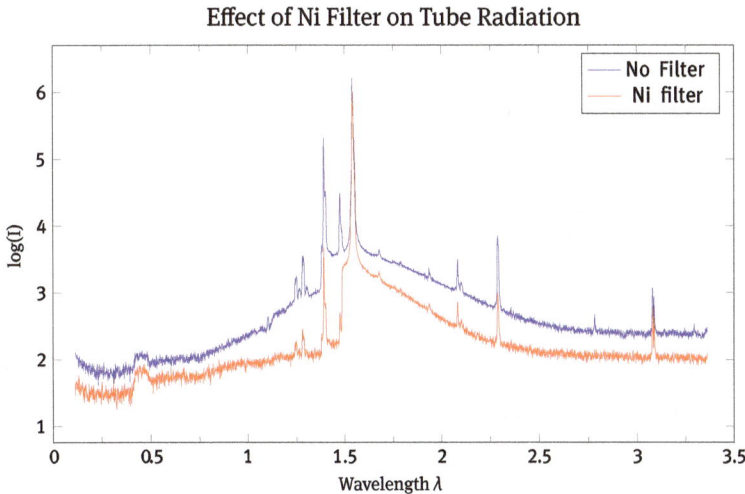

Figure 6.5: Radiation from an aged copper tube at 40kV without (blue) and with (red) Ni Filter analyzed by a Ge (111) crystal.

In the case of copper radiation, nickel is a suitable filter material that attenuates efficiently the Cu$K\beta$ radiation, since the Ni absorption edge is locaated between the Cu $K\alpha$ and $K\beta$ energies. The thickness of the nickel filter is small, of the order of 20 μm, to achieve attenuation of the Cu$K\beta$ line by a factor of 500, while passing the $K\alpha$ radiation.

The unfiltered radiation emanating from the focal spot of the tube at a voltage of 40 kV is attenuated below the nickel absorption edge of 1.4879 Å, reducing the intensity of the copper $K\beta$ radiation by more than an order of magnitude, while the copper $K\alpha$ intensity is attenuated by about a factor of 2.

The short wavelength spectrum intensity is also reduced by the nickel filter, giving a reduced background intensity. The attenuation of the $K\beta$ radiation by a filter is often sufficient for qualitative work. For samples where diffracted intensities are high, the $K\beta$ radiation will be visible in the pattern. The attenuation of the unwanted $K\beta$ emission line can exceed a factor of 10, and taking into account that the intensity ratio between the $K\beta$ to $K\alpha$ emission is of the order of 1:10, an overall attenuation of about 1:500 is easily achieved.

The choice of using a filter or a monochromator depends on the experiment: for powder diffraction, where a signal to noise ratio (S/N ratio) of 10^3 is desired, a filter may be sufficient. In the case of molybdenum radiation, a zirconium foil is usually used to attenuate the $MoK\beta$ radiation. However, for samples that fluoresce, a monochromator placed in the diffracted beam path may be needed to eliminate the unwanted background due to the fluorescence. In particular, copper radiation is sufficiently energetic to generate strong X-ray fluorescence in manganese, iron and cobalt. If no monochromator can be used, iron containing samples may be studied with $FeK\alpha$ radiation instead, avoiding the fluorescence background.

6.3 Monochromatization of X-rays

The emission spectrum of a tube is often conditioned to contain only a narrow wavelength range, in particular including a characteristic X-ray emission line. A monochromator can be used to pick this wavelength range via a Bragg reflection from a well-known lattice plane of the monochromator material. In many cases, the crystals used for these purposes are close to perfect, with a reflection coefficient for X-rays that can approach values of the order of 0.3 to 0.5, but only over a very small wavelength range. More often, however, the reflectivity of a monochromator is around 0.1 to 0.2, and may include a wider wavelength range. Such losses need to be offset by either increased power levels and brilliance, or through focusing methods that include X-ray optical elements such as multilayer mirrors, capillary optics or curved (focusing) monochromator crystals. An often used monochromator material is highly oriented graphite, providing good reflectivity for both the $K\alpha_1$ and $K\alpha_2$, while eliminating the $K\beta$ radiation. By using a monochromator for the diffracted beam, fluorescence can be eliminated at the cost of the intensity reduction.

A different approach involves an energy dispersive detector that is tuned to the desired wavelength range, acting as an energy bandpass similar to a monochromator. In this case, the intensity penalty is reduced, allowing a faster data acquisition. Modern energy dispersive X-ray detectors such as Si-PIN diodes and silicon drift detectors can have a band width of the order of 130 to 150 eV, sufficient to eliminate the Kβ radiation and pass all of the Kα radiation.

A thin film of an element with a high electron number Z has a total reflection angle of the order of 0.5° for copper radiation. It is therefore possible to build reflective optical elements that can focus X-rays. This is further enhanced if multilayer systems are employed, where light and heavy elements alternate, using Bragg reflection geometry of a 1-dimensional stack. Such X-ray mirrors (Göbel mirrors) will reflect at higher angles and can collect a substantial solid angle of radiation. If a parabolic multilayer mirror is used, then a parallel beam will be produced. An elliptical multilayer mirror can be set up so that the focal spot of an X-ray tube is in one of the focus points of the ellipse, with a convergent X-ray beam propagating from the mirror. With multilayer mirrors now available, X-ray beams can be conditioned in many ways. Furthermore, X-ray mirrors attenuate the Kβ lines emanating from an X-ray tube, and suppress background radiation. They function as band pass systems similar to monochromators. For multilayer mirrors, the surface and interface roughness needs to be minimized, since roughness will reduce the intensity.

Thin, hollow glass capillaries can be used to collect X-rays from a focal spot. The bending radius of the capillaries is such that internal total X-ray reflection conditions are satisfied, and the X-rays propagate inside the capillary. With the capillaries tapered, a high intensity is achieved at the capillary exit. Such capillary optics find application where a small spot illumination of a sample is needed, for instance, in microdiffraction applications.

6.4 Absorption coefficient

Every material absorbs X-rays, with an absorption coefficient μ that has units of inverse distance. The intensity of a propagating electromagnetic wave in a medium can be described by

$$I = I_0 e^{-\mu t}$$

with t the path length in the material, and μ its linear absorption coefficient. The linear absorption coefficient is again characteristic for each element, and it is a function of the energy (wavelength). The mass absorption coefficient for an element is defined as $\mu_{\text{element}}/\rho_{\text{element}}$, with ρ the density. The total mass absorption coefficient of a compound is obtained by the weighted average of the mass absorption coefficients of the constituents, with the weight fractions w_{element} according to the composition. The mass absorption coefficient for a compound with n elemental weight fractions w_i is expressed as

$$\frac{\mu}{\rho} = \sum_i w_i \left(\frac{\mu}{\rho}\right)_i, \quad i = 1\ldots n$$

The linear mass absorption coefficients for the elements are tabulated versus energy (wavelength) in the *International Tables of Crystallography*. Additionally, such information can be found online in the *X-ray Data Booklet* (http://xdb.lbl.gov). This publication further contains information on the interactions between X-rays and matter.

Optimal crystal size

For a single crystal measurement in transmission mode, the sample size may be chosen in regard to the absorption coefficient. The diffracted intensity is proportional to the volume of the crystal, but is attenuated by the X-ray beam traveling through the sample. The total diffracted intensity for a spherical sample can therefore be approximated by

$$I = I_0 \frac{4\pi}{3} r^3 e^{-\mu 2r}$$

with r the radius of the sample and μ the absorption coefficient. For maximal intensity, the derivative of I with respect to r is zero, giving

$$\frac{dI}{dr} = 0 = 4\pi r^2 e^{-\mu 2r} - \frac{4\pi}{3} r^3 2\mu e^{-\mu 2r}$$

leading to

$$r = \frac{3}{2} \mu^{-1}$$

or for the sample diameter, $d = 3\mu^{-1}$. Since in a single crystal measurement, the incident beam is collimated to about 0.8 mm to 0.5 mm diameter, a sample should not substantially exceed a diameter of 0.3 mm to ensure sample immersion in the plateau region of the incident beam. Therefore, weakly absorbing samples can be quite large, up to a diameter of 300 μm, whereas intermetallic and ceramic samples often have optimal sizes of less than 100 μm. If a strongly absorbing sample is too large, then the center of a reflection may have a lower intensity than the edges. Therefore, checking the reflection profiles is good practice to ensure that the data is of high quality.

Bibliography

[8] M. H. Menenhall, A. Henins, L. T. Hudson, C. I. Szabo, D. Windover, and J. P. Cline. High-precision measurement of the x-ray Cu Kα spectrum. J. Physics B, At. Mol. Opt. Phys., 80:115004, 2017.

7 Single crystal methods

7.1 Single crystal diffractometer

The discussion of the Ewald construction revealed that single crystal methods using a monochromatic radiation (single wavelength) require motion of the crystal to ensure that the Laue equations can be satisfied. This is achieved with a positioning device, a diffractometer, a combined system of rotation axes to align a reciprocal vector in respect to the incoming wave, and to position a detector at the proper 2θ angle to be in reflection position. In addition, the sample is mounted on a goniometer head that has at least x, y and z linear adjustments to position the sample in the center of the diffractometer. Goniometer heads with two additional arcs can also be used, allowing further adjustments of the sample. Figure 7.1 shows two types of goniometer heads, one for linear adjustments, and the other with additional orthogonal arcs that are located below the translation axes.

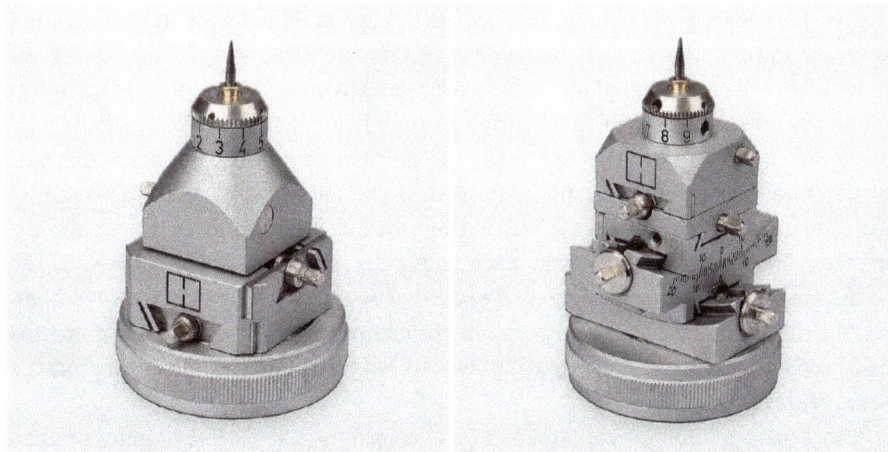

Goniometer head with linear translations.   Eucentric goniometerhead with two arcs.

Figure 7.1: Images courtesy of HUBER Diffraktionstechnik GmbH & Co. KG.

The crystal is glued on the top of a thin glass fiber that is potted into a 3 mm thick brass holder that fits into the top of the goniometer head. Magnetic mounting studs are also available, often equipped with cryoloops, where a thin fiber is mounted in a loop that allows to pick up a crystal with either a drop of solvent or oil. The crystal is immobilized by cooling the cryoloop below the temperature where the solvent/oil solidifies. This provides an excellent mount that exerts minimal amounts of strain. The goniometer is then mounted on the diffractometer and the crystal is positioned in the center of the diffractometer using the goniometer translation axes.

https://doi.org/10.1515/9783110610833-007

A modern 4-circle diffractometer is under control of a computer, equipped with stepper or servo motors, enabling positioning and scanning of all the axes simultaneously, with angular positioning accuracy of the order of $0.003°$. Dedicated systems can have higher angular positioning accuracy, but mechanical stability may impose limits. A 4-circle diffractometer can operate with a point detector, where a single diffracted intensity profile is registered by scanning one or two axes, or with an area detector, where simultaneously, integrated intensities of a number of different reflections are registered while the crystal is rotated. In the case of a point detector system, the diffractometer uses the 2θ and θ axes to satisfy the bisecting reflection condition ($\theta_{in} = \theta_{out}$), while two more angles χ and ϕ are used to align a reciprocal vector such that the Ewald construction is satisfied. The two axes to orient a crystal, χ and ϕ, are realized in a system that can be implemented using the convention by Euler. Such a system consists of a circle with its axis oriented horizontally providing the χ-angle, combined with a ϕ angle rotation shaft that can be positioned at any χ-angle. This system is called an *Euler cradle*. A general 4-circle diffractometer is therefore realized using an Euler cradle (χ- and ϕ-axes) mounted on an θ-circle that, with the detector mounted in the 2θ-axis, provides four degrees of freedom. To satisfy to Bragg condition for an arbitrary scattering vector, the ϕ-axis is used to rotate the vector into the χ-plane, the χ-rotation aligns the vector in the horizontal plane, the θ-axis is used to bring the reciprocal lattice vector into the reflection position against the primary beam, while the 2θ-axis moves the detector to the proper position to register the diffracted intensity. The scattering vector therefore is oriented in the plane formed by the incoming beam to the center of the diffractometer, and from the center to the detector setup. A simple 4-circle diffractometer is shown in Figure 7.2, and a schematic drawing is shown in Figure 7.6. The diffractometer axes are positioned at $2\theta = 30°$, $\theta = 15°$, $\phi = 0°$ and $\chi = 45°$. In addition, the diffractometer has graphite monochromators for the incident and diffracted beam, and is used for the analysis of highly textured thin films and to measure rocking curves of single crystals.

There are a number of variations to this design, such as partial χ-circles or open Euler cradles, avoiding the shielding of high angle reflections by the χ-circle. Offset χ-circles increase access to the sample volume and, for instance, allow mounting environmental chambers for high and low temperature measurements. An example of such a system is shown in Figure 7.3, where the schematics of an offset open Euler cradle is depicted. The ϕ-axis is positioned at $\chi = 180°$ and is designed to accommodate a cryostat or heater system.

In principle, only three axes are required to bring a reciprocal lattice vector into a reflection position to satisfy the Ewald construction. With four degrees of freedom provided by the 4-circle diffractometer, an additional constraint is needed. The *bisecting condition*, setting the reflection plane such that the angles between the reflection plane and the incoming and the diffracted beam are equal provides the additional constraint. The flexibility of the 4-circle diffractometer, however, allows other constraints to be imposed, depending on the sample geometry, or sample environmental

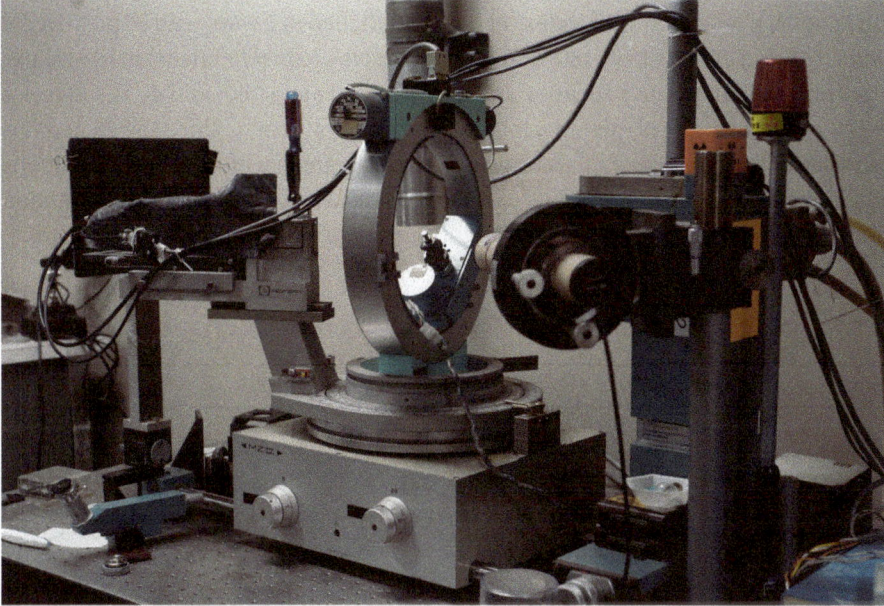

Figure 7.2: Small general purpose 4-circle diffractometer with Euler cradle and point detector: the axes are positioned at $2\theta = 30°$, $\theta = 15°$, $\phi = 0°$ and $\chi = 45°$. The system has a graphite incident beam monochromator and diffracted beam analyzer, and is used for texture, rocking curves and thin film diffraction.

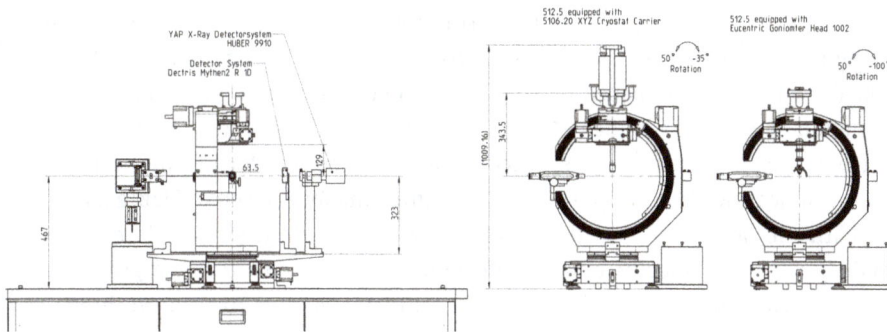

Figure 7.3: Schematic view of a Huber 4-circle system with Euler cradle, consisting of a Huber 424 $2\theta/\theta$ base and a Huber 512 open Euler cradle. Image courtesy of HUBER Diffraktionstechnik GmbH & Co. KG.

requirements that may limit the motion range of a particular axis. With a full χ-circle, the same crystal plane can be oriented in more than one way, and by convention, the value of χ is usually between 0 and 90°. Strictly speaking, only a quarter of the χ-circle is therefore needed. For a given orientation of an (hkl) reflection plane with angles χ and ϕ, the opposite vector $(\bar{h}\bar{k}\bar{l})$ has the angles $-\chi$ and $\phi + 180°$. For a full circle

Euler cradle, a reflection with index (hkl) can be aligned in 2 ways with 2θ positive: at $0 < \chi < 90°$ and ϕ, and at $-\chi+180°$ and $\phi+180°$. The related ($\bar{h}\bar{k}\bar{l}$) (Friedel pair) also has two positions with positive 2θ: at $-\chi$ and $\phi + 180°$, and at $\chi + 180°$ and ϕ. The detector can be positioned at $\pm2\theta$, giving a total of four positions each. Since the 2θ angle is identical for the Friedel pairs, accurate angle measurements of these eight positions for a number of (hkl)s give absolute 2θ values that can be used for the determination of lattice parameters. Table 7.1 lists the eight equivalent positions for a reflection (hkl).

Table 7.1: Equivalent angle setting for a reflection (hkl).

Setting	Index	2θ	ω	χ	ϕ
1	hkl	$+2\theta$	$+\omega$	$+\chi$	$+\phi$
2	$\bar{h}\bar{k}\bar{l}$	-2θ	$+\omega$	$+\chi$	$+\phi$
3	hkl	$+2\theta$	$-\omega$	$-\chi + 180°$	$+\phi + 180°$
4	$\bar{h}\bar{k}\bar{l}$	-2θ	$-\omega$	$-\chi + 180°$	$+\phi + 180°$
5	hkl	-2θ	$-\omega$	$+\chi + 180°$	$+\phi$
6	$\bar{h}\bar{k}\bar{l}$	$+2\theta$	$-\omega$	$+\chi + 180°$	$+\phi$
7	hkl	-2θ	$+\omega$	$-\chi$	$+\phi + 180°$
8	$\bar{h}\bar{k}\bar{l}$	$+2\theta$	$+\omega$	$-\chi$	$+\phi + 180°$

A different realization of a 4-circle diffractometer was pioneered by the company Enraf–Nonius®, where a κ-axis diffractometer was developed. In this geometry, the Euler cradle is replaced by a combination of 2 rotation axes: the κ-axis is positioned at an angle of the order of 50 to 60° to the θ-axis while the ϕ-axis is mounted on the κ-axis arm. The χ motion is replaced by a combination of the κ- and θ-axes motions, with maximum equivalent χ angles $\pm100°$ for a κ of 50°. The κ geometry has found wide acceptance as the sample is easily accessible from the top, for instance for sample cooling or heating. An example of a κ-axis diffractometer is given in Figure 7.4.

With the advent of area detectors, the χ-axis, while desirable, is not strictly needed, since the detector records intensities over a χ and 2θ range given by the detector area configuration. Therefore, the diffractometer will only need three axes, simplifying the construction. For a large area detector, it is even possible to only use one axis, the ϕ-axis, to rotate the crystal, since the 2θ-, and χ-positions are determined from the position of the reflection on the area detector. In this case, care must be taken that a unique set of reflections is collected in this geometry, as an accidental alignment of a high symmetry axis close to the ϕ rotation axis may restrict the accessible reciprocal space.

An example of a modern diffractometer with two microfocus sources and area detector and a 4-circle goniometer is shown in Figure 7.5. The two microfocus X-ray sources are on the left, the area detector to the right, the goniometer in the middle with a quarter χ-circle.

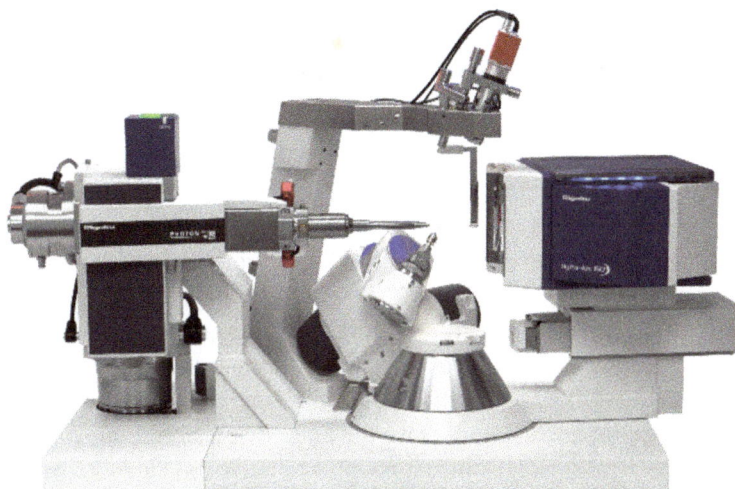

Figure 7.4: Rigaku κ-axis diffractometer with HiPix area detector. X-ray source is to the left, detector to the right. The κ-axis is positioned to the left of the θ-axis. Image courtesy of Rigaku Corporation.

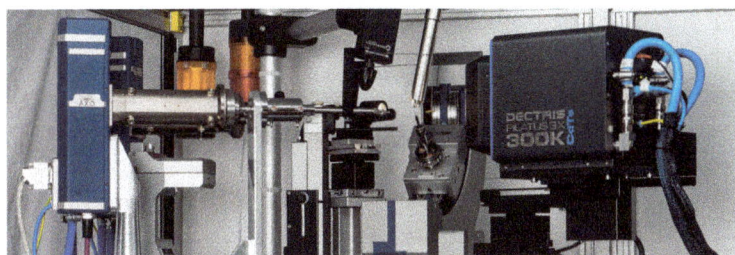

Figure 7.5: Dual microfocus source diffractometer with area detector and gonimeter with a quarter χ-circle. Image courtesy of STOE & Cie GmbH.

Data acquisition for a typical small molecule crystal can now be achieved in a matter of hours in a laboratory setting. This compares favorably to a system with point detector, where such a data collection took several days of measurements. Data acquisition times are even shorter at a synchrotron beamline, where full data sets can be obtained in a fraction of an hour.

7.1.1 Orientation matrix

A crystal mounted on a goniometer head will generally have its unit cell axes oriented in an arbitrary way. To explore the reciprocal space of this crystal, this orientation needs to be determined and mapped onto the coordinates of the diffractometer. For this, a number of conventions have been developed that define the setup of the diffractometer coordinates, and the mathematical relationships between unit cell axis directions and the diffractometer axes have been established. A short introduction is presented below.

Diffractometer angles and coordinates
A diffractometer shall be arranged in the following way: with the 2θ-axis vertical, a plane horizontal to it is defined: the primary beam will pass from the source through the center of the crystal to the detector (for $2\theta \neq 0°$) in this horizontal plane. The instrument angles are adjusted so that the primary and diffracted beam are in this horizontal plane. The plane of the Euler cradle is perpendicular to the primary beam for the case of $\theta = 0°$. In the special case where $\chi = 0°$, the 2θ-, θ and ϕ-axes are all collinear.

Moving the 2θ-axis by an angle 2θ will move the Euler cradle and sample by an angle of θ around the vertical axis. It is further possible to move the Euler cradle independently by an additional angle ω via the same axis. In this way, the reciprocal lattice vector bisects the angle between the primary beam and the diffracted beam. The ϕ rotation is supported by the Euler cradle that provides the χ-axis rotation. If the angle $\theta = 0$, then the central axis of the Euler cradle is aligned with the primary beam. From the 2θ, θ, χ and $\phi = 0$ position, the sense of the angles can be defined. The zero-point of the ϕ-axis can be chosen arbitrarily; per convention, it is set that with all other diffractometer angles at zero, the ϕ-axis is initialized with the position of the goniometer alignment pin parallel to the primary beam. The convention defined by Busing and Levy is shown in Figure 7.6.

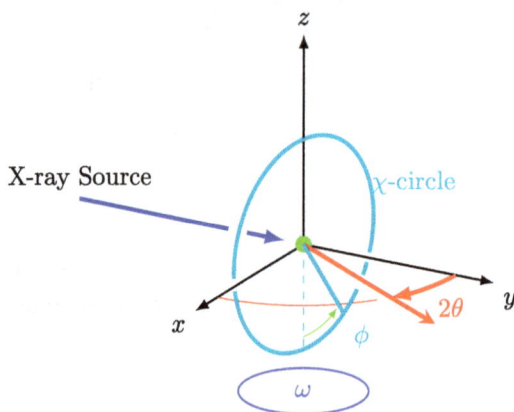

Figure 7.6: Diffractometer angle definition by the Busing and Levy convention.

Diffractometer coordinate system
A single crystal with an arbitrary orientation needs to be referenced in respect to the coordinate system of the diffractometer and an orientation matrix is determined to re-

late the two systems so that the Euler angles χ and ϕ (or κ-angles) for any index (hkl) (a reciprocal lattice vector) can be calculated. The *International Tables of Crystallography* contain a chapter on the mathematics of the orientation matrix, and the publication by *Busing* and *Levy* [9] gives details on how to determine the orientation matrix using three independent non-coplanar reflections, or two noncollinear reflections and the unit cell. The findings of Busing and Levy are briefly discussed. Several right-handed coordinate systems need to be defined. First, a reciprocal lattice vector $\mathbf{v}^*$ is defined by the three reciprocal vectors $\mathbf{b}_1^*, \mathbf{b}_2^*$ and $\mathbf{b}_3^*$, the reciprocal unit cell axes. The vector $\mathbf{v}_c$ is described in terms of crystal Cartesian axes that are attached to the reciprocal lattice (equation (7.1)). This Cartesian coordinate system is chosen such that the x-axis is parallel to $\mathbf{b}_1^*$, the y-axis is in the plane of $\mathbf{b}_1^*$ and $\mathbf{b}_2^*$, and the z-axis is perpendicular to this plane. It follows that

$$\mathbf{v}_c = \mathbb{B}\mathbf{v}^*$$

with

$$\mathbb{B} = \begin{pmatrix} b_1 & b_2 \cos\beta_3 & b_3 \cos\beta_2 \\ 0 & b_2 \sin\beta_3 & -b_3 \sin\beta_2 \cos\alpha_1 \\ 0 & 0 & 1/a_3 \end{pmatrix} \tag{7.1}$$

with a_i and α_i the direct, and b_i and β_i the reciprocal lattice parameters.

The ϕ-axis coordinate system is a set of Cartesian axes that is thought to be attached to the ϕ-axis of the instrument. With all instrument angles set to zero, its x-axis will point along the scattering vector, its y-axis will point in the direction of the primary beam, and its z-axis will be along the vertical instrument axis.

The matrix $\mathbb{U}$ is now an orthogonal matrix relating the ϕ-axis system to the crystal Cartesian system so that

$$\mathbf{v}_\phi = \mathbb{U}\mathbf{v}_c$$

The matrix $\mathbb{U}$ is called the *orientation matrix* since it relates the particular way the crystal is mounted to the ϕ-axis system of the instrument.

In a similar way, three additional Cartesian systems are defined, each attached to the χ-, ω- and θ-axes, respectively, for the case the instrument angles are all zero. It is possible to define transformation matrices for each of these coordinate systems:

$$\mathbf{v}_\chi = \mathbb{P}\mathbf{v}_\phi$$
$$\mathbf{v}_\omega = \mathbb{X}\mathbf{v}_\chi$$
$$\mathbf{v}_\theta = \mathbb{W}\mathbf{v}_\omega$$

with the matrices as

$$\mathbb{P} = \begin{pmatrix} \cos\phi & \sin\phi & 0 \\ -\sin\phi & \cos\phi & 0 \\ 0 & 0 & 1 \end{pmatrix}$$

$$X = \begin{pmatrix} \cos\chi & 0 & \sin\chi \\ 0 & 1 & 0 \\ -\sin\chi & 0 & \cos\chi \end{pmatrix}$$

and

$$W = \begin{pmatrix} \cos\omega & \sin\omega & 0 \\ -\sin\omega & \cos\omega & 0 \\ 0 & 0 & 1 \end{pmatrix}$$

Two more coordinate systems need to be defined, a laboratory system that is fixed with respect to the primary beam, and a 2θ system that is attached to the detector axis. These two systems will coincide with the previously defined ϕ coordinate system if the instrument angles are all set to zero. A vector $\mathbf{v}$ is transformed as follows:

$$\mathbf{v}_{lab} = \mathbb{T}\mathbf{v}_\theta = \mathbb{N}\mathbf{v}_\omega$$
$$\mathbf{v}_{2\theta} = \tilde{\mathbb{T}}\mathbf{v}_\theta = \mathbb{M}\mathbf{v}_\omega$$

with

$$\mathbb{T} = \begin{pmatrix} \cos\theta & \sin\theta & 0 \\ -\sin\theta & \cos\theta & 0 \\ 0 & 0 & 1 \end{pmatrix}$$

$$\mathbb{N} = \mathbb{T}W = \begin{pmatrix} \cos\nu & \sin\nu & 0 \\ -\sin\nu & \cos\nu & 0 \\ 0 & 0 & 1 \end{pmatrix}$$

with $\nu = \omega + \theta$, and

$$\mathbb{M} = \tilde{\mathbb{T}}W = \begin{pmatrix} \cos\mu & \sin\mu & 0 \\ -\sin\mu & \cos\mu & 0 \\ 0 & 0 & 1 \end{pmatrix}$$

with $\mu = \omega - \theta$. With the exception of χ, all angles are left-handed rotations around their respective axes.

It is now possible to write the *basic diffractometer equations* that govern the angle setting for a given reflection (hkl). For this, it is assumed that the diffractometer is perfectly aligned, the sample is at the center of the Euler cradle (χ-circle) and is an infinitely small perfect scatterer, and the primary beam emanates from a monochromatic point source. Since an actual instrument is a reasonable approximation of these conditions, the calculations are considered valid for real instruments. For a reflection $\mathbf{h} = (hkl)$ to be observed, it is necessary for θ to satisfy the Bragg equation and that the reciprocal lattice vector is along the x-axis of the θ-coordinate system. The length q of the scattering vector is easily evaluated in one of the Cartesian coordinate systems:

$$q = (h_{c1}^2 + h_{c2}^2 + h_{c3}^2)^{1/2}$$

$$\mathbf{h}_c = \mathbb{B}\mathbf{h}$$

with the Bragg equation giving

$$\sin\theta = \frac{\lambda q}{2}$$

For diffraction to occur in the horizontal plane, the vector $\mathbf{h}_\theta$ has to be of the form

$$\mathbf{h}_\theta = (q\,0\,0)$$

with the relation

$$\mathbf{h}_\theta = \mathbb{W}\mathbb{X}\mathbb{P}\mathbb{U}\mathbb{B}\mathbf{h}$$

In this way, the values for the diffractometer angles are obtained via the above condition that orients the reciprocal lattice vector of the crystal system parallel to the x-axis of the θ-coordinate system, and moves the detector to the 2θ position. The four equations given above constitute the *basic diffractometer equations*. It is, however, clear that there are an infinite number of angle combinations that satisfy the basic diffractometer equations, and it is necessary to introduce additional constraints to arrive at a small number of solutions.

Bisecting geometry

With the matrix $\mathbb{U}\mathbb{B}$ known, it is always possible to calculate

$$\mathbf{h}_\phi = \mathbb{U}\mathbb{B}\mathbf{h}$$

for any set of indices $\mathbf{h}$. In the case of a 4-circle instrument, it is customary to constrain the angles to *bisecting geometry* where the angle ω is constrained to zero so that the χ-plane (Euler cradle) bisects the angle defined by the primary and diffracted beams. Physically, this means that the vector $\mathbf{h}_\phi$ is brought into the scattering position by first rotating the angle ϕ to bring the vector into the χ-plane, and then moving the χ-angle to bring the vector into the horizontal plane. In this way, the (hkl) plane is in reflection condition. This procedure leads to the following angles:

$$\phi = \arctan\!\left(\frac{h_{\phi2}}{h_{\phi1}}\right)$$

and

$$\chi = \arctan\!\left(\frac{h_{\phi3}}{(h_{\phi1}^2 + h_{\phi2}^2)^{1/2}}\right)$$

These equations will give angles in the range $-90° \le \chi \le 90°$ since the square root is taken positive. Alternate setting of the angles are given

$$\phi' = 180° + \phi$$
$$\chi' = 180° - \chi$$

which is equivalent to a rotation about the scattering vector by 180°. The full set of equivalent positions is given in Table 7.1.

Evaluation of the orientation matrix
Knowledge of the crystal system (lattice parameters) was implicitly assumed for orienting a reflection. If three noncollinear reflections with their indices are known, the UB matrix can be determined. Given the angles of three reflections $2\theta_i$, ω_i, χ_i and ϕ_i for reflection i, the scattering vector in the ϕ-axis system can be computed:

$$\mathbf{h}_{i\phi} = \frac{\lambda}{2} \sin \theta_i \mathbf{u}_{i\phi}$$

with the vector $\mathbf{u}_{i\phi}$ given as

$$\mathbf{u}_\phi = \begin{pmatrix} \cos \omega \cos \chi \cos \phi - \sin \omega \sin \phi \\ \cos \omega \cos \chi \sin \phi + \sin \omega \cos \phi \\ \cos \omega \sin \chi \end{pmatrix}$$

For each of these three reflections, the matrix UB performs the transformation

$$\mathbf{h}_{i\phi} = \text{UB}\mathbf{h}_i$$

with $\mathbf{h}_i$ the index vector. Therefore, the matrix $\mathbb{H}_\phi$ constructed from the three column vectors $\mathbf{h}_{i\phi}$ and the matrix $\mathbb{H}$ constructed from the three index column vectors $\mathbf{h}_i$ are related by

$$\mathbb{H}_\phi = \text{UB}\mathbb{H} \tag{7.2}$$

and, therefore,

$$\text{UB} = \mathbb{H}_\phi \mathbb{H}^{-1} \tag{7.3}$$

In the case of an unknown unit cell, a larger number of reflections need to be found, so that a minimal direct unit cell can be determined. This means that the indices for three reflections need to be determined, with the additional reflections serving to check the index choices. A number of routines in direct and reciprocal space have been implemented to find the unit cell, with the routines usually part of the control software of a diffractometer. A typical way of determining the unit cell of a crystal is based on the search for a number (up to 20) of reflections, followed by an indexing routine to achieve integer (*hkl*) values for all reflections. This is usually followed by additional exploration of reciprocal space for additional reflections that may have fractional indices.

Area detectors are very efficient in recording a large number of reflections in a short time, thus it may be more advantageous to collect a large volume of reciprocal space before indexing all the reflections. Such measurements rotate one axis (ω- or ϕ-axis) by an angle of 0.3° to a few degrees while collecting a frame. Therefore, the positions of recorded reflections have well determined 2θ and χ-angles; however, the ω or ϕ angle is not always well determined. To find the unit cell, similar techniques as discussed above are used, with the added consideration that one angle may only be approximately known.

Once a valid orientation matrix is defined, it is possible to explore reciprocal space and align any (*hkl*) Bragg reflection, even with fractional indices. Scans in any direction in reciprocal space can be executed, and intensity recorded. It is therefore important to define the diffraction geometry: The incoming beam is collimated to reduce the angular spread (beam divergence), and the slits that define the diffracted beam control the diffracted beam acceptance angle. If a graphite monochromator is used, the incoming beam divergence is at least of the order of 0.2° due to the mosaic of the graphite. A crystal with a mosaic spread smaller than this will show an apparent reflection peak width that is defined by the mosaic of the monochromator and other diffractometer conditions. For instance, collimation using variable slits and beam collimators can reduce the beam divergence, and consequently, change the observed reflection profile.

7.1.2 Intensity measurement

For structure determinations, the position and the intensity at a reciprocal lattice point has to be measured. The ideal diffractometer setup provides an incident monochromatic planar wave that interacts with an ideal crystal where the reciprocal lattice points are δ functions, and the diffracted intensities are measured with high accuracy over a large dynamic range. It is obvious that such a system does not exist, and different compromises are made to achieve an accurate intensity measurement.

In a real crystal, the reciprocal lattice points are not δ-functions due to the interference function that is determined by the number of repetitions of the unit cell along the crystal dimensions. The mosaic structure of a crystal affects the reciprocal lattice by giving rise to an assembly of slightly misaligned domains distributed along a circle with the radius of the d-spacing of the (*hkl*). Any strain present in the crystal will broaden this circle, as the d-spacing has a not a single value. A laboratory X-ray source is polychromatic and may be monochromatized, and contains a range of wavelengths, usually centered around an element specific characteristic wavelength. In addition, the incoming beam has an angular divergence that is controlled by the beam collimation, providing a cone of different incoming beam directions. These effects are folded into the reciprocal lattice, where each reciprocal lattice point occupies an extended volume in reciprocal space. The wavelength range of the incoming beam can be modeled either as an assembly of Ewald spheres with different radii, or as an additional radial extension of the reciprocal lattice points. Since the reciprocal lattice "point" occupies a finite volume, integration over this volume is needed. It is therefore necessary to move the crystal through a number of positions (angular scans) while acquiring intensity measurements to obtain the integrated intensity that is then used to derive $|F(\mathbf{S})|$ values. Scans can therefore involve multiple axes in the case of a point detector, or a single axis in the case for an area detector. In both cases, an integrated intensity is obtained, whereas the detailed procedure depends on the type of instrument and the analysis software used.

The strategy for this measurement depends on the particulars of the diffractometer setup and the detector type that is used. In the following, a brief description of different detectors is given. Due to the rapid development of detector technology, this discussion will give a necessarily incomplete overview.

7.1.3 Detectors

A radiation detector measures, in some way, the number of Xray photons per unit time that enter through its aperture. An ideal detector therefore registers every X-ray photon, preferably with infinite spatial resolution and infinite dynamic range, without any dark count. While such a detector does not exist, there are different detector technologies that approximate an ideal detector. There are two basic types of detectors, point detectors and area detectors, which will be discussed briefly.

Area detectors
In the case of a 2-dimensional detector or X-ray sensitive film, the positional resolution is defined by the pixel of the detector itself, or the size of the silver halide grains. Therefore, readout of a 2-dimensional detector produces a 2-dimensional map of the number of photons registered. A modern 2-dimensional detector can register photons directly, and the resolution is determined by the pixel size. The detector has a dynamic range, where the number of photons per second is limited due to physical phenomena, such as dead-time, a time window when the detector is processing one event and cannot register another photon, readout where the detector does not process any incoming photons, the maximum number of events that can be stored, etc. In some cases, the response to the accumulated photons is digitized, with an analog-to-digital conversion (DAC) system. Depending on the bit-resolution of the DAC, the total number of photons that can be counted in a given time interval may be limited.

A charge coupled device (CCD) detector, where the CCD chip is sensitive to photons in the visible range may be combined with a fluorescence layer, where the X-ray photon interacts with the atoms to produce light that the CCD chip can register. In this case, one X-ray photon may produce many "down converted" photons, thus enhancing the detector sensitivity. CCD chips tend to have a small area, insufficient to cover the large area desired for fast data acquisition. This was overcome through a combination of a large fluorescence layer and a fiber-optic taper, a system of a well-ordered optical fiber bundle that is drawn in a way to match the cross-section at the large area of the fluorescence layer to the small area of the CCD chip. Such detectors need a geometrical correction to establish the correspondence of the fluorescent screen location to the CCD pixel location as the optical fiber taper tends to have some irregularities in the arrangement of the fibers. To reduce the inherent thermal noise of the CCD chip, a cooling system based on Peltier elements is often applied to bring the CCD to a temperature of about 230 K.

A 2-dimensional image plate detector uses a layer of barium sulfide doped with europium (BaS:Eu), or europium doped barium fluoride bromide (BaFBr: Eu). The interaction of X-ray photons with the layer will knock electrons from the europium atoms, in effect ionizing them and exciting the europium atom to a higher energy state. While some electrons immediately recombine with the europium atoms, other electrons get trapped at defects in the material, leaving the europium atoms ionized. The energetic depth of the trap is shallow so that a low energy photon (about 2 eV) can give the electron enough energy to leave the trap. A detrapped electron will eventually recombine with the ionized europium atom and the energy upon recombination is released via emission of a photon. This europium fluorescence has a well defined wavelength that is different than the detrapping light, and is registered with a photon detection system. Spacial resolution is achieved by scanning the image plate point-by-point using a laser light source to detrap the electrons and a fast photodetector (for instance, a photomultiplier tube) to register the europium fluorescence signal emanating from the illuminated spot. With the trap density sufficiently high, the image plate can register X-ray photons over a large dynamic range with high position accuracy. The image plate detector has thus replaced the traditional wet X-ray film that worked via sensitizing silver halide grains, followed by chemical fixation of the sensitized areas. An example of an image plate system is given in Figure 7.7, where a compact diffractometer setup is shown.

A recent development is a hybrid pixel detector based on silicon technology, where each pixel converts the impinging X-ray photons to an electronic charge that is read by a readout circuit. The readout electronic for each pixel is based on an application specific integrated circuit, and is fabricated independently of the active photon detection semiconductor. The photon detectors are silicon (Si) or cadmium telluride (CdTe) based, and are bonded to the readout circuit, providing direct readout for each pixel. These detectors can be operated in single-photon mode, where individual photons are counted within a preset time interval. They have low dark counts and

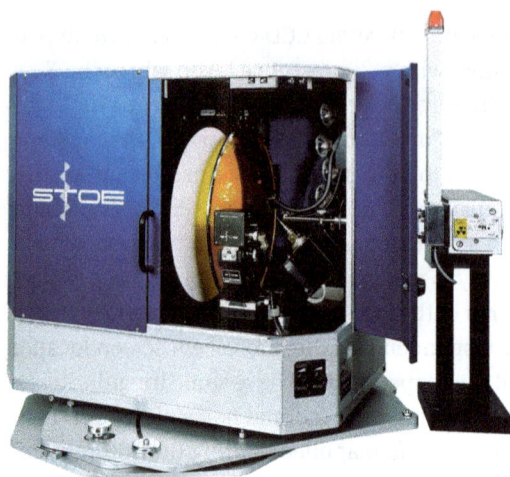

Figure 7.7: STOE IPDS-II image plate system, with the X-ray tube source on the right. Image courtesy of STOE & Cie GmbH.

a high dynamic range, combined with excellent spatial resolution. In addition, the detectors do not need to be cooled to low temperature, and can provide some amount of energy discrimination. They are finding application at synchrotron beam lines and in laboratory set-ups, and are now replacing CCD detectors. Examples of laboratory systems with hybrid pixel detectors are shown in Figures 7.4 and 7.5.

Point detectors

The scintillation detector registers X-ray photons via a scintillation process. The X-ray photon interacts with the atoms in a thallium doped NaCl crystal (NaCl:Tl), producing light pulses as the X-ray photon is slowed to a stop in the crystal. The light pulses are subsequently converted to an electrical signals using a photo multiplier or a photo diode. The stopping power of the NaCl:Tl crystal for different energy X-ray photons is known, and the conversion process is well understood. Since the X-ray photon is stopped in the crystal, the amount of light for a given energy X-ray photon is proportional to its energy. The integrated pulse strength (pulse height) of the electrical pulse is thus proportional to the energy of the incoming X-ray photon. The detector can therefore be tuned to accept photon energies in an energy window matched to the X-ray energy of the source. Unfortunately, the energy resolution of a scintillation detector is not very high, so that different wavelength are registered, such as the β-line, and possibly fluorescence from the sample. For Cu radiation, fluorescence is pronounced when Fe, Co and Ni atoms are present in the sample; atoms that can be excited by the Cu radiation to emit their characteristic radiation. The robustness of the scintillation detector, comparably low cost and low dark count make them useful for many applications.

Recent developments in detector technology have improved the energy resolution, most notably in silicon drift detectors and silicon PIN diodes, allowing to observe diffracted/scattered radiation for a narrow energy window, thus eliminating fluorescence effects that will give a high background. These detectors find application in powder diffraction, where energy discrimination is important to eliminate sample fluorescence for a high signal-to-background ratio. Their main application, however, is in energy dispersive spectroscopy for the analysis of sample composition, and they are often incorporated in a scanning electron microscope to provide elemental mapping capabilities as well as composition/stoichiometry determination. Energy resolution is of the order of 130 eV with count rates that can reach up to 10^6 counts per second. This energy resolution is sufficient to eliminate the $K\beta$ radiation. Furthermore, linear, semiconductor based detectors covering an extended 2θ range are incorporated in many diffractometer setups.

Crystal motion

The measurement of an integrated intensity requires different types of scans for different diffractometer setups. If a point detector is used, the 2θ and θ angles may be scanned over an angle range. A radial scan, a coupled $2\theta - \theta$ scan maintains bisecting geometry. In reciprocal space, such a scan is along a radial direction, going through the center of the reciprocal lattice point (hkl) and the origin. Enough of the background on the low as well as the high angle side of the reflection is usually measured so that the observed intensity is well-defined, and a profile fit may be used to subtract the background intensity. Using a laboratory tube source, the presence of the $K\alpha_1$ and $K\alpha_2$ radiation requires that the scan range is increased with the diffraction angle to ensure that the intensity is integrated over both wavelengths. A different scan type only scans the ω (θ)-axis. In this scan, the mosaic spread of the crystal is explored, and an integrated intensity from a crystal with a large mosaic structure can be measured. In such a scan, the 2θ-axis is stationary, usually with a slit opening in front of the detector that defines a 2θ-angle range that is sufficient for both the $K\alpha_1$ and $K\alpha_2$ radiation to be registered. Both types of scans give integrated intensities. It is therefore necessary that the orientation matrix is well-defined so that the scans are centered at the proper reciprocal space position. A $2\theta/\theta$ scan gives an intensity $I(\mathbf{S}) \propto |F(\mathbf{S})|^2$, which is an *integrated intensity*, capturing all the diffracted radiation from the total volume of the crystal.

If an area detector is used, the detector registers intensity over a 2θ range, as well as over a χ range. A ϕ- or ω-scan can is then used to obtain integrated intensities.

Since motion of the crystal is required, a fixed scan speed (angular velocity) is generally used. This has a geometrical effect in that the reciprocal lattice points further from the origin and perpendicular to the rotation axis travel faster and, therefore, spend less time in the diffraction position than reciprocal lattice points closer in. This effect can be visualized using the Ewald construction and is discussed further.

Lorentz factor

Measuring an integrated intensity of a reflection requires that the crystal be moved through the Ewald sphere while the intensity at each position is recorded. The measured intensity therefore needs to be normalized by the angular velocity of the crystal motion. A crystal of volume Δ is rotated with an angular velocity ω across a reflection position, with the rotation axis contained in the reflection plane hkl. These conditions are valid for all reflections that are in the equatorial plane, bisecting the Ewald sphere. The intensity of the primary beam impinging on the crystal is given by I_0 photons per unit time (seconds) and area (mm^2). A detector placed at a distance r from the crystal will receive all the diffracted intensity during the rotation, given as I photons. If the primary beam is polarized, the amplitude $\xi(\mathbf{S})$ of the scattered wave in direction of the scattering vector $\mathbf{S}$ is given by

$$\xi(\mathbf{S}) = \xi_0 \frac{1}{4\pi\varepsilon_0} \frac{e^2}{mc^2} \frac{1}{r} G(\mathbf{S}) \sin\phi; \quad \mathbf{S} = \mathbf{s} - \mathbf{s}_0 \tag{7.4}$$

with ϕ the angle between the direction of the polarization of the incident wave and $\mathbf{s}$. For an unpolarized wave, the scattered intensity in direction $\mathbf{s}$ is therefore $I(\mathbf{s}) = I_e|G(\mathbf{S})|^2$, with I_e the number of photons scattered by a classical electron comprising the polarization factor P. To obtain the total intensity I, an integration of the angular coordinates Ω of the interference function $J^2(\mathbf{aS})J^2(\mathbf{bS})J^2(\mathbf{cS})$ by using

$$I = \int_{\theta-\delta\theta}^{\theta+\delta\theta} d\theta \int_{1 \text{ refl.}} Id(\text{surface}) = r^2 |F(hkl)|^2 \frac{I_e}{V_n} \int_{\text{angles}} J_M^2 J_N^2 J_P^2 \, d\Omega \tag{7.5}$$

The term v_n is a function of ω and θ and represents the speed of the passage of a reciprocal lattice point traversing the Ewald sphere. Since the integration is carried out in angular coordinates, a modified Ewald construction is used (Figure 7.8): the sphere radius is normalized to 1 and the reciprocal lattice is multiplied by λ.

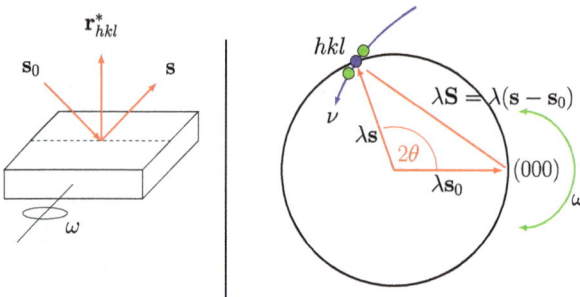

Figure 7.8: Lorentz factor corresponding to a crystal rotation around an axis perpendicular to the reciprocal vector $\mathbf{r}^*_{hkl}$.

Turning the crystal with an angular velocity ω, the reciprocal lattice point hkl moves with a velocity of $v = \lambda r^* \omega$. The transit velocity across the Ewald sphere is therefore the projection of $\mathbf{v}$ onto the direction of the scattering vector $\mathbf{s}$, and $v_n = v \cos\theta$. The length of the vector λr^* is $2\sin\theta$ and, therefore, $v_n = 2\omega \sin\theta \cos\theta = \omega \sin 2\theta$. The integrated intensity is reduced as the velocity v_n is increased. The quotient $L(\theta) = \omega/v_n$ is called the *Lorentz factor*. For the geometric condition shown in Figure 7.8, the Lorentz factor is

$$L(\theta) = \frac{1}{\sin 2\theta}$$

The integrated intensity is increased the slower the reciprocal lattice point moves across the Ewald sphere. By reducing the scan speeds for weak reflections, their integrated intensity can be measured more precisely. As the (000) reflection is always located on the Ewald sphere and is independent of the orientation of the crystal, its Lorentz factor is $L(0) = \infty$. Additionally, the Lorentz factor depends on the experimental conditions; a different diffraction geometry will use a different Lorentz factor. For instance, if the crystal rotation axis is not in the reflection plane (perpendicular to the reciprocal lattice vector), the velocity v is smaller. As an example, the Lorentz factor for a powder diffraction setup is often given as

$$L(\theta) = \frac{1}{\sin^2\theta \cos\theta}$$

The integration over a reciprocal lattice point can be approximated, and becomes the following:

$$\int_{\text{angles}} J_M^2 J_N^2 J_P^2 \, d\Omega = \int_{\text{max}} J_M^2 J_N^2 J_P^2 \, d^3(\lambda S) = \lambda^3 \int_{\text{max}} J_M^2 J_N^2 J_P^2 \, d^3(S) = \frac{\lambda^3 \Delta}{V_{\text{unit cell}}^2}$$

It is now possible to express the total integrated intensity obtained from a moving crystal (or an assembly of crystallites, powder) as

$$\frac{I(hkl)\,\omega}{I_0} = \left(\frac{1}{4\pi\varepsilon_0}\right)^2 \left(\frac{e^2}{mc^2}\right)^2 \lambda^3 \Delta \frac{1 + \cos^2 2\theta}{\sin 2\theta} \frac{|F(hkl)|^2}{V_{\text{unit cell}}^2} = KL(\theta)P(\theta)|F(hkl)|^2 \qquad (7.6)$$

It should be noted that the diffracted intensity depends on the unit cell volume, indicating that a large unit cell has relatively smaller diffraction intensities. This is due to the fact that a large unit cell will produce a large number of possible reciprocal lattice points that all receive some intensity, while the total scattering power is proportional to the electron density. For instance, organic molecules with large van der Waals distances have large unit cells and commensurate low electron density. Therefore, they do not diffract as strongly as for instance a metallic crystal with small unit cell and high electron density. Furthermore, the diffracted intensity is proportional to λ^3 (equation (7.6)). For the two most common wavelengths, Cu and Mo radiation, the difference is of the order of a factor of 2. Therefore, the same sample will diffract by almost an order of magnitude stronger using Cu radiation than Mo radiation. This is also a reason that for laboratory based systems, Cu radiation is often used for organic crystals to enhance the overall diffracted intensities, whereas Mo or Ag radiation are preferred for inorganic and intermetallic compounds. Laboratory diffractometers for protein crystallography use Cu radiation almost exclusively. In addition, absorption and extinction corrections need to be added to the expression (7.6), allowing to extract the structure factor $|F(hkl)|$ from the observed intensity.

Friedel law

The fact that the structure factor of a reflection $(\bar{h}\bar{k}\bar{l})$ is the complex conjugate $F^*(hkl)$ of $F(hkl)$ makes the observed intensities of the two reflections (hkl) and $(\bar{h}\bar{k}\bar{l})$ equal

$$|F(hkl)|^2 = |F(\bar{h}\bar{k}\bar{l})|^2 \tag{7.7}$$

Equation (7.7) represents the Friedel law and states that even for a non-centrosymmetric crystal, the two intensities are equal. Furthermore, the Friedel law explains why diffraction methods only allow to classify a crystal according to the 11 Laue classes, and not according to the 32 crystal classes. However, the Friedel law only holds approximately since it is valid for kinematical scattering and diffraction only. This is based on the assumption that the atomic form factor $f(\sin\theta/\lambda)$ is a function that gives real numbers. This is not strictly the case, since atoms do have an internal electronic structure that results in dispersion effects, where the X-ray photons interact with the electrons of the elements. These dispersion effects introduce an imaginary component to the atomic form factor. In particular, dispersion effects become prominent if the energy of the incoming X-ray radiation is close to the absorption edge of an atom. In the case of a centrosymmetric structure, there is always a symmetry related atom present that offsets this interaction. In a non-centrosymmetric crystal, this is not the case, and the structure factors no longer strictly obey the Friedel law, making $|F(hkl)| \neq |F(\bar{h}\bar{k}\bar{l})|$. This effect is used to determine the absolute configuration of a non-centrosymmetric structure, for example, it can be used to distinguish between a left- and a right-handed screw axis. Generally, the modification to the Friedel law is small, and intensities need to be measured with high accuracy to ensure that dispersion effects are significant. For an absolute configuration determination of a non-centrosymmetric crystal, the differences in the Friedel pairs, reflections $I(hkl)$ and $I(\bar{h}\bar{k}\bar{l})$ need to be determined. Application of the dispersion correction depends on the wavelength used, with the values for common wavelengths for each element listed in the *International Tables for Crystallography, Volume C* (https://it.iucr.org/C).

7.2 Space group determination

7.2.1 Crystal system and Laue class

The diffractogram, the total 3-dimensional diffraction pattern comprised of all accessible reflections from a single crystal, permits the classification of the crystal according to the 11 Laue classes. For example, for a crystal of the Laue class $4/m$, the reflections (hkl), $(\bar{k}hl)$, $(\bar{h}\bar{k}l)$, $(k\bar{h}l)$, $(\bar{h}\bar{k}\bar{l})$, $(k\bar{h}\bar{l})$, $(hk\bar{l})$, $(\bar{k}h\bar{l})$ are equivalent and show the same intensity. The Laue class $4/m$ comprises the crystal classes 4, $\bar{4}$ and $4/m$, and a projection of the diffractogram onto a plane perpendicular to the 4-fold axis is tetragonal.

For a crystal of the Laue class $4/mmm$ comprising the crystal classes $4mm$, 422, $\bar{4}2m$ and $4/mmm$, there is an additional symmetry compared to $4/m$; in this case, the intensities of the (hkl) and (khl) reflections are also equivalent. Therefore, in the general case (hkl) with $h \neq k$, $h \neq 0$, $k \neq 0$ and $l \neq 0$, there are 16 equivalent reflections. A projection of the diffractogram on a plane perpendicular to the 4-fold axis possesses the symmetry $4mm$.

A question now arises: Is it possible to obtain further symmetry information from a diffractogram? Some intensities in diffraction patterns are sometimes observed to be very weak or even zero since their structure factors are either very small or zero. If the indices of these absent reflections follow certain parity rules, then these missing intensities are called *systematic absences*. These *systematic absences* indicate the presence of translations, reflection glide planes, screw axes or in general, symmetry elements without fixed points. Three types of systematic absences are observed: *global, zonal* and *axial* absences. Global absences concern all reflections (hkl), while zonal absences concern reflections in a reciprocal lattice plane including the origin, for instance reflections of type $(hk0)$, and axial absences concerning a reciprocal lattice direction passing through the origin, for example, reflections of the type $(h00)$. Table 7.2 lists conditions for systematic absences.

7.2.2 Global absences, centered unit cells

The rules for systematic absences are deduced using the equation for the structure factor $F(hkl)$. For a C-centered unit cell, the vector $(\mathbf{a} + \mathbf{b})/2$ is a lattice translation: An atom in position (x, y, z) has an equivalent atom at the position $(\frac{1}{2} + x, \frac{1}{2} + y, z)$. The contribution of these two atoms to the structure factor $F(hkl)$ is given by

$$[f_m]_t e^{2\pi i(hx+ky+lz)}\{1 + e^{\pi i(h+k)}\}$$

The following conditions apply:

$$1 + e^{\pi i(h+k)} = 0 \text{ for } h + k = \text{odd}$$
$$= 2 \text{ for } h + k = \text{even}$$

Table 7.2: Systematic absences.

Reflection	Condition	Lattice or symmetry element	Translations
hkl	$h + k + l = 2N$	lattice I	$\frac{1}{2}(\mathbf{a} + \mathbf{b} + \mathbf{c})$
	$h + k = 2N$	lattice C	$\frac{1}{2}(\mathbf{a} + \mathbf{b})$
	$h + l = 2N$	lattice B	$\frac{1}{2}(\mathbf{a} + \mathbf{c})$
	$k + l = 2N$	lattice A	$\frac{1}{2}(\mathbf{b} + \mathbf{c})$
	hkl all even or all odd	lattice F	$\frac{1}{2}(\mathbf{a} + \mathbf{b}); \frac{1}{2}(\mathbf{a} + \mathbf{c}); \frac{1}{2}(\mathbf{b} + \mathbf{c})$
	$-h + k + l = 3N$	lattice R (reverse)	$\frac{1}{3}(2\mathbf{a} + \mathbf{b} + \mathbf{c}), \frac{1}{3}(\mathbf{a} + 2\mathbf{b} + 2\mathbf{c})$
	$h - k + l = 3N$	lattice R (obverse)	$\frac{1}{3}(\mathbf{a} + 2\mathbf{b} + \mathbf{c}), \frac{1}{3}(2\mathbf{a} + \mathbf{b} + 2\mathbf{c})$
0kl	$k = 2N$	plane b (100)	$\frac{1}{2}\mathbf{b}$
	$l = 2N$	plane c (100)	$\frac{1}{2}\mathbf{c}$
	$k + l = 2N$	plane n (100)	$\frac{1}{2}(\mathbf{b} + \mathbf{c})$
	$k + l = 4N$	plane d (100)	$\frac{1}{4}(\mathbf{b} + \mathbf{c})$
h0l	$h = 2N$	plane a (010)	$\frac{1}{2}\mathbf{a}$
	$l = 2N$	plane c (010)	$\frac{1}{2}\mathbf{c}$
	$h + l = 2N$	plane n (010)	$\frac{1}{2}(\mathbf{a} + \mathbf{c})$
	$h + l = 4N$	plane d (010)	$\frac{1}{4}(\mathbf{a} + \mathbf{c})$
hk0	$h = 2N$	plane a (001)	$\frac{1}{2}\mathbf{a}$
	$k = 2N$	plane b (001)	$\frac{1}{2}\mathbf{b}$
	$h + k = 2N$	plane n (001)	$\frac{1}{2}(\mathbf{a} + \mathbf{b})$
	$h + k = 4N$	plane d (001)	$\frac{1}{4}(\mathbf{a} + \mathbf{b})$
hhl	$l = 2N$	plane c ($1\bar{1}0$)	$\frac{1}{2}\mathbf{c}$
	$2h + l = 2N$	plane n ($1\bar{1}0$)	$\frac{1}{2}(\mathbf{a} + \mathbf{b} + \mathbf{c})$
	$2h + l = 4N$	plane d ($1\bar{1}0$)	$\frac{1}{4}(\mathbf{a} + \mathbf{b} + \mathbf{c})$
h00	$h = 2N$	helices $2_1, 4_2, [100]$	$\frac{1}{2}\mathbf{a}$
	$h = 4N$	helices $4_1, 4_3, [100]$	$\frac{1}{4}\mathbf{a}$
0k0	$k = 2N$	helices $2_1, 4_2, [010]$	$\frac{1}{2}\mathbf{b}$
	$k = 4N$	helices $4_1, 4_3, [010]$	$\frac{1}{4}\mathbf{b}$
00l	$l = 2N$	helices $2_1, 4_2, 6_3, [001]$	$\frac{1}{2}\mathbf{c}$
	$l = 3N$	helices $3_1, 3_2, 6_2, 6_4, [001]$	$\frac{1}{3}\mathbf{c}$
	$l = 4N$	helices $4_1, 4_3, [001]$	$\frac{1}{4}\mathbf{c}$
	$l = 6N$	helices $6_1, 6_3, [001]$	$\frac{1}{6}\mathbf{c}$
hhl	$h = 2N$	helix $2_1, [110]$	$\frac{1}{2}(\mathbf{a} + \mathbf{b})$

Since this relationship is valid for all atoms in the unit cell, the conclusion is that these conditions apply to all reflections (hkl), in this case, all reflections with $(h + k) = 2N + 1$ (odd) have intensity zero and are therefore systematically absent. For each of the centered lattices, type A, B, C, F, I, R, a rule for systematic absences can be constructed, whereas the primitive lattice P does not give any global systematic absences.

It is instructive to examine these rules in view of coordinate choices for the unit cell. Each centered unit cell can be transformed into a primitive unit cell, albeit without the symmetry apparent in the metric of the unit cell. In the case of a C-centered unit cell, the transformation given by

$$\mathbf{a'} = \frac{1}{2}(\mathbf{a} - \mathbf{b})$$
$$\mathbf{b'} = \frac{1}{2}(\mathbf{a} + \mathbf{b})$$
$$\mathbf{c'} = \mathbf{c}$$

The indices transform covariantly, given by

$$h' = \frac{1}{2}(h - k)$$
$$k' = \frac{1}{2}(h + k)$$
$$c' = c$$

Since the Laue equations are valid for integers only, the indices h' and k' are also integer numbers, and $(h + k)$ is therefore even. Indices with $h + k = 2N + 1$ (odd) are thus not part of the reciprocal lattice.

7.2.3 Zonal absences, glide planes

Given the existence of a reflection glide plane $\mathbf{a}$ perpendicular to the lattice vector $\mathbf{c}$ and passing through the origin, an atom of type m at position (x, y, z) has a symmetry equivalent atom at position $(\frac{1}{2} + x, y, -z)$ (or for the glide mirror $\mathbf{a}$ located at $\frac{c}{4}$: $(\frac{1}{2} + x, y, \frac{1}{2} - z)$). The contribution of these two atoms to the structure factor is

$$[f_m]_t [e^{2\pi i(hx+ky+lz)} + e^{2\pi i(hx+ky-lz)} e^{\pi i h}]$$

Therefore, indices belonging to the zone $(hk0)$ show intensities that are zero for $h = 2N + 1$ or odd integers. Therefore, glide mirror planes generate zonal absences characteristic with their orientation.

The origin of the zonal absences stems from the fact that the periodicity of the structure projected onto the glide mirror plane has a periodicity that is different from the structure. The Fourier transform of the projection, in this example taken along **c** onto the plane ($00l$), has the structure factor

$$F(hk0) = V_{\text{unit cell}} \int_0^1 dx \int_0^1 dy \int_0^1 dz \langle \rho(xyz) \rangle e^{2\pi i(hx+ky+0z)}$$

$$= \frac{V_{\text{unit cell}}}{c} \int_0^1 dx \int_0^1 dy \left[\int_0^1 d(cz) \langle \rho(xyz) \rangle \right] e^{2\pi i(hx+ky)}$$

$$= S_{ab} \int_0^1 dx \int_0^1 dy \langle \rho'(xy) \rangle e^{2\pi i(hx+ky)}$$

Here, S_{ab} is the surface area of the projection of the unit cell [**a** × **b**], $\langle \rho'(xy) \rangle$ is the projection of $\langle \rho(xyz) \rangle$ onto the plane **a**. However, this projection has a periodicity of $\mathbf{a}' = \mathbf{a}/2$. It follows that $h' = h/2$ is an even integer for all reflections with indices ($hk0$). A zonal absence indicates that a projection of the structure on a plane has a periodicity that is different from the original 3-dimensional structure due to a glide mirror plane (reflection followed by a translation).

7.2.4 Axial absences, helical axes

The existence of a screw axis 2_1 parallel to the axis **c** passing through the origin implies that an atom of type m in position x, y, z has a symmetry equivalent atom at the position $(-x, -y, \frac{1}{2} + z)$. If the axis does not pass through the origin but through a point (p, q) in the (**a**, **b**)-plane, the symmetry equivalent atom position is at $(p - x, q - y, \frac{1}{2} + z)$, with $(p, q) = (0, \frac{1}{2})$, $(\frac{1}{2}, 0)$, or $(\frac{1}{2}, \frac{1}{2})$. The contribution of these two atoms to the structure factor is given by

$$[f_m]_t [e^{2\pi i(hx+ky+lz)} + e^{2\pi i(-hx-ky+lz)} e^{\pi i l}]$$

For an axial reflection ($00l$), the following is obtained:

$$[f_m]_t e^{2\pi i l z} [1 + e^{\pi i l}]$$

For the second expression in the square brackets, the following relations hold:

$$1 + e^{\pi i l} = 2; \quad \text{for } l = 2N \text{ even integer}$$
$$= 0; \quad \text{for } l = 2N + 1 \text{ odd integer}$$

Therefore, the observed intensities along the reciprocal $\mathbf{c}^*$-axis are zero for l odd.

The origin of these absences can again be interpreted by the projection of the structure, but in this case, on the helical (screw) axis. By analogy, the structure factor is given by

$$F(00l) = c \int_0^1 dz \langle \rho''(z) \rangle e^{2\pi i l z}$$

with $\langle \rho''(z) \rangle$ the projection of $\langle \rho(xyz) \rangle$ onto the **c**-axis. If the direction of **c** coincides with an axis of type 2_1, 4_2 or 6_3, the projection is invariant via a translation $\mathbf{c}' = \mathbf{c}/2$. In this case, the index $l' = l/2$ is an even number for all reflection $(00l)$. A systematic axial absence indicates that the periodicity of the structure projected onto a line is a fraction of the periodicity of the 3-dimensional structure ($1/2$, $1/3$, $1/4$ or $1/6$), due to the presence of a helical (screw) axis.

7.2.5 Rotations and roto-inversions

Rotations and roto-inversions do not generate any systematic absences. This is unfortunate, as the presence of a center of symmetry greatly simplifies the mathematical procedures for calculations. Since diffraction methods do not always allow an unambiguous determination for the presence of a center of symmetry, other methods may be employed. For instance, a transparent crystal may be checked for its ability to double the frequency of light propagating though it. This effect, second harmonic generation (SHG) of light, is only possible if a center of inversion is *absent*. Therefore, positive SHG rules out the presence of an inversion center, while absence of SHG is a strong indication that the crystal possesses a center of inversion.

7.2.6 Deduction of the space group from the diffraction symbol

From an observed diffraction pattern, the Laue class can be determined, giving the crystal system information. Then, global absences are checked, followed by zonal and axial absences. This information allows to form the diffraction symbol that is used to identify possible space groups. The International Tables for Crystallography A contains tables that list the diffraction symbols and possible space groups, for all potential unit cell setups. There are 101 unique extinction symbols and 122 diffraction symbols for the 230 space groups. Therefore, an unambiguous space group determination is not always possible using the diffraction symbols. In only 50 cases is it possible to arrive at a unique space group, whereas the remaining 72 diffraction symbols indicate several possible space group symmetries.

As an example, a diffraction pattern displays the Laue symmetry *mmm*, indicating orthorhombic symmetry. No global absences are found, thus the unit cell is primitive *P*, with Laue group *Pmmm*. Further inspection finds the following zonal absences: (0*kl*): none, (*h0l*): $h + l = 2N$ (*n*-glide associated with the b-axis) and (*hk*0) : $k = 2N$ (*b*-glide associated with the c-axis). In addition, axial absences are (*h*00) : $h = 2N$, (0*k*0) : $k = 2N$, and (00*l*) : $l = 2N$. This gives the diffraction symbol *P–nb*. The possible space groups are thus $P2_1nb$ (#33) and *Pmnb* (#62), the former non-centrosymmetric with point group *2mm*, the latter centrosymmetric, with point group *mmm*. Space group #33 in the standard setting is $Pna2_1$. A unit cell transformation with $\mathbf{a}' = \mathbf{b}$, $\mathbf{b}' = \mathbf{c}$, $\mathbf{c}' = \mathbf{a}$ will arrange the unit cell for the standard setting. Space group #62 in the standard setting is *Pnma*, and the unit cell transformation $\mathbf{a}' = \mathbf{b}$, $\mathbf{b}' = \mathbf{a}$, $\mathbf{c}' = -\mathbf{c}$ will set up the unit cell for the standard space group setting.

In this example, the diffraction symbol led to two possible space groups. The presence or absence of the center of inversion needs to be established for an unambiguous symmetry assignment.

7.2.7 Calculation of structure factors for a diffraction pattern

An algorithm can now be described for the calculation of the diffraction pattern for a given structure. The structural information, usually given in a *Crystallographic Information File* (CIF), contains at a minimum, the unit cell dimensions, symmetry, atomic position and, if available, atomic displacement parameters. These data are used for the calculation of the structure factors $F(hkl)$. To obtain intensities, the geometrical factors need to be included, as well as source parameters, such as polarization of the incident beam, type of sample rotation, beam paths in relation to the sample, etc. First, the calculation of the structure factors will be discussed. An algorithm will be sketched, describing several steps, that can easily be implemented using a spread sheet calculator, MatLab, Mathematica, Octave or other computational tools. In addition, the web holds a number of useful applets, programs and tools for such calculations.

Generation of (*hkl*)

The unit cell information is used to generate a list of indices (*hkl*). For this, a cutoff needs to be defined, usually in 2θ or, if none given, the maximum possible for a given wavelength. A routine to produce a list of d-spacings $d(hkl)$ uses the unit cell information together with the symmetry information. The symmetry information will inform of systematic absences (global, zonal, axial), and the constraints and multiplicities of the generated (*hkl*). The d-spacing calculations are given according to:

Triclinic:

$$\frac{1}{d^2} = \frac{1}{V^2}(S_{11}h^2 + S_{22}k^2 + S_{33}l^2 + 2S_{12}hk + 2S_{13}hl + 2S_{23}kl)$$

where the following abbreviations are used:

$$S_{11} = b^2 c^2 \sin^2 \alpha$$
$$S_{22} = a^2 c^2 \sin^2 \beta$$
$$S_{33} = a^2 b^2 \sin^2 \gamma$$
$$S_{12} = abc^2 (\cos \alpha \cos \beta - \cos \gamma)$$
$$S_{13} = ab^2 c (\cos \alpha \cos \gamma - \cos \beta)$$
$$S_{23} = a^2 bc (\cos \beta \cos \gamma - \cos \alpha)$$

Monoclinic:

$$\frac{1}{d^2} = \frac{1}{\sin^2 \beta} \left(\frac{h^2}{a^2} + \frac{k^2 \sin^2 \beta}{b^2} + \frac{l^2}{c^2} - \frac{2hl \cos \beta}{ac} \right)$$

Orthorhombic:

$$\frac{1}{d^2} = \frac{h^2}{a^2} + \frac{k^2}{b^2} + \frac{l^2}{c^2}$$

Tetragonal:

$$\frac{1}{d^2} = \frac{(h^2 + k^2)}{a^2} + \frac{l^2}{c^2}$$

Rhombohedral:

$$\frac{1}{d^2} = \frac{(h^2 + k^2 + l^2) \sin^2 \alpha + 2(hk + hl + kl)(\cos^2 \alpha - \cos \alpha)}{a^3 (1 - 3 \cos^2 \alpha + 2 \cos^3 \alpha)}$$

Hexagonal:

$$\frac{1}{d^2} = \frac{4}{3} \left(\frac{h^2 + hk + k^2}{a^2} \right) + \frac{l^2}{c^2}$$

Cubic:

$$\frac{1}{d^2} = \frac{(h^2 + k^2 + l^2)}{a^2}$$

hkl list

The *Bragg equation* can now be used to calculate a $\sin \theta / \lambda$ for a given (hkl) index triple. A list of (hkl)'s an be generated by an algorithm where, for instance, the index l is varying fastest, index k follows l and h varies the slowest. Starting at (000), changing the index l will give (001), (002),..., until the Bragg equation sets a limit. The procedure is repeated with k incremented (010), then (011), (012),..., etc. Using the symmetry information, negative indices will have to be added where necessary. The list of hkl triples can subsequently be sorted according to increasing $\sin \theta / \lambda$.

Atomic form factor calculation

With the list of the hkl and $\sin\theta/\lambda$, the atomic form factor $f(\sin\theta/\lambda)$ can now be calculated for each hkl. If needed, the anomalous contribution to the form factor can be included as well. The form factor calculation is based on the parametrization given earlier:

$$f\left(\frac{\sin\theta}{\lambda}\right) = \sum_{i=1}^{4} a_i e^{-b_i(\frac{\sin\theta}{\lambda})^2} + c$$

Structure factor calculation

The structure factor can now be calculated, using the atomic form factors at every $\sin\theta/\lambda$, and the atomic positions in the unit cell. The anomalous effects can be included using the f' and f'' corrections by modifying the form factor

$$f_{atom} = f_0 + f' + if''$$

Here, f_0 is the kinematic form factor, and f' and f'' are wavelength dependent corrections for dispersion. For non-centrosymmetric structures, the differences in $F(hkl)$ and $F(\bar{h}\bar{k}\bar{l})$ need to be taken into account, and the $F(hkl)$ and $F(\bar{h}\bar{k}\bar{l})$ are listed separately. Atomic displacement factors are included at this point, since individual displacement parameters may be given. Anisotropic displacement parameters require additional computations that can be easily set up in subroutines.

Intensity calculations

With the structure factor $F(hkl)$ calculated, a raw intensity $I(hkl)$ is now obtained as $|F(hkl)|^2$ that includes all structural parameters. Experimental parameters are added for a particular diffraction geometry that includes geometrical factors, such as the *Lorentz* factor in addition to the *polarization* factor. Furthermore, normalization of the highest intensity to a value to 1000 can be added using a *scale* factor.

Several programs that perform this calculation are freely available. For instance, the program *VESTA* (https://jp-minerals.org/vesta/en/) can display structures and perform calculations of structure factors.

Bibliography

[9] W. R. Busing and H. A. Levy. Acta Cryst., 22:457, 1967.

8 Powder diffraction methods

8.1 Introduction

The power of powder diffraction methods is in their simplicity and ease of application. Unlike single crystal diffraction, where the crystal must be carefully oriented to
observe a diffraction intensity, the random orientation of many crystallites ensures
that there are always a number of small crystals in an orientation that satisfies the
Ewald construction. The reciprocal lattice concept is useful to understand the origin
of powder diffraction patterns. Starting with the reciprocal lattice and the Ewald construction where a single crystal is shown with its distinct reciprocal lattice points, a
large number of crystals with different orientations will now be considered. If all orientations are equally likely (with a probability of $1/4\pi$), the reciprocal lattice points are
now distributed on the surface of a sphere with the distance from the origin equal to
$1/d_{hkl}$. The surface generated for a (hkl) reflection will always intersect with the Ewald
sphere and ensures that the Bragg condition is satisfied. The intersection of the Ewald
sphere with the sphere of all reciprocal lattice points results in a circle that defines a
diffraction cone. Figure 8.1 shows this condition for one particular (hkl), together with
the resulting diffraction cone that can be recorded on a 2-dimensional detector.

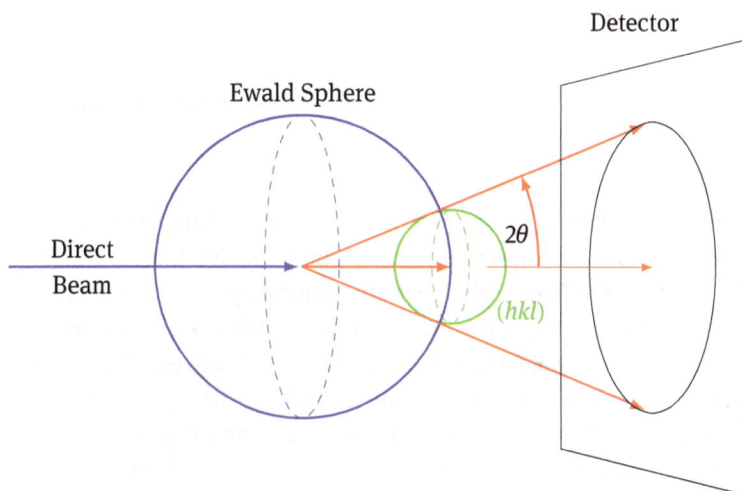

Figure 8.1: Ewald construction for a powder sample: the incident beam is shown in blue, the
d-spacing for (hkl) is given by a sphere (green); the diffracted cone (red) is recorded on a
2-dimensional detector.

The probability that a crystallite is found in an orientation that is consistent with the
Ewald construction, approaches unity and, therefore, the conditions for diffraction

https://doi.org/10.1515/9783110610833-008

will be fulfilled for an incoming wave with wave vector s_0. Since all possible orientations are present, a powder sample will produce a cone of diffracted intensity for each d-spacing present in the lattice. On a flat 2-dimensional detector intercepting the cone, circular or elliptical diffraction patterns are observed, depending on the orientation of the detector in respect to the incident beam and the sample. (see also Figure 8.2).

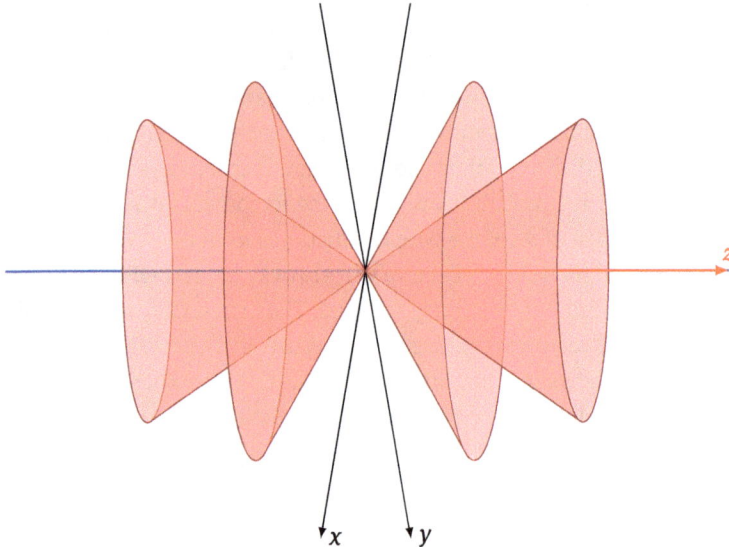

Figure 8.2: Debye cones from a powder diffraction experiment: The incident beam is along the z-axis (horizontal).

In the case of a simple cubic lattice, the reciprocal lattice points belonging to the same (hkl)'s have the same distance from the origin. A multiplicity factor is therefore needed to account for the number of different (hkl)'s with the same d-spacing. This increases the probability of finding a crystallite orientation that satisfies the Ewald construction. In Figure 8.3, a 2-dimensional cut through the reciprocal lattice is shown, with circles representing the random orientation of the reciprocal lattice points. Furthermore, each circle includes multiple reciprocal lattice points, and depending on the symmetry of the lattice, multiplicity factors are calculated to account for the different (hkl)s of the same family of reflections.

For the 3-dimensional simple cubic lattice, a given (hkl) has symmetry equivalent (hkl)s that have the identical d_{hkl}-spacing. All possible (hkl)'s are considered with their multiplicities based on all permutations and $(\pm h, \pm k, \pm l)$ possibilities. For example, the multiplicity of the (h00) reflections is 6 (three permutations and two $\pm h$), of the (hh0) reflections 12 (three permutations and four $\pm h$), for (hk0) reflections 24 (six permutations and four $\pm h, \pm k$), for (hhh) reflections 8 (eight $\pm h$), etc. The multiplicity for a

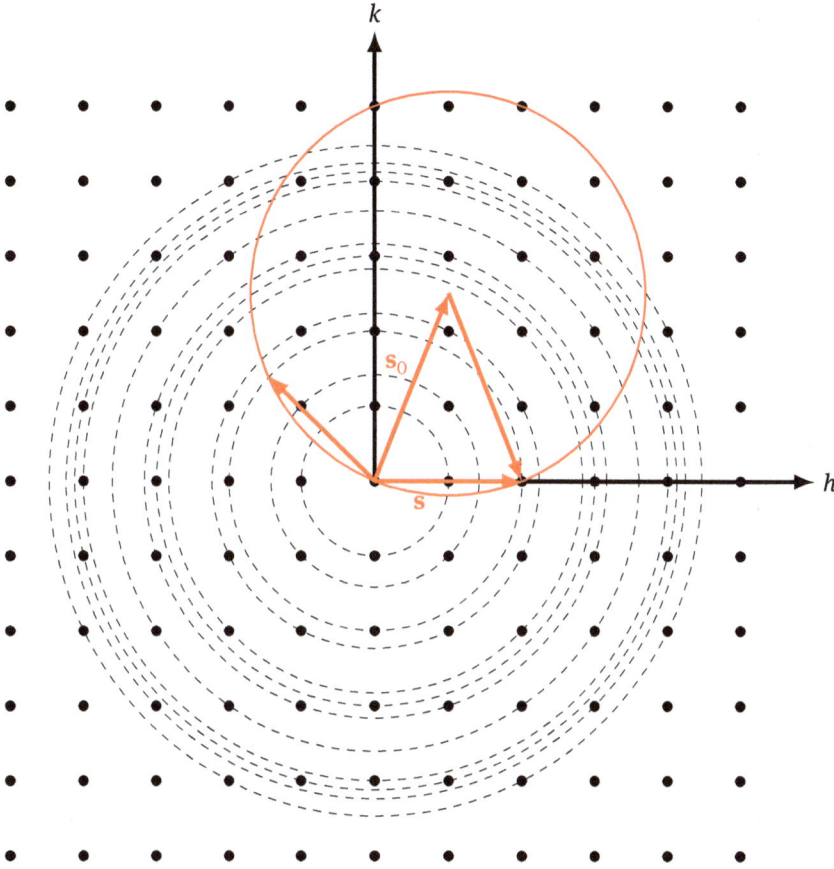

Figure 8.3: Two-dimensional reciprocal lattice with the first 12 spheres of random orientations represented by circles. Only the in-plane (*hk*0) triples are shown. The Ewald sphere is shown in red, with
scattering vectors **s** shown as red arrows. The two possible reflections with index (200) are shown.
Reflection are possible for each intersection of the Ewald sphere with the dashed circles.

general (*hkl*) in a cubic system (for instance, (123)) is therefore 48 (six permutations
and eight ± possibilities). For the calculation of intensities, the multiplicities for the
reflections need to be included since a larger multiplicity increases the probability of
finding a crystallite in reflection position.

The random orientation of many crystallites is required to ensure that the Ewald
construction is satisfied. While the complexity of a powder diffractometer is much reduced since no positioning systems such as an Euler cradle or κ-axis is required, the
sample preparation is more involved, as a perfect random packing of a large number
of small crystallites is desired. This has led to different sample preparation methods
to ensure that a *perfect random crystallite distribution* is achieved, without preferred
orientation of the crystallites. Sample movement during data acquisition is often used

to increase the crystallite sampling over different orientations. For many samples with a fine grain structure that combine a large number of small crystallites, this condition is naturally met. Examples include metals and alloys quenched from the melt, or mineralogical rock samples that have been cooled rapidly, and therefore contain a large number of small grains in random orientation. These materials approximate the perfect random distribution of crystallites and give excellent powder patterns, with even intensity distribution along the Debye cones. In the following, different implementations of powder diffraction techniques will be discussed briefly.

8.2 Debye–Scherrer camera

A simple implementation of a powder diffraction system is the *Debye–Scherrer/Hull* camera. The use of the term "camera" survived the era of photographic film that was used to record the diffraction intensities. The powder sample is filled into a capillary made from Lindemann Glass (low X-ray absorption glass) with diameter of the order of 0.3 mm or less, and wall thickness of the order of 10 μm. Care has to be taken that the capillary is well packed without preferentially aligning the powder grains. The sample in the capillary is then placed at the center of a cylinder that defines the position of the film/detector. For a laboratory setup, the preferred diameter of the cylinder is 114.59 mm (360 mm/π) so that 1 mm on the surface of the cylinder will by equivalent to 1°. A collimator defining a round beam delivers the X-rays to the sample from an X-ray tube mounted in point focus geometry. Opposite of the collimator is a beam stop that eliminates stray radiation to reduce the background due to air scattering. The beam diameter is matched to the capillary diameter, and the sample is rotated to increase the orientational distribution of the crystallites. Due to the light sensitivity of the photographic film, the cylinder is closed light tight. While the system is remarkably simple, the read-out of the reflection positions and intensities is more involved. For the read-out, a light box is used with a micrometer-coupled optical lens system to measure the reflection positions and estimate its intensity. More recently, optical scanning techniques were used to produce digitized diffraction data.

For a number of applications, Debye–Scherrer cameras are very useful, for instance, in the case when only very small amounts of a powder is available, or if a qualitative phase analysis is needed.

In the Debye–Scherrer camera, the diffraction cones intersect with the cylindrical radiation detector (photographic film, 2-dimensional detector), resulting in partial rings with different curvature that are recorded on the film (see Figure 8.4). For a 2θ of 90°, the partial ring is recorded as a straight line, whereas diffraction cones at very low (close to 0°) and very high (close to 180°) angles are recorded as full rings. Several methods to eliminate systematic errors were developed, such as the asymmetric positioning of the film strip in the camera to measure potential film shrinkage during processing, and measurement of the same reflection on both sides of the collima-

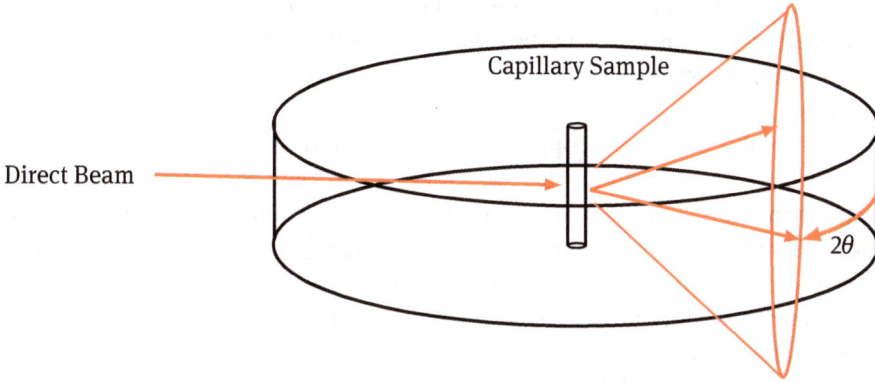

Figure 8.4: Schematic of a Debye–Scherrer camera: the sample is the cylinder in the middle, the incident beam comes from the left (red) and creates Debye cones (only one shown) that are recorded by the 2-dimensional detector (film).

tor/beam stop to accurately define the zero point of the angular measurement (Straumanis method).

The resolution in the Debye–Scherrer geometry is related to the diameter of the camera, as the separation of two closely spaced reflections increases with the camera radius. This is clearly a function of the collimation of the incoming beam that defines the incident beam divergence. The collimator design is optimized to achieve optimal illumination of the sample under condition of a well-defined incident beam direction. For the Bragg equation written as

$$\lambda = 2d \sin \theta$$

the full differential under condition that $d\lambda = 0$ gives

$$\frac{d\theta}{dd} = -\frac{\tan \theta}{d}$$

The separation D of two close reflections on the detector is proportional to the radius R of the camera and is given as $D = \theta \times 2R$. Therefore, the differential for the angle is

$$d\theta = \frac{dD}{2R}$$

Combining the two relationships gives the resolving power $\frac{d}{\Delta d}$ of the camera as

$$\frac{d}{\Delta d} = -\frac{2R}{\Delta D} \tan \theta \qquad (8.1)$$

where d is the mean d-spacing for two close reflections and Δd their difference. For two close reflections to be clearly identified, their minimum separation is given as

ΔD and is a function of the collimation and the sample. The resolution of the system therefore increases linearly with the camera radius R. Equation (8.1) also shows that the resolution increases with the tangent of the diffraction angle θ; therefore, high resolution is found in the back scattering region, where θ is larger than 45°.

In general applications, copper or iron radiation from a standard X-ray tube in point focus orientation and respective $K\beta$ filters are employed. If the sample fluoresces, then a different radiation wavelength will have to be used since the photographic film does not discriminate the energy of the diffracted radiation. Exposure times required for a Debye–Scherrer pattern were of the order of one to several hours for a copper tube with a power setting of 0.8kW and a double coated X-ray film, and a camera diameter of 114.59 mm.

The basic geometry of the Debye–Scherrer camera system has been implemented in a modern diffractometer, where the traditional silver halide film is replaced by a curved position sensitive detector spanning an extended angular range of the order of $2\theta = 120°$. Depending on the signal processing and digitization, the detector readout may be based on up to 4096 channels, resulting in a position resolution of 0.03°, sufficient for reflection line widths that are of the order of 0.15° full width at half maximum (FWHM).

8.2.1 Sample preparation

A cylindrical powder sample is needed for the Debye–Scherrer geometry, with crystallite size of the order of 40 µm (about 325 mesh). This can be achieved by filing of a metal, or by grinding a material that is not too ductile. Grinding is achieved using a (agate) mortar and pestle, and the resulting powder may be sifted through a 325 mesh sieve. The powder can then be attached to the surface of a glass fiber, using a small amount of adhesive, or filled into capillaries made from Lindemann Glass (borosilicate glass). Standard capillary diameters are 0.3 mm, but capillaries with diameter of 0.1 mm care also available, and larger diameters (up to 1 mm) may be utilized for samples with small absorption coefficients. However, the smaller diameter capillaries are more difficult to pack with a powder, and smaller powder particles are desired. Generally, 40 µm particles will pack well into a capillary with a diameter of 0.3 mm and wall thickness of 10 µm. Care has to be taken to achieve a dense packing of the powder in the capillary, and compacting the powders by various means, such as controlled dropping of the capillary, may be applied.

Materials with a large absorption coefficient, such as tungsten, absorb almost all of the X-rays in transmission mode. This leads to reflections that are split and offset to higher angles. To counteract such effects, the samples can be diluted with a material that does not show X-ray signatures, such as corn starch. However, dilution may reduce the number of powder particles to the point where the particle orientations are no longer random, affecting the quality of the reflections distributed in the Debye cone. One way to increase the orientation randomness is to rotate the sample around its cylindrical axis so that the orientations of the powder particles are sampled more extensively. For most Debye–Scherrer systems, a small motor to rotate the sample is integrated into the camera design.

8.3 Focusing systems

Diffractometers with focusing or parafocusing geometry can achieve higher resolution than Debye–Scherrer systems while retaining high intensity for powder samples. Figure 8.5 shows the parafocusing geometry, where rays emanating from a point source impinge on the sample that is curved along an arc from point A to point D. All angles, from point S (source) to point F (focus) via A, B, C and D all have the same angle between incoming and diffracted beam. A system may therefore use a point source with divergent beam that illuminates the sample mounted on the arc from A to D, and use the detector positioned in point F. Since the angles at points A, B, C and D have the value of 2θ, the angle between the source and focus is 4θ. The system has therefore twice the resolution than a Debye–Scherrer system, with the resolving power

$$\frac{d}{\Delta d} = -\frac{4R}{\Delta S}\tan\theta \qquad (8.2)$$

with S measured along the Rowland circle.

8.3.1 Seemann–Bohlin camera

This parafocusing system was described by Seemann and by Bohlin, and is therefore known as a Seemann–Bohlin camera. Instead of a point source, the system is cylindrical and a line source is used, as generated by an X-ray tube in line focus. The sample is now extended along the cylinder axis, and in the focus F, another slit matched to the source will be used. A 2-dimensional detector (film, image plate, etc.) can be placed on the Rowland circle, starting from D to close to the source (Figure 8.5). While the Seeman–Bohlin system has superior resolution compared to the Debye–Scherrer system, it is not possible to cover the same angular range as easily. This is due to the almost grazing incidence of the diffracted intensities close to $2\theta = 0°$ and $2\theta = 180°$. For an asymmetric setup, an angular range of about $2\theta = 120°$ can be achieved, with the detector covering $240°$. The line source will have a divergence in the vertical orientation that produces broadening of the reflections away from the equatorial position. This effect can be reduced by, for instance shortening the line source, or using additional incident beam conditioning elements that reduce the vertical divergence.

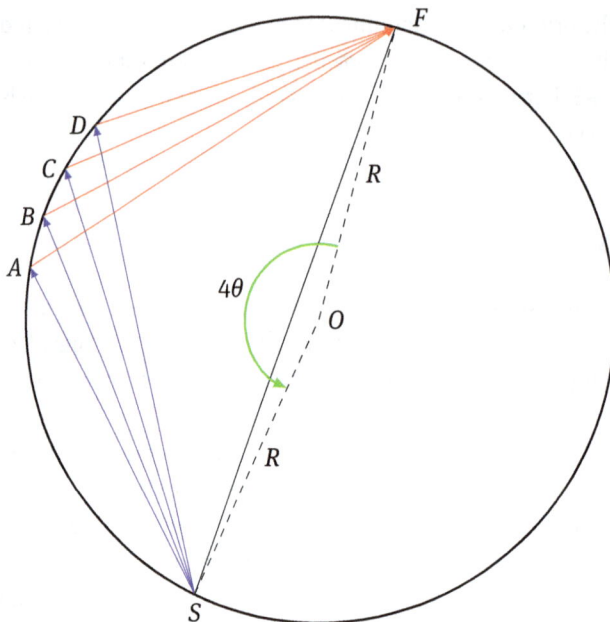

Figure 8.5: Parafocusing geometry.

8.3.2 Guinier camera

The Guinier camera utilizes a focusing incident beam monochromator and focusing geometry similar to the Seemann–Bohlin geometry (Figure 8.6). To achieve focusing, a crystal monochromator is bent and cut. If the monochromator is symmetrically cut and bent, the distances between source and focus are equal. For an asymmetrically cut and bent crystal (Johannson type), the distances between source and focus are unequal, and allow more flexibility in the construction of the system. The asymmetric cut is used for mounting the monochromator on the tube enclosure that necessitates a short distance from the tube focal line and a longer distance to the camera. The monochromator is a narrow pass filter that is adjusted so that only the respective $K\alpha$ lines, or even only the $K_{\alpha 1}$ line pass. The background in the pattern is thus strongly reduced, and results in a larger signal-to-noise ratio for the intensities. Unless the sample is fluorescing, the Guinier system will give a superior pattern compared to the Debye–Scherrer system.

The Guinier system modifies the intensities due to the monochromator that produces a partially polarized beam. With the diffraction angle of the monochromator given as α, the total polarization factor becomes

$$P = \frac{1 + \cos^2 2\alpha \cos^2 2\theta}{1 + \cos^2 2\alpha}$$

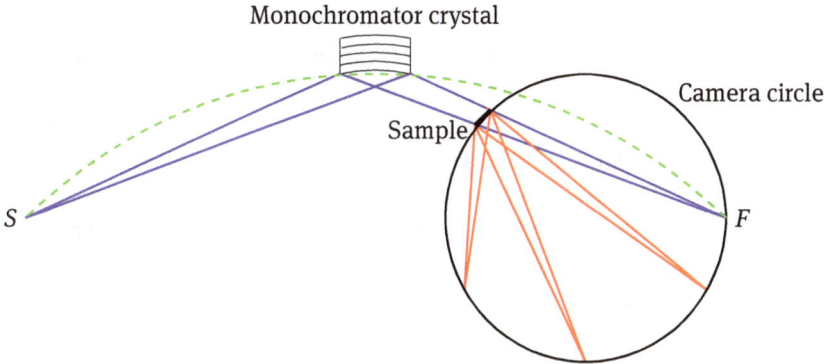

Figure 8.6: Schematic of Guinier system with monochromator crystal. The tube source S is imaged by the monochromator to the focus F (Rowland circle is shown as dashed green line); the sample is located on the camera circle. Diffracted intensities are shown in red. The detector is on the camera circle and registers 4θ.

The Guinier system has a number of different implementations that were given hyphenated names, such as the *Guinier–de Wolff* camera, the *Guinier–Hägg* system, or the *Guinier–Jagodzinski* system, all based on silver halide film detectors. Due to the nonperpendicular incidence of the reflections onto the film detector, single coated films were generally used, necessitating longer exposures than the Debye–Scherrer system.

A modern implementation of the Guinier systems uses an image plate in place of the film, with a self-contained read-out system, giving digitized intensity and position data (see for instance https://www.xhuber.com/en/products/2-systems/23-x-ray-cameras/g670). With the sample in transmission mode, thin samples are required to reduce attenuation by heavily absorbing elements. To offset the reduced number of crystallites, the sample is oscillated to illuminate a larger number of grains.

Guinier cameras use the line focus of a tube source, which presents a line width of the order of 40 µm under a take-off angle of 6°, indicating that under optimal focusing conditions, the FWHM could be of the order of 0.05° for a detector radius of 114.59 mm (equivalent to 360 mm circumference). FWHM values of 0.1° or better are easily obtained from well-crystallized sample, with small variations of the FWHM over the whole angular range. However, similar to the Seemann–Bohlin geometry, the vertical divergence from the line focus source tends to broaden the reflections away from the equatorial plane. This is usually addressed by adding a Soller slit system, a number of parallel plates of a heavy absorber material such as molybdenum or tantalum, to reduce the vertical divergence. The generally nonperpendicular incidence of the diffracted intensities leads to asymmetries in the reflection profiles, with the most symmetrical line profile at perpendicular incidence onto the detector.

Guinier system are very compact diffractometer setups with high resolution, that use a small amounts of material. They give excellent low background diffraction patterns due to the use of a crystal monochromator tuned to the $K_{\alpha 1}$ radiation. The standard angular range is of the order of 2θ from 5 to 100°, due to the fact that the detector registers 4θ, but different positions of the sample and detector are possible so that the full range, from forward to back scattering, is covered. Calibration of the system uses a National Institute of Standards and Technology (NIST) standard such as silicon or lanthanum hexaboride to obtain accurate peak positions. Often, such a standard is added to the powder for an *in-situ* calibration. The high resolution, low background and single wavelength, make the Guinier system well suited for studying complex powder patterns.

8.4 Powder diffractometer

The powder diffractometer is currently the most widely used system for the acquisition of powder diffraction patterns. There are a number of manufacturers of diffraction instruments that produce modern systems using recent developments in X-ray source and detection technologies. The diffractometer is a versatile instrument that can perform a number of tasks, such as accurately measuring the intensity of X-ray reflections with a high angular resolution, or analyze an incoming beam by diffraction off a well-known crystal. In the first case, the term "diffractometer" is warranted since a diffraction pattern of the substance is recorded, whereas in the second case, the wavelength (energy) distribution of the incoming X-ray beam is recorded. The latter is therefore also called a "spectrometer" for wavelength dispersive X-ray spectroscopy.

Similar to the Debye–Scherrer system, the diffractometer consists of an X-ray source, sample mount, and a detection system. In contrast to the Debye–Scherrer system, the sample is a flat plate, which, together with the X-ray source and the detector defines a plane, the *Diffractometer plane*. However, the parafocusing condition is not met for all diffraction angles for a flat sample. For parafocusing conditions to be satisfied, the positions of source and detector would need to be continually adjusted during an angle scan, while the sample would need to be bent, with the sample curvature adjusted according to the diffraction angle. In Figure 8.7, a schematic of a diffractometer is shown. The central piece is a goniometer that combines two vertical axes, the θ axis for the sample, and the 2θ-axis for the detector. The flat sample is mounted vertically in the center of the goniometer and can be rotated about the θ-axis that coincides with the vertical sample direction. The incoming beam together with the diffracted beam define the diffractometer plane. The source and the receiving slit in front of the detector are both located on the goniometer/diffractometer circle, with the distance between the source and the θ-axis of the order of 200 to 250 mm. Smaller goniometers are in use for compact systems, whereas larger goniometers are often used at synchrotron beam lines and for applications that require a heavy load. Several scatter slits are used to define the incoming and diffracted beams, and monochromator/analyzer crystals may be included. It is desirable that the illuminated

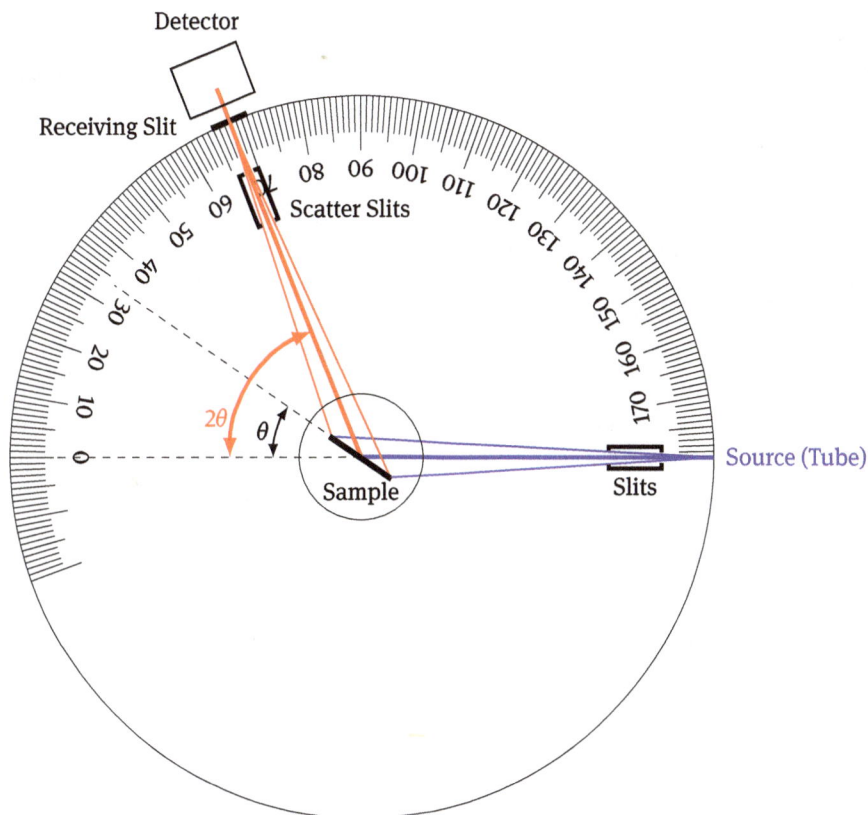

Figure 8.7: Schematic of a diffractometer: a flat sample is mounted in the center of the diffractometer where the θ-axis is also located, and a detector is traveling around the sample at twice the angular speed as the sample, ensuring the $2\theta - \theta$ relationship. The source and the receiving slit in front of the detector are located on the diffractometer circle.

surface of the sample is identical to the surface that is seen by the detector and the scatter slits. With the sample and detector moving at ratio of 2 : 1, a $2\theta - \theta$ motion of the detector and sample is achieved, and the angle of beam incidence is always equal to the exit angle of the diffracted beam. In this way, the source and detector receiving slit will always be on the focusing circle in the diffractometer plane. Using a flat plate sample, this focusing condition is somewhat weakened, but the surface of the sample is always kept tangent to the focusing circle.

With a tube take-off angle of 6°, the effective source size from a fine focus tube is 8 mm × 0.04 mm (12 mm × 0.04 mm for long fine focus). The diffractometer is imaging the source via the sample onto the receiving slit. It follows that receiving slits smaller than the projected source line width do not increase the resolution. Receiving slits of the order of 0.1 to 0.3 mm are often in use and are a good compromise between angular resolution and diffracted intensity.

The source emits radiation not only parallel to the diffractometer plane, but also in directions with a component along the line focus of the tube. This vertical divergence of the incoming X-ray beam needs to be reduced, usually by means of Soller slits, which consist of horizontal parallel plates of a heavy absorber material such as molybdenum or tantalum. These slits can reduce the intensity by a factor of 2 or more, but increase the resolution of the diffractometer by reducing the vertical divergence to a few degrees in addition to reducing the reflection tails. High resolution horizontal Soller slits, with angular acceptance of the order of 0.5° or smaller can be used in place of slits if sample displacements are expected. Additional slits define the illumination of the sample (see Figures 8.7 and 8.8) and ensure that the sample area illuminated by the source is the same area from where the detector will receive diffracted intensities. In this way, scattering from areas not containing the sample will not get registered by the detector, thus improving the background. Due to the rotation of the sample, the illuminated area changes with the θ angle. In some implementations, a variable slit system is employed to keep the illuminated sample volume constant for different angles.

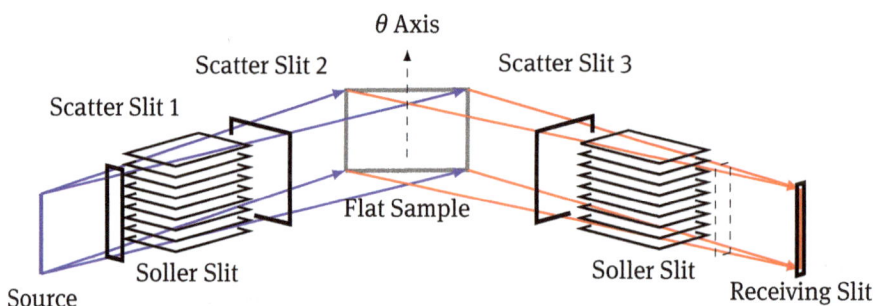

Figure 8.8: Beam path from the source to the detector receiving slit: The vertical line emanating from the tube illuminates the sample, and the detector receiving slit only accepts intensity emanating from the sample.

Several implementations of modern diffractometers are available: the $2\theta - \theta$ goniometer can be oriented horizontally or vertically, depending on the desired foot print of the system.

The vertical goniometer arrangement has become popular for its smaller foot print and for the easy access to the sample holder. Additionally, the sample is less likely to loose powder particles in this geometry, However, a drawback is that symmetric measurements of the diffracted beams at both the $\pm 2\theta$ positions are not easily accessible. Such measurements may for instance be used for an accurate determination of the 2θ angle of a reflection.

A variation of the vertical diffractometer is the $\theta - \theta$ goniometer, where the sample remains stationary, but both the source and the detector move at the same speed, realizing again the condition of equal angle of incidence and exit from the sample. This arrangement allows investigation of liquid samples, which need to remain horizontal at all time. The $\theta - \theta$ geometry has become popular with additional sample environment controls and *in-situ* studies, such as temperature, humidity, etc., which benefit from a stationary sample.

The horizontal goniometer setup follows the traditional spectrometer setup, with source, sample and detector in a horizontal plane (see Figure 8.7). The weight of the goniometer and tube tower can be distributed over a larger surface area. Often, the tube source and goniometer are mounted on a common plate to produce a rigid mechanical coupling, improving mechanical stability maintaining alignment.

The goniometer itself can be compact, with typical source to sample distance of 200 to 250 mm. The flexibility of adjusting the diffractometer circle allows easy addition of Xray optical elements and beam conditioning elements such as slits, monochromators and X-ray mirrors.

Diffractometers with a mechanical linkage between the 2θ- and θ-axes providing the $2:1$ speed ratio require a single motor only and allow compact goniometers with small footprints, suitable for small desktop devices. For standard powder diffraction, independent motion control for the sample is not needed, and such systems give excellent powder diffraction data at potentially lower cost.

In contrast, most modern powder diffractometers have independent motors for each axis, giving additional flexibility for measurements. While the additional motion control increases the complexity of the goniometer, the independent control over both rotation axes enables additional scan types and increases the flexibility of the system. For instance, in θ scans with stationary detector, sample texture information can be obtained. In grazing incidence geometry where the sample is stationary and the incident beam at a shallow angle, a thin sample such as a polycrystalline thin film can be measured. For this case, the asymmetry between angle of incidence and the angle of exit results in a higher intensity at the cost of broader reflections.

8.4.1 Diffractometer alignment

Achieving optimal intensity and resolution of a diffractometer requires that several elements need to be aligned properly. Figures 8.7 and 8.8 show the diffractometer and details of the source-sample-receiving slit geometry. For optimal alignment, the focal line of the X-ray tube (radiation source) and the receiving slit need to be located on the diffractometer circle. The incoming beam needs to pass through the center of the goniometer, where both, the 2θ and θ axes are located. The focal line and the receiving slit have to be vertically aligned parallel to the goniometer axes. The sample surface has to coincide with the goniometer axes, and the slits should be arranged symmetrically around the incident and the diffracted beam. The Soller slits need to be horizontally aligned to allow a maximum of the intensity to pass through and minimize the vertical divergence. Any alignment procedure is iterative and will need to be checked several times. With a diffractometer circle of the order of 200 to 250 mm, the nominal angular range of the source is of the order of 200 μrad. Similarly, a receiving slit of the order of 0.2 mm has an angular range of 1mrad or 0.0573°. A well-aligned system will resolve the copper $K\alpha$ doublet at diffraction angles of less than 40°. A possible alignment procedure may include the following steps:

- The tube tower and the goniometer are leveled to ensure that the goniometer axes and the tube focal line are parallel.
- The scatter slits and Soller slit assembly are mounted on the tube tower and adjusted so that the slits are parallel to the tube focal line, and the Soller slits are horizontal. Set screws will allow fine adjustments of these elements.
- The distance from the tube focal spot to the goniometer center is adjusted to the desired diffractometer radius.
- With suitable slits defining the incident beam, the tube assembly is adjusted so that the incident beam is passing though the goniometer center.
- If an alignment jig such as a beam tunnel or a pin hole is available, it can be mounted in place of the sample. For the case of a beam tunnel, the θ axis is now moved until intensity is observed passing through. The tube can then be translated again to optimize the intensity. The beam tunnel is moved around the θ axis to further optimize the intensity, and the adjustment is repeated until no intensity increase is observed. This alignment also defines the zero point of the θ axis.
- The soller slit for the diffracted beam is adjusted using the direct beam to ensure that the Soller slits are parallel to each other. The scatter slits for the diffracted beam are adjusted so that they are symmetrical.
- The receiving slit is placed on the diffractometer circle and is adjusted to ensure parallel orientation with the goniometer axes and tube focal line. The position of maximum intensity defines the zero point of the 2θ axis.

Depending on the diffractometer type, the details of the alignment process will vary. For instance, instead of a single tube translation, the tube may have a range of possible motions, such as travel tangential to the diffractometer circle and along the incident beam, and it may also include a limited range of rotations. The incident beam slit assembly generally has options for rotation/tilt to align the slits parallel to the focal line, and a translation to center the slit around the focal line. Similarly, the diffracted beam slit assembly will have adjustments for the Soller slit and the scatter slits. Manufacturers have begun to include automated alignment capabilities, with motorized stages to carry out the alignment and facilitate configuration changes for different types of measurements.

Filters, monochromator and analyzer
In the simplest setup, the tube will emit a broad spectrum of radiation with the high energy cutoff given by the generator acceleration voltage, usually of the order of 40kV

to 50kV, and tube current of 20 to 40 mA (power load of 1 to 2 kW). Since the diffraction will use the elemental characteristic radiation, for instance the CuKα radiation, the CuKβ radiation will also be present, together with the Bremsstrahlung spectrum. The broad spectrum will therefore add to the background, and the Kβ lines will also be present in the diffraction pattern. Since the Kβ intensity is of the order of 10 % of the Kα radiation, it needs to be attenuated. This is achieved by either adding a filter into the incident beam, or using a monochromator crystal. An effective method is the use of filters with an absorption edge chosen to strongly attenuate the Kβ radiation while passing the Kα radiation. Some tube towers integrate a filter wheel that contains a number of different filters for the most common tube based wavelengths. For a copper target, a nickel filter is employed, which attenuates the CuKβ intensity by a factor of about 10. The nickel filter also attenuates the higher energy Bremsstrahlung and, therefore, reduces the overall background. However, the filter is also affecting the Kα radiation, reducing the overall intensity of the incident beam. The attenuation of the Kβ line to less than 1/100 of the Kα intensities is often enough for a qualitative powder pattern. However, if strong reflections are present in the diffraction pattern, with count rates exceeding a few thousand counts per second, then it is expected that the corresponding Kβ reflections will still be visible in the diffraction pattern.

Incident beam monochromator

If a filter is not sufficient for the attenuation of the Kβ line in a tube spectrum, then an incident beam monochromator may be considered. The monochromator crystal, often pyrolithic graphite or lithium fluoride (LiF) is used to provide a narrow wavelength window centered around the Kα radiation wavelength. Lower and higher energy photons do not get reflected at the same Bragg angle (with the exception of higher harmonics, e. g., $\lambda/2$, $\lambda/3$, etc.), which reduces florescence as well as background. The monochromator, however, strongly reduces the incident beam intensity, and is therefore often used for measurements of highly reflective samples, such as single crystalline thin films. In the case of a large single crystal sample, such as a silicon wafer, an incident beam monochromator is necessary to reduce spurious reflections due to characteristic emission of elements other than the target (for instance, tungsten L radiation, Fe Kα radiation, etc.), elements that are used in the construction of the X-ray source.

Diffracted beam monochromator

The receiving slit of the diffractometer will not only receive diffracted intensities, but also fluorescence radiation from the sample itself in addition to diffraction from all the wavelengths present in the spectrum of the incident beam. It is therefore desirable that only a narrow wavelength range is accepted at the detector. In this way, the background will be strongly attenuated, whereas the diffracted intensities will be accepted by the detector, improving the signal to noise (signal to background) ratio. One way to achieve this is to add a diffracted beam monochromator after the receiving slit, with the analyzer crystal oriented to produce a Bragg reflection for the characteristic $K\alpha$ radiation. Therefore, a second diffractometer system is mounted on the 2θ arm of the goniometer that is set up so that only the $K\alpha$ radiation will reach the detector. It therefore includes slits and an adjustable θ axis for the analyzer crystal, whereas the detector will be in a fixed position corresponding to the 2θ angle of the chosen Bragg reflection. Analyzer crystals may use the lithium fluoride (LiF) (200) reflection, or more common, the (002) reflection of highly oriented (pyrolithic) graphite. The mosaic spread of the highly oriented graphite is of the order of 0.2°, with a reflection power of the order of 20 %. In this way, the $K\alpha$ radiation is attenuated by a factor of 5, whereas the background may be attenuated by a factor of 10 to 20, giving a better signal to noise ratio even though the overall intensity is reduced.

Analyzer setup

An analyzer crystal (LiF, pyrolithic graphite, etc.) needs to be integrated into a diffractometer setup for a single Bragg reflection. Alignment of the analyzer crystal is therefore similar to the alignment of the diffractometer itself and involves setting the Bragg angle of the analyzer crystal to optimize the intensity of the $K\alpha$ radiation. If a LiF crystal is used, the mosaic spread is much smaller, of the order of 0.02°, and a fine adjustment of the analyzer crystal is required. This is often achieved with a simple tangential arm geometry that allows highly controlled movement of the analyzer over a small angular range, usually set in the vicinity of the analyzer Bragg angle. An analyzer crystal represents a narrow band pass filter for the diffracted spectrum, and will only pass a small wavelength region of the spectrum centered around the $K\alpha$ radiation into the detector. This means that the fluorescence radiation emanating from the sample upon illumination with a high energy incident beam will not reach the detector.

Incident beam monochromator and analyzer combination

A combination of an incident beam monochromator and diffracted beam analyzer promises excellent attenuation of the $K\beta$ wavelength and good reduction of the background intensities. However, the diffracted intensities are also strongly attenuated, increasing data acquisition times. For highly reflecting samples, such a combination is often used to evaluate stress/strain relationships. Furthermore, if the monochromator and analyzer are matched and set up in a nondispersive geometry, excellent diffraction reflection profiles can be obtained over a large 2θ range.

X-ray mirrors

X-ray mirrors have found use in many diffractometer designs, either in focusing or parallel beam geometries. An X-ray Bragg mirror consists of a large number of alternating layers, such as molybednum and silicon, or nickel and boron, layers that are deposited on an elliptical or parabolic surface. An elliptical mirror will therefore form an image of the tube source at the second focal spot, and will collect a larger solid angle of intensity. The intensity delivered to the sample is therefore increased over beam conditioning based on slits. Due to the fact that a small angle Bragg reflection is used, the mirror acts as an energy (wavelength) band pass filter. The $K\beta$ line is attenuated by at least an order of magnitude, and shorter X-ray wavelengths are eliminated. The reflectivity of X-ray mirrors exceeds the reflectivity of monochromator crystals, but the energy band pass width of the mirror is larger than that of a crystal. Combining an X-ray mirror with a incident beam monochromator gives a powerful beam conditioning system that uses the larger angular range of the X-ray flux emanating from the tube and directs it to the monochromator crystal, combining excellent flux with a narrow X-ray energy spread.

With a combination of two parabolic X-ray mirrors, one at the incident beam, the other for the diffracted beam, a parallel beam system can be set up. The first mirror converts the divergent incident beam to a parallel beam that is wider than the actual source. This parallel beam is then diffracted from the sample, resulting again in a parallel diffracted beam. A second parabolic mirror then receives the diffracted parallel beam and images it onto a receiving slit. This geometry reduces any effects due to sample displacements and roughness of the sample surface and can be used for high resolution powder diffraction. The parallel beam geometry may necessitate additional sample movement to produce a better sampling of the statistical distribution of the powder grains of the sample, since the divergence of the incident beam is very small.

8.5 Detectors

For diffractometer applications, a single channel detector is often the system of choice. High sensitivity to X-ray photons is required, and a photon detection efficiency of 100 % is desired. A *scintillation* detector, consisting of a thallium doped sodium iodide (Tl:NaI) single crystal and a photomultiplier is a robust detection system. The number of visible photons that are created by the scintillator is proportional to the energy of the incoming X-ray photon, and the number of electrons that can be collected is therefore proportional to the energy of the incoming X-ray photon. Linear amplifiers then convert the charge pulse from the photomultiplier into a pulse signal that is a measure of the energy of the incoming X-ray photon. A single channel analyzer is then used to discriminate the pulse heights of the pulses generated, rejecting detector pulses outside the energy window. In this way, the background level in a powder pattern can be reduced for a well-tuned amplifier—single channel analyzer chain. A scintillation detector can be set up to reach a quantum efficiency of 100 %, so that every photon entering the detector is registered. The dark count, the counts when no X-ray photons from the source are generated, can be as low as 1 count per second, so that longer count times are possible without unduly increasing the background.

Modern Peltier cooled solid state detectors allow setting energy band pass widths of the order of 130 eV, while accepting count rates of 10^4 to 10^5 counts per second. The advantage of such a detector is that only photons in the narrow energy range are counted, reducing background and potential fluorescence from elements present. For single channel counting, such detectors are desirable as they simplify the diffractometer design while maintaining the general band pass characteristics of an analyzer crystal. The first generation of energy dispersive detector based on lithium drifted silicon or germanium required cooling to liquid nitrogen temperatures and incorporated a small liquid nitrogen dewar on the detector arm of the goniometer. While these detectors provided the energy resolution, they had limitations on the count rate, and have been superseded by silicon PIN diodes and silicon drift detectors (SDDs).

Linear detectors, such as silicon strip detectors, combine linear position detection with high count rates and energy discrimination, reducing data acquisition times. For best resolution, the linear detector should follow the diffractometer circle, but a small angular range detector can be linear, since the errors produced by the deviation from the a curved detector are small and can be corrected. Angular ranges of the order of 5° (or 87.5 mrad) are often used to increase the data acquisition rates. Linear detector also do have energy band pass capabilities and give data with low background and excellent signal-to-noise ratio.

Statistical treatment of counter detectors

In general, emission and diffraction of an X-ray photon is a stochastic process, and photon counting will follow statistical processes as well. Measuring the same intensity at short time intervals will show count rate fluctuations that follow a statistical model. It is therefore desirable to accumulate a large number of X-ray photons to reduce these statistical fluctuations. The statistical deviation σ from a certain number of observed X-ray photons N is a function of N only

$$\sigma = \sqrt{N}$$

and it follows that the relative standard deviation σ_{rel} is given by

$$\sigma_{rel} = \frac{\sqrt{N}}{N} = \sqrt{N}$$

Measuring an intensity of a diffracted reflection to a certain degree of accuracy, the count rate of the reflection must therefore be taken into account. For instance, to achieve a 1% accuracy, the total number of counts required is therefore 10000, so that $\sqrt{10000} = 100$, and the ratio $\sqrt{10000}/10000 = 0.01$. If a reflection with a count rate of

100 counts per second needs to be measured to an accuracy of 3 %, then the counting time has to be 11.11 seconds, resulting in 1111 counts, with a standard deviation of 33 counts. It also follows that a high background will reduce the overall accuracy. If it is assumed that a reflection is giving 4000 counts (standard deviation 63 counts) on a background of 1000 counts (standard deviation of 32 counts), then the total number of counts is 5000, with a total standard deviation of

$$\sigma_{tot} = (\sigma_{Peak}^2 + \sigma_{Background}^2)^{\frac{1}{2}}$$

or about 71 counts. However, the percentage standard deviation becomes larger, since the background is also fluctuating, and the individual deviations have to be taken into account. If the number of counts in the background are N_B and the number of counts of a reflection is N_R, then the relative standard deviation is given by

$$\sigma_P = \frac{(N_R + N_B)^{\frac{1}{2}}}{N_R - N_B} \tag{8.3}$$

For the example, according to equation (8.3) the relative standard deviation is 0.023, or 2.3 %, larger than the relative standard deviation of 1.5 % for the peak of 4000 counts. For a diffractometer measurement to give accurate intensity data, the background intensities should therefore be as low as possible so that weak reflections can still be measured with a reasonable accuracy. What needs to be kept in mind is the fact that doubling the data acquisition time will improve accuracy by a factor of $\sqrt{2} = 1.41$. If a powder pattern should be measured with an intensity range of 10^3, with the strongest reflection giving 10000 counts per second, then the weakest reflection will have 10 counts per second, with a standard deviation of about 3 counts per second. This weak reflection is therefore barely above background. Assuming the background count is of the order of 1 count per second, then counting for 10 seconds at each position will give about 100 counts for the weakest reflection with a standard deviation of $\sigma = 10$, on a background of about 10 counts. The relative standard deviation is then 0.12, with the weak reflection about 8 times above the background. In contrast, the strongest reflection will now be 100000 counts plus 10 counts background, and the relative standard deviation is 0.003. For a laboratory source, extending the counting time can very quickly require long data acquisitions. For instance, a scan from 5° to 90° at a step width of 0.02° will produce 4251 data points. For a counting time of 1 second per data point, the scan will take about 1.2 hours. Increasing the data acquisition time by a factor of 10 will improve the average relative accuracy by about a factor of 3, but requires an approximate 12 hour scan. It follows that linear detectors that collect angular ranges simultaneously will reduce data acquisition times. Depending on the requirements, the combination of X-ray source, incident and diffracted beam elements and detector allow many different configurations to optimize the powder data measurement.

8.6 Sample preparation

An ideal powder pattern should have a dense and even distribution of intensity over the Debye cone. The sample therefore needs to have (1) a sufficiently large number of crystallites that (2) are randomly oriented. Condition (1) requires that the powder is fine enough to ensure even intensity distribution in a Debye cone. For a flat plate sample that is stationary, a small crystallite size is required for reproducible intensities. Crystallite sizes less than 10 μm are desired, with 5 μm a good size. Grinding the sample, followed by passing the powder through a sieve to remove larger crystallites, produces a sample with a sufficient number of small crystallites to produce uniform coverage of the Debye cones.

The powder sample needs to be mounted on a sample holder/support. A sample holder may be fabricated from an aluminium plate, with a hole or a depression to accept the powder sample. A sample thickness of 2 mm is sufficient to ensure that most materials satisfy the condition of optimal sample volume, with the exception of samples consisting of light elements, such as organic and polymeric systems. Alternatively, microscope slides with curved or straight-walled cavities with sufficient well depth can also be used. The powder may be mixed with a small amount of binder, such as collodion or an easily soluble grease to ensure that the powder is not released from the sample mount for different sample orientations. The powder is packed into the cavity and lightly compressed, followed by flattening the surface. This ensures that the sample surface is well-defined and is placed at the proper position relative to the θ-axis. Care has to be taken that the powder does not develop a nonrandom orientation distribution during this procedure. In addition, the sample holder should be constructed in a way that it does not contribute any intensity to the powder pattern.

If only qualitative intensities are needed, a quick way to prepare a sample consists in using a flat microscope slide cut to fit into the sample holder of the diffractometer. A thin layer of grease is applied, and the powder is dropped onto the slide from a height of a few centimeters. In this way, the powder assumes a randomized distribution during the fall, and adheres to the slide in a reasonable random orientation. The disadvantage is that the sample is usually not thick enough, resulting in a less than optimal scattering volume and intensity ratios that are not exact. Using a slide with a well can improve the sample preparation, with thicker samples providing improved intensity ratios. For a maximum diffracted intensity, the effective thickness t of the sample depends on the powder packing efficiency, the density of the material, as well as the absorption coefficient of the sample

$$t \geq \frac{3.2}{\mu} \cdot \frac{\rho}{\rho'} \sin \theta \tag{8.4}$$

with μ the absorption coefficient, ρ the density of the material, ρ' the effective density of the material as a powder sample, and θ the scattering angle. A steep angle of incidence θ requires a thicker sample than a shallow angle of incidence. If a pattern is acquired up to $2\theta = 120°$, then at the highest angle θ the maximum of $\sin\theta = \sqrt{3}/2 \approx$ 0.866. The overall powder density may be of the order of 50 to 80 %, requiring a sample thickness of about twice of $\frac{3.2}{\mu}$. The absorption coefficient of an inorganic material for copper radiation may be of the order of 100 mm^{-1}; for example, the CuKα absorption coefficient of BaTiO$_3$ is 146.95 mm^{-1}. If the packing density of the powder is of the order of 80 %, then the sample thickness may be of the order of 25 μm. Materials with a smaller absorption coefficient will require thicker samples, whereas highly absorbing materials can remain quite thin. It is therefore clear that using a simple flat slide as the sample support is often adequate if the sample has a high absorption coefficient and is densely packed. If a sample holder has a well depth of the order of 2 mm, then most materials satisfy the condition given in (8.4).

8.7 Texture

Powder diffraction assumes a random orientation of a large number of crystallites so that the Debye cones have an even intensity distribution. Deviations from random orientation are called *texture*. To fulfill the condition of random packing of crystallites in sufficient number, the sample preparation is crucial. Deviations from random packing may be introduced by the crystallite shape, which affects their packing in the sample holder. Grinding the sample to produce a suitable powder may induce preferred cleavage, resulting in a powder with crystallites of nonrandom shapes. The mechanical process of mounting the powder in the sample holder may also produce a preferred orientation. The Debye cones emanating from such a sample may be smooth, but with intensity variation along the circumference of the cone. In the case of an insufficient number of crystallites, the Debye cones will become "spotty," with a low number of crystallites contributing to the intensity. This results in measured intensities that deviate strongly from the ideal case of random packing.

Analysis of the intensity variations along the Debye cone allows the determination of the preferred alignment axis. For instance, a layered material may form thin platelets, which will preferentially pack parallel to the platelet plane. The diffraction pattern will therefore have a preferred orientation perpendicular to the platelet plane, but a random orientation perpendicular.

Similarly, materials that are extruded or drawn into a wire show texture, since the crystallites are aligned by the application of mechanical stress. In the case of an extruded polymer fiber, the polymer molecules are stretched and aligned along the extrusion axis, but randomly packed perpendicular to the extrusion axis, resulting in a *fiber texture*. During the wire drawing process, the shear forces tend induce slip in the material along the planes of high atomic density, resulting in a reorientation of the crystallite grains. For example, a drawn copper wire has a "fiber" texture with the [111] direction preferred.

8.8 Examples of powder diffractometer system

Manufacturers of complete diffractometer systems have achieved a combination of X-ray source and detector that allows the collection of a powder pattern in a short time, with narrow line profiles and low background. Systems with high resolution incident beam monochromators can pick the Kα_1 radiation, giving powder patterns with a single wavelength; such pattern are suitable for full pattern refinement.

Figure 8.9 shows the Rigaku SmartLab diffractometer that can be configured with a high-intensity rotating anode source for Cu Kα radiation. A hybrid pixel array detector with energy discrimination capability is mounted on the diffracted beam side, making this diffractometer an excellent multipurpose high resolution system with high sample throughput.

Figure 8.9: Rigaku SmartLab diffractometer: Vertical Goniometer, θ - θ geometry. The high-intensity rotating anode X-ray source is on the left side, the detector assembly on the right side. Image courtesy of Rigaku Corporation.

Figure 8.10 shows the Bruker D8 diffractometer equipped with a compact microfocus X-ray source. The diffractometer has the $\theta - \theta$ geometry and the sample is mounted on

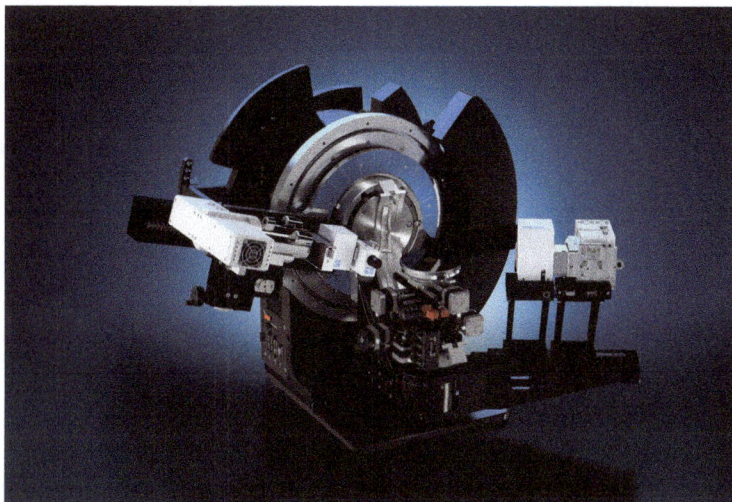

Figure 8.10: Bruker D8 Diffractometer: Vertical Goniometer, θ - θ geometry. The high-brilliance microfocus X-ray source is on the left side, the quarter-circle Euler cradle in the middle and the detector assembly on the right side. The detector arm has an additional degree of freedom and can be rotated out of the scattering plane. Image courtesy of Bruker Corporation.

the $xyz\phi$-stage that is connected to a χ-quarter circle. The detector arm has an additional degree of freedom that allows rotation out of the diffraction plane for investigations of thin film samples in grazing incidence.

Figure 8.11 shows a Stoe Stadi $2\theta - \theta$ powder diffractometer with high resolution monochromator. Three linear detectors are mounted on the detector arm. The beam path is indicated in color: red for the beam emanating from the tube, green for the monochromatic beam from the monochromator to the sample and blue the diffracted beam. The sample is mounted in a capillary.

8.9 Example of a powder diffraction pattern

A powder pattern of the mineral fluorite, CaF_2, has been prepared on a zero-background plate, with a small amount of grease to ensure adherence of the powder to the sample holder. The powder was dropped onto the plate from a height of approximately 5 cm, and subsequently tapped to distribute the powder over an area of about $1\,cm^2$. The as prepared powder was mounted on a Scintag PAD-V vertical $2\theta - \theta$ powder diffractometer with a standard copper fine focus tube, graphite analyzer and a single channel scintillation detector. The slits at the tube side included a Soller slit, and vertical slits of 0.5 and 3 mm. The slits on the detector side used a scatter slit of 3 mm, and a receiving slit of 0.2 mm, as well as a Soller slit. The tube power settings were 45 kV and 20 mA. The powder scan used the step scan mode, with step width of 0.02° and

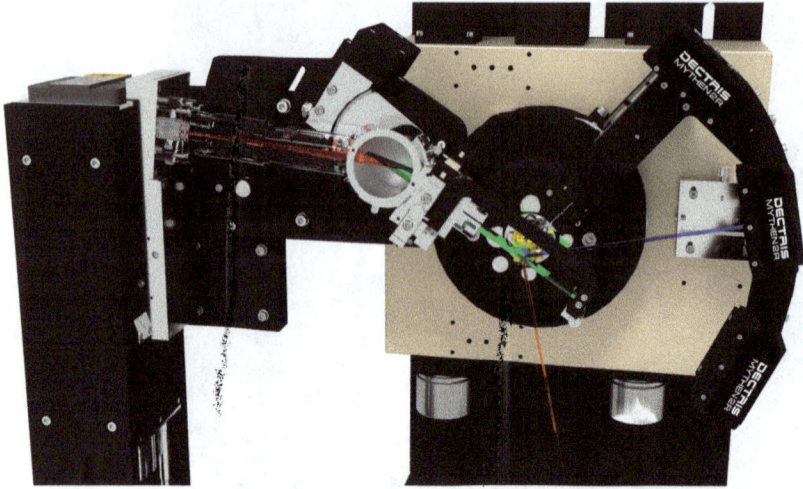

Figure 8.11: Stoe $2\theta-\theta$ diffractometer: Vertical Goniometer with stationary tube and monochromator. Three linear detectors are mounted on the 2θ arm. The beam path is indicated in color: red shows the beam emanating from the tube, green shows the monochromatized beam from the monochromator to the capillary holding the sample, blue the diffracted beam. Image courtesy of Stoe & Cie, GmbH.

counting time of 10 seconds/step in the angle range of 15° to 120°. The resulting powder pattern is shown in Figure 8.12, showing well resolved $CuK\alpha_1$ and $CuK\alpha_2$ peaks at higher angles. The vertical intensity scale is linear.

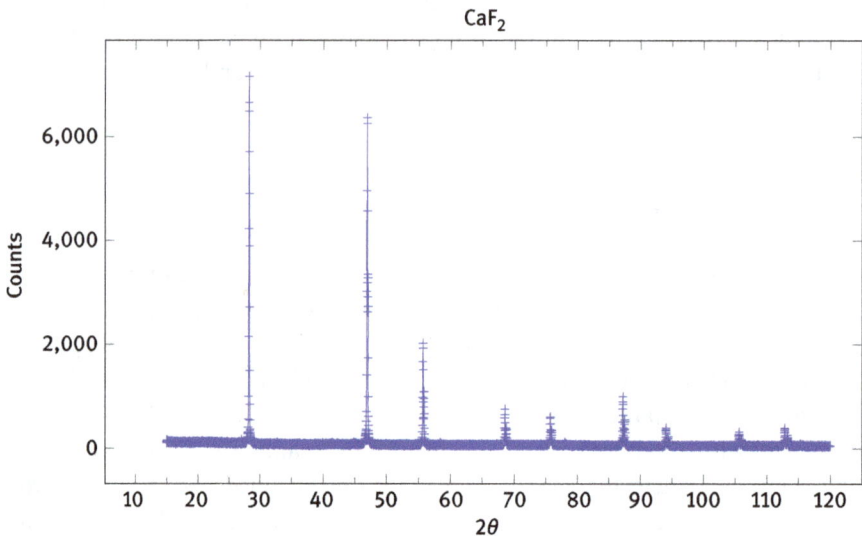

Figure 8.12: Powder pattern of CaF_2 with $CuK\alpha$ radiation.

The resolved $CuK\alpha_1$ and $CuK\alpha_2$ reflections with index (311) at a position of 55.68° are easily recognized, while the (222) reflections at 58.39° is very weak, but clearly above background. The full width at half maximum (FWHM) is of the order of 0.07° of the reflections, a reasonable compromise for this system in terms of intensity and resolution. Modern systems may give powder patterns with smaller FWHM and higher count rates, depending on the particular setup. To show the weak reflections, a pattern may also be presented in a semilog plot, with the intensity axis logarithmically scaled. If the peak shape of the reflections can be approximated by a Gaussian, then the peaks will have a parabolic shape. Deviations from a Gaussian shape can therefore be easily seen, where the peaks no longer have a parabolic shape. A semilog plot of the powder pattern is shown in Figure 8.13, over the whole angular range of 15° to 120°. The weak reflections, where the calcium ion and the two fluorine ions are interfering destructively are emphasized more. There are also a number of weak reflections at lower angles that are satellites to the strong reflections. These weak reflections are $CuK\beta$ lines that have a smaller wavelength than the $CuK\alpha$ radiation. They pass through the diffracted beam analyzer system due to their close energy to the copper radiation, but are strongly attenuated. A tighter collimation at the diffracted beam analyzer system could eliminate these reflections. However, their intensities are low enough not to affect the overall pattern.

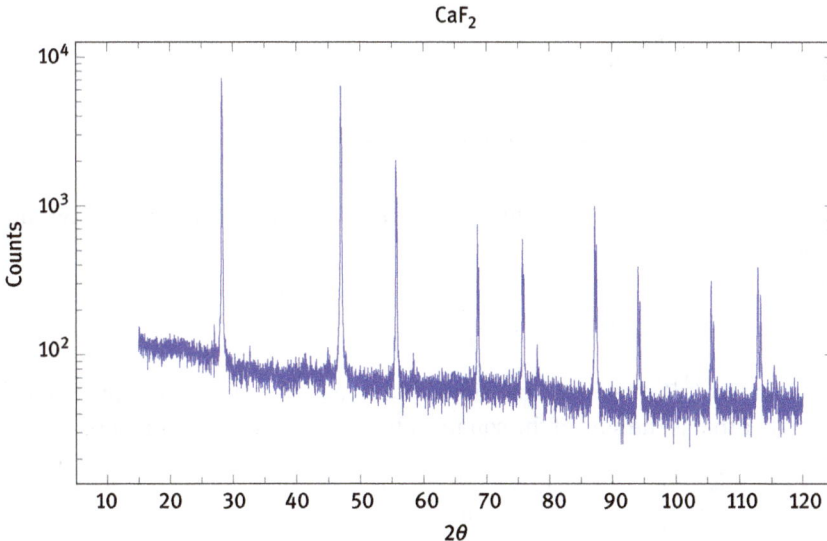

Figure 8.13: Powder pattern of CaF_2 shown in a semi-log plot, emphasizing the weak reflection intensities.

8.10 Indexing of a powder pattern

In contrast to a single crystal pattern that separates each (*hkl*) in 3-dimensional space, the powder pattern collapses this 3-dimensional pattern of (*hkl*)s into one single dimension. While it is always possible to calculate the powder pattern for a given crystal structure, with known lattice parameters and the atomic positions, it is a much harder problem to find a set of (*hkl*) values that explain the positions of all reflections that are observed in an unknown pattern. For the CaF_2 pattern given in Figure 8.12, a Table 8.1 of observed reflections and their respective peak heights can be set up.

Table 8.1: Observed reflection positions and integrated intensities for CaF_2.

2θ	Observed Intensity	Normalized Intensity
28.232	6890	963.4
32.733	23	3.2
46.945	7152	1000.0
55.687	2258	315.7
58.403	31	4.3
68.560	793	110.9
75.743	749	104.7
78.085	96	13.4
87.210	1204	168.3
94.061	522	73.0
105.618	402	56.2
112.842	533	74.5
115.334	63	8.8

Indexing this pattern, assuming a cubic unit cell, is not difficult. The Bragg law can be rewritten as

$$\sin^2 \theta = \frac{\lambda^2}{4} \times \frac{1}{a^2} \times (h^2 + k^2 + l^2)$$

and by dividing by $\lambda^2/4$, the right-hand side contains a prefactor $1/a^2$ multiplied by an integer number that is the sum of the square of the individual indices of (*hkl*),

$$\frac{4}{\lambda^2} \sin^2 \theta = \frac{1}{a^2}(h^2 + k^2 + l^2)$$

For each observed reflection, an index (*hkl*) needs to be found, under the condition that *h*, *k* and *l* are small integers. A new table can therefore be constructed (for instance, using a spread sheet) that lists the values of $\frac{4}{\lambda^2} \sin^2 \theta$. To find integer values for the sum of h^2, k^2 and l^2, ratios are taken be dividing all the $\frac{4}{\lambda^2} \sin^2 \theta$ values (λ = 1.54056 Å) by the first value in the table. If it is assumed that the first reflection has a low index, such as (100), (110), (111) or (200), the ratios should indicate possible integer values. The next column executes the division of the $\frac{4}{\lambda^2}\sin^2\theta$ values by the value for the first line. The new column is searched for patterns that indicate a possible multiplier to give integer values. The new column shows that in relation to the first reflection, some integer multiples are present, as well as fractional multiples that divide into two groups only, with 1/3 and 2/3 fractions within errors. Therefore, a multiplier of 3 will produce integer values for the sums of $(h^2 + k^2 + l^2)$ for all observed reflections in the CaF_2 pattern. Indexing can now proceed by assigning (hkl)s to every observed reflection.

Table 8.2 shows all columns as well as the assigned indices for the observed reflections. A unit cell can now be determined, with a value of 5.47 Å. A more precise determination of the unit cell can then follow, based on the fact that most errors in positions are vanishing for $2\theta = 180°$. With this unit cell, the powder pattern is reevaluated to ensure that all reflections that are observed are indexed and no reflection of observable intensity is omitted. The small unit cell found makes this an easy problem to solve, and a spread sheet calculation can be set up for larger and more complicated unit cells.

Table 8.2: Indexed Pattern of CaF_2.

2θ	θ	$\frac{4}{\lambda^2}\sin^2\theta$	$\frac{\sin^2\theta}{\sin^2\theta_1}$	×3	(hkl)
28.234	14.117	0.10025	1.000	3	(111)
32.716	16.358	0.13369	1.335	4	(200)
46.943	23.472	0.26738	2.667	8	(220)
55.685	27.843	0.36766	3.668	11	(311)
58.394	29.197	0.40117	4.002	12	(222)
68.565	31.196	0.53467	5.333	16	(400)
75.732	37.866	0.63517	6.339	19	(331)
78.065	39.033	0.66872	6.671	20	(420)
87.239	43.620	0.80168	7.997	24	(422)
94.061	47.031	0.90238	9.001	27	(333)/(511)
105.612	52.806	1.06957	10.669	32	(440)
112.836	56.418	1.16983	11.669	35	(531)
115.325	57.663	1.20329	12.003	36	(600)/(442)

8.11 General powder pattern indexing

The random orientation of crystallites results in a uniform intensity distribution on the Debye cones, and with it, the loss of the 3-dimensional information present for a single crystal. However, the 3-dimensional information can be recovered from the 1-dimensional pattern, because the peak position is a function of the unit cell dimensions, whereas the integrated intensities are a function of the content of the unit cell.

In addition, the peak shapes carry information on size and strain properties of the material, information that is usually not used for indexing purposes. Indexing a powder pattern, assigning (hkl)s to every observed reflection, is a necessary first step to extract structural information from the pattern.

The Ewald construction informs that for diffraction to occur, the scattering vector has to be a member of the reciprocal lattice. Therefore, the position of a reflection will give the absolute length d_{hkl} of this reciprocal lattice vector. Based on the geometry of a general lattice, the square of the length of a reciprocal lattice vector $\mathbf{r}^* \cdot \mathbf{r}^*$ is given by

$$\mathbf{r}^*_{hkl} \cdot \mathbf{r}^*_{hkl} = (h\mathbf{a}^* + k\mathbf{b}^* + l\mathbf{c}^*) \cdot (h\mathbf{a}^* + k\mathbf{b}^* + l\mathbf{c}^*)$$

This expression can be written out using the metric matrix of the reciprocal space as

$$(\mathbf{r}^*_{hkl})^2 = \frac{1}{d^2_{hkl}} = h^2\mathbf{a}^{*2} + k^2\mathbf{b}^{*2} + l\mathbf{c}^{*2} + 2hk\mathbf{a}^*\mathbf{b}^* \cos\gamma^* + 2hl\mathbf{a}^*\mathbf{c}^* \cos\beta^* + 2kl\mathbf{b}^*\mathbf{c}^* \cos\alpha^*$$

Using Bragg's law to express $1/d^2_{hkl}$, one finds

$$\frac{1}{d^2_{hkl}} = \frac{4}{\lambda^2} \sin^2\theta = Q_{hkl}$$

The quantity Q_{hkl} may be given with a prefactor of 10^4 for convenience. The components of the reciprocal metric matrix can now be used to define the following expression:

$$Q_{hkl} = h^2 a_{11} + k^2 a_{22} + l^2 a_{33} + hk a_{12} + hl a_{13} + kl a_{23}$$

For a given powder pattern, depending on the symmetry, a set of 1 to 6 coefficients a_{ij} need to be found, under condition that h, k and l are small integers. The coefficients a_{ij} describe the reciprocal lattice parameters and, therefore, the direct space unit cell can be determined. Table 8.3 gives the quadratic forms for the different crystal systems. However, a measured 2θ value does have an associated error, requiring that the quadratic form falls within $Q \pm \Delta$, with Δ a function of 2θ for the case of uniform accuracy of the reflection position. High accuracy of the reflection positions in the powder pattern is therefore paramount. Choosing very small values for the six a_{ij}, indexing of a pattern may be achieved, but it is not necessarily a correctly indexed pattern. Given that the set of (hkl)'s are expected to be small integers, a criterion is needed to assess the quality of the indexing. A useful measure is the *de Wolff figure of merit* [10], which is defined as

$$M_{20} = \frac{Q_{20}}{2\langle Q \rangle N_{20}}$$

with N_{20} the number of calculated Q-values up to Q_{20}, with Q_{20} the Q-value of the 20th observed reflection that is indexed. The number $\langle Q \rangle$ is the average deviation of the

Table 8.3: Quadratic forms and crystal systems.

Crystal system	Quadratic form
triclinic	$Q = h^2 a_{11} + k^2 a_{22} + l^2 a_{33} + hka_{12} + hla_{13} + kla_{23}$
monoclinic	$Q = h^2 a_{11} + k^2 a_{22} + l^2 a_{33} + hla_{13}$
orthorhombic	$Q = h^2 a_{11} + k^2 a_{22} + l^2 a_{33}$
tetragonal	$Q = (h^2 + k^2)a_{11} + l^2 a_{33}$
hexagonal	$Q = (h^2 + hk + k^2)a_{11} + l^2 a_{33}$
cubic	$Q = (h^2 + k^2 + l^2)a_{11}$

calculated Q from the observed Q-values. The figure of merit is large for an indexed pattern where the averaged deviation $\langle Q \rangle$ is small, and the number of calculated possible Q-values based on the derived unit cell is small. If an indexed pattern uses a unit cell that is too large, then the number N_{20} will be large, reducing the value of M_{20}. In general, indexation of the first 20 observed lines in a pattern that results in $M_{20} > 10$ is expected to be reasonable. However, this does not necessarily mean that the unit cell obtained is actually correct.

An alternate figure-of-merit was defined by Smith and Snyder [11], as

$$F_N = \frac{1}{|\Delta 2\theta|} \times \frac{N}{N_{possible}}$$

Here, N is the number of indexed lines, and $N_{possible}$ is the number of lines allowed by the unit cell, up to the Nth line. The term $|\Delta 2\theta|$ is the average absolute difference between the observed and calculated 2θ positions. For instance, a F_{20} figure-of-merit would include the first 20 observed reflections in a pattern. If possible, N should be the number of observed lines, but a low symmetry pattern may have an exceedingly large number of possible calculated 2θ positions, so that a potentially valid index (hkl) exists. As a compromise, the first 30 observed reflections are taken, defining F_{30} as the figure-of-merit, or the actual number of reflections if there are fewer observed reflections.

While these two figure-of-merit values allow an assessment of the indexing of a powder pattern, it is not an absolute proof that the indexing is correct. If both, the M_{20} and F_{30} figure-of-merit are large, then the obtained pattern indexing is likely correct. For indexing with a figure-of-merit $M_{20} < 10$, the solution for the unit cell should be rejected. For values $M_{20} > 10$, the solution should be checked, and the pattern needs to be evaluated further to ensure that all reflections are properly indexed. A subsequent unit cell refinement with a different program than the indexing program should be carried out to further assess the quality of the proposed unit cell.

Powder indexing would be simpler if experimental errors and ambiguities could be eliminated. It is thus desirable to have an ideal sample that is measured with an ideal diffractometer. For instance, an internal standard may be used to get a calibration of the angular positions of the reflections from the unknown pattern. Thus, zero-point offsets and other systematic errors can be determined and corrected. Sample problems include weak peaks that may not have been observed, or the presence of an impurity that gives additional reflections that were included in the unit cell search algorithm. Contamination of the X-ray source by other elements, such as tungsten (*L* lines), may further add complications.

High quality powder diffraction data can be obtained via the mail-in service, for instance at the Advanced Photon Source at Argonne National Laboratory. The synchrotron powder data from a high quality sample will have very narrow peak profiles, with FWHM of the order of 0.05° or better, a single wavelength of the order of 0.5 Å, and an angular range of up to 50° in 2θ, giving a $\sin\theta/\lambda \approx 0.845$. The large dynamic range in intensity makes weak peaks observable, that may be missed in a laboratory source system. With the sample mounted in Kapton capillaries and rotated during data acquisition, texture is virtually eliminated, giving an "ideal" powder pattern. As a counterpoint, small amounts of impurities will also be observed, whereas they may not be noticeable in patterns acquired using a laboratory system.

Indexing programs

There are a number of powder pattern indexing programs that are freely available. Some of the most widely used programs are
- TREOR [12]
- ITO [13]
- DICVOL [14]

A more extensive discussion of programs for powder indexing is given in [15], where indexing results from a number of programs, including the ones listed above, are presented.

8.12 Calculation of a powder pattern

The calculation of a powder pattern requires the unit cell and its content, and the details of the diffractometer. The intensity of a reflection with index (*hkl*) is given by

$$I(hkl) = K \times M(hkl) \times L \times P \times |F(hkl)|^2 \times T(hkl)^2$$

where *K* is a scale factor, *M*(*hkl*) is the multiplicity factor for a reflection with index (*hkl*). The Lorentz factor *L* for a Bragg–Brentano diffractometer is given by

$$L(\theta) = \frac{1}{\sin^2\theta\cos\theta}$$

the polarization factor P for a Bragg–Brentano diffractometer is given by

$$P(\theta) = \frac{1 + \cos^2 2\theta}{2}$$

If additional monochromators/analyzers are used, their respective polarizations need to be included in the polarization factor. The structure factor $F(hkl)$ is the sum over all the atomic form factors of the atoms in the unit cell with their respective phases, and the displacement factor $T(hkl)$ takes into account that atoms are displaced from the ideal position by thermal fluctuations.

As an example, the pattern for CaF_2 is calculated. The unit cell is cubic, with space group $Fm\bar{3}m$, and lattice parameter $a = 5.4456$ Å. The atomic positions are given in Table 8.4, and the atomic form factor parameters are listed in Table 8.5.

Table 8.4: Atomic position of CaF_2.

Atom	Wyckoff	x	y	z
Ca	4a	0	0	0
F	8c	$\frac{1}{4}$	$\frac{1}{4}$	$\frac{1}{4}$

Table 8.5: Atomic Form Factors for Ca^{2+} and F^{1-}.

El	a_1	b_1	a_2	b_2	a_3	b_3	a_4	b_4	c
Ca^{2+}	15.6348	−0.00740	7.95180	0.608900	8.43720	10.3116	0.853700	25.9905	−14.875
F^{1-}	3.63220	5.27756	3.51057	14.7353	1.26064	0.442258	0.940706	47.3437	0.653396

The CaF_2 structure contains 4 calcium atoms and 8 fluorine atoms. Since the symmetry is face centered cubic, the systematic extinction rules only allow (hkl)s with unmixed indexes. The structure factor calculation reduces therefore to the sum over 1 calcium atom and 2 fluorine atoms, multiplied by the factor $1 + \exp(2\pi i(h/2 + k/2)) + \exp(2\pi i(h/2 + l/2)) + \exp(2\pi i(k/2 + l/2))$. For the two fluorine positions in the unit cell, $(\frac{1}{4}, \frac{1}{4}, \frac{1}{4})$ and $(-\frac{1}{4}, -\frac{1}{4}, -\frac{1}{4})$ are used, resulting in

$$F(hkl) = f_{Ca}e^{2\pi i 0} + f_F[e^{2\pi i(h/4+k/4+l/4)} + e^{-2\pi i(h/4+k/4+l/4)}],$$

which can be reduced to

$$F(hkl) = f_{Ca} + 2f_F \cos 2\pi(h/4 + k/4 + l/4)$$

This means that only the calcium atom is contributing to the (111) reflection, and the calcium and fluorine atoms are interfering destructively for the (200) reflection, whereas constructive interference is found for the (220) reflection. Therefore, the (111) and the (220) reflections will be strong, whereas the (200) reflection will be very weak, as seen in the pattern shown in Figure 8.12. The (200) reflection is almost zero; this is due to the fact that the 2 fluorine atoms have almost an equal number of electrons as the calcium atom. The calculation can be set up in a spread sheet or using Matlab®, Mathematica® or another computational environment. Using copper radiation, the pattern calculation gives the results in Table 8.6. It is clearly seen that the polarization factor P has the minimum at $2\theta = 90°$, whereas the Lorentz factor L has the minimum around $2\theta = 110°$. The form factor is falling off quite fast, with the effective number of electrons for fluorine for the (511)/(333) reflection only equivalent to about 3 electrons, and for calcium equivalent to about 8.5 electrons. In addition, the displacement parameter is adding a further attenuation to the intensities, reducing the intensities by about 40 % for the high diffraction angles. The combined LP factor will counteract the drop in the form factor and the effect of the displacement, eventually increasing the observed intensity in the back scattering region closer to $2\theta = 180°$. The agreement with the observed pattern is acceptable, since preferred orientation has not been considered, assuming a perfect random distribution. Qualitatively, the reflections that are calculated to be strong are observed strong in the pattern, and the reflections calculated as weak are not observed. The apparent dynamic range in intensity of the diffractometer used to measure the CaF$_2$ powder pattern is not quite 3 orders of magnitude, making reflections with a normalized intensity below 10 difficult to observe. Higher dynamic intensity ranges are possible with linear detectors that can count for commensurately longer times, increasing the statistical significance of the weak reflections. Furthermore, any reduction of the background counts will increase the dynamic range of the measurement. However, by modeling the background of the pattern, the apparent dynamic range in intensity can be increased. Higher dynamic intensity ranges are routinely achieved at synchrotron beam lines optimized for powder diffraction.

It should be noted that the measured CaF$_2$ pattern contains both the Cu Kα_1 and Kα_2 radiation, with an intensity ratio of 2:1. Therefore, normalization of the intensities of the pattern has to include separation of the Cu Kα_1 and Kα_2 radiation. Since the apparent second strongest reflection, the (220) reflection, has already a very noticeable separation of the Cu Kα_1 and Kα_2 reflections, it is actually the strongest reflection in the pattern for a single wavelength. This is consistent with the calculated pattern that shows the (220) reflection with strongest intensity and the observed intensities that were extracted under consideration of the Cu Kα_1 and Kα_2 splitting (see Figure 8.12 and Tables 8.1 and 8.6).

Table 8.6: Calculated powder pattern for CaF_2.

h k l	2θ	M	L	P	f_{Ca}	f_F	F(hkl)	T(hkl)	I(hkl)	I	I_{obs}
1 1 1	28.364	8	17.18	0.887	15.536	7.781	62.143	0.961	434811	954.13	963.4
2 0 0	32.867	6	13.04	0.853	14.868	7.262	1.900	0.948	114	0.25	3.2
2 2 0	47.167	12	6.817	0.731	12.756	5.760	97.101	0.899	455712	1000.00	1000.0
3 1 1	55.957	24	5.145	0.657	11.605	4.997	46.421	0.864	130371	286.08	315.7
2 2 2	58.681	8	4.778	0.635	11.283	4.787	6.835	0.853	823	1.81	4.3
4 0 0	68.916	6	3.788	0.565	10.226	4.106	73.750	0.808	45598	100.06	110.9
3 1 1	76.132	24	3.341	0.529	9.622	3.719	38.488	0.776	62806	83.10	104.7
4 2 0	78.480	24	3.227	0.520	9.449	3.608	8.929	0.766	1885	4.14	13.4
4 2 2	87.730	24	2.888	0.501	8.865	3.233	61.323	0.727	68906	151.21	168.3
5 1 1	94.614	24	2.730	0.503	8.518	3.009	34.072	0.698	38277	40.93	
3 3 3	94.614	8	2.730	0.503	8.518	3.009	34.072	0.698	6217	13.64	73.0
4 4 0	106.291	12	2.604	0.539	8.062	2.713	53.949	0.653	20923	45.91	56.3
5 3 1	113.614	48	2.608	0.580	7.841	2.561	31.364	0.628	28142	61.76	74.5
6 0 0	116.141	6	2.625	0.597	7.774	2.528	10.875	0.619	1112	0.94	
4 2 2	116.141	24	2.625	0.597	7.774	2.528	10.875	0.619	4448	3.74	8.8
6 2 0	126.916	24	2.796	0.680	7.535	2.377	49.154	0.587	38029	83.45	
5 3 3	136.122	24	3.110	0.760	7.380	2.280	29.518	0.564	15725	34.51	
6 2 2	139.527	24	3.284	0.789	7.331	2.251	11.317	0.557	2469	5.42	
4 4 4	157.040	8	5.232	0.924	7.152	2.146	45.774	0.528	22571	49.53	

8.13 Particle size and strain

8.13.1 Particle size

The reflection positions in a powder pattern are defined by the unit cell, and their intensity by the content of the unit cell. Additional information is contained in the reflection profile itself, in the actual shape of the reflections. Based on the interference function defined earlier, a finite crystal of M, N and P unit cells along the **a**-, **b**- and **c**-axes has a diffraction envelope described by

$$I \propto |F(hkl)|^2 J_M^2(\mathbf{aS})J_N^2(\mathbf{bs})J_P^2(\mathbf{cS})$$

For large numbers M, N and P, the interference function approaches a delta function. In contrast, for finite numbers M, N and P, the maxima are reduced in height and become broadened, with minor maxima in between the main maxima. The interference function normalized for N planes in one direction is given by

$$I(S) = N^2 = \frac{\sin^2(2\pi NaS/2)}{\sin^2(2\pi aS/2)}$$

with S the scattering vector. The width of a peak n at position S_n at half maximum, the FWHM, will be used, with $I(S_n \pm \Delta s) = N^2/2$, assuming that a reflection profile is symmetric.

$$I(S_n + \Delta s) = \frac{\sin^2(2\pi Na(S_n + \Delta s/2)/2)}{\sin^2(2\pi a(S_n + \Delta s/2)/2)} = \frac{N^2}{2}$$

Taking the square root results in

$$\frac{\sin(2\pi Na(S_n + \Delta s)/2)/2}{\sin(2\pi a(S_n + \Delta s/2)/2)} = \frac{N}{\sqrt{2}}$$

Using the fact that $\sin(\alpha + \beta) = \sin \alpha \cos \beta + \cos \alpha \sin \beta$ and that S_n is a reciprocal lattice vector with $\cos(\pi aS) = 1$, the ratio becomes

$$\frac{\sin(\pi Na\Delta s)/2}{\sin(\pi a\Delta s)/2} = \frac{N}{\sqrt{2}}$$

Since Δs is small and N not too small, then the sine function can be approximated by $\sin x = x$. The equation then becomes

$$\sin(\pi Na\Delta s/2) = \frac{(\pi Na\Delta s/2)}{\sqrt{2}}$$

This nonlinear equation of the type $\sin x = x/\sqrt{2}$ has a solution for $x = 1.3916$, which can be obtained numerically. Therefore, the thickness t of a slab with N the number of Bragg planes is linked to the FWHM in s as

$$t = Na = \frac{2 \times 2.783}{\pi \Delta s}$$

Since the scattering vector is given by $s = (2/\lambda) \sin \theta$, an approximation for the peak FWHM is given by $\beta = 2\Delta s/(ds/d\theta) = 2\Delta s/((2/\lambda) \cos \theta)$. With this, it is possible to link the thickness of the slab to the FWHM of a reflection at 2θ by

$$t = Na = \frac{5.5662\lambda}{2\pi\beta \cos \theta} = \frac{0.88\lambda}{\beta \cos \theta} = \frac{K\lambda}{\beta \cos \theta} \tag{8.5}$$

Note that β is in radians. Equation (8.5) is called the Scherrer equation, with K a constant that depends on the model, and has values of the order of 1.0, in this particular case, $K = 0.88$. The derivation in one dimension neglects details of the crystallite shape and overall size, but illustrates that based on the interference function, a finite FWHM can be obtained for diffraction from a finite slab. If a crystallite of silicon ($a = 5.431$ Å has a diameter of 5 μm, then the slab contains of the order of 9000 planes. The broadening of the (111) reflection at $2\theta = 28.442°$ is of the order of 0.0018°, and would not be noticed on a diffractometer with a FWHM of 0.05°. A silicon nanoparticle of average diameter of 50 nm would produce a FWHM of about 0.20°, clearly larger than a reflection width from a perfect sample of about 0.05°. Application of the Scherrer equation, however, requires a good knowledge of the optical setup (slits, monochromator and analyzer crystals and mirrors, etc.) of the diffractometer, so that the reflection profiles and profile widths are well known over the desired angular range. In particular, the intrinsic profile width b obtained using a NIST Standard Reference Material (for instance, LaB_6, silicon), needs to be established. The two profiles that are often used for this type of analysis are the Gaussian and the Lorentzian (Cauchy) functions:

$$I(\epsilon) = I_0 e^{-k^2 \epsilon^2} \quad \text{and} \quad I(\epsilon) = \frac{I_0}{1 + k^2 \epsilon^2}$$

In the case of Gaussian profiles, the total breadth B of the profile, either at FWHM or the integral breadths are given by

$$B^2 = b^2 + \beta^2$$

whereas for Lorentzian (Cauchy) profiles, the total breadth is given by

$$B = b + \beta$$

Since the actual reflection profiles are usually neither pure Gaussian nor pure Lorentzian in shape, the calculation of the profile broadening due to the finite crystal size effects needs to take this into account; hence, the profile function that best describes the actual diffractometer profiles needs to be known. Furthermore, the crystallite size broadening is proportional the $(\cos \theta)^{-1}$.

8.13.2 Strain

Strain (compressive or tensile) will affect the d-spacing in a sample, either increasing or decreasing it. If the strain is *homogenous*, then a displacement of the reflections will occur. For small changes in the d-spacing (Δd), the Bragg law can be written as

$$\sin(\theta + \delta\theta) = \frac{\lambda}{2(d - \Delta d)}$$

Since the strain $\epsilon = (\Delta d)/d$, differentiating Bragg's law gives

$$\delta d = \frac{\lambda}{2} \frac{\cos \theta}{\sin^2 \theta} \delta\theta$$

and

$$\frac{\delta d}{d} = \frac{1}{\tan \theta} \delta\theta$$

Since homogenuous strain results in a small displacement, strain broadening of the reflections requires different strains in different grains; thus the strain is *inhomogeneous*. Relating the inhomogeneous strain to the fraction $\delta d/d$ and applying the fact that small changes in the d-spacing translate to a change in diffraction angle θ with a $(\tan \theta)^{-1}$ behavior gives the relationship

$$\beta_\epsilon = C_\epsilon \tan \theta$$

where C_ϵ is of the order of 4 to 5, depending on the assumption describing the inhomogeneous strain. The reflection broadening due to the inhomogeneous strain therefore follows a different relationship with the diffraction angle θ, in this case, it is proportional to the $\tan \theta$.

8.13.3 Size and strain

Often, small size and strain are present in a material. For instance, small grains in cold worked metals are highly strained by the process. Therefore, using reflection profile widths to determine size and strain needs an approach that separates the two distinct behaviors in 2θ for the size and strain behavior. If the total width produced by the two effects is additive, then

$$\beta_{tot} = \beta_{size} + \beta_\epsilon = \frac{K\lambda}{\tau \cos \theta} + C_\epsilon \tan \theta$$

This equation can be modified by multiplication of $\cos \theta$, resulting in

$$\beta_{tot} \cos \theta = \frac{K\lambda}{\tau} + C_\epsilon \sin \theta \tag{8.6}$$

In a plot that takes the total reflection widths multiplied by $\cos \theta$ versus $\sin \theta$ should result in a straight line, with the slope given by C_ϵ and the y-axis intercept as $K\lambda/\tau$. This plot representing equation (8.6) is called the *Williamson–Hall plot* and allows separating the two contributions of the line broadening. In general, the Williamson–Hall plot works quite well for processed metals, but deviations from linear behavior is often observed for other materials. Even negative intercepts are possible, which do not represent any physical quantity. Therefore, the Williamson–Hall plots need to be regarded with caution, but the method may give good results for relative measurements, where the same material is processed under different conditions, and relative changes are followed using this method.

Bibliography

[10] P. M. de Wolff. J. Appl. Cryst., 1:108, 1968.
[11] G. S. Smith and R. L. Snyder. J. Appl. Cryst., 12:60–65, 1979.
[12] P. E. Werner, L. Eriksson, and M. Westdahl. J. Appl. Cryst., 18:367–370, 1985.
[13] J. W. Visser. J. Appl. Cryst., 2:89, 1969.
[14] A. Boultif and D. Louër. J. Appl. Cryst., 37:724, 2004.
[15] J. Bergmann, A. Le Bail, R. Shirley, and V. Zlokazov. Z. Kristallographie., 219:783–790, 2004.

9 Qualitative and quantitative evaluation of powder patterns

Powder diffraction is a very versatile and easy method for phase analysis and forensic investigations. Diffraction from a powder requires two axes at most, with the implementation in Bragg–Brentano geometry as a well-known example. Integrating intensities over the Debye cone by using sample motion and linear or area detectors, close to ideal powder patterns can be obtained. Powder patterns can therefore be analyzed quantitatively, giving sample information such as phase fractions, stoichiometry, stress/strain, grain size, etc. A qualitative powder pattern analysis may involve phase identification (what phases are present), refinement of unit cell parameters, in contrast, quantitative analysis needs to describe the sample as well as the diffractometer experiment.

9.1 Qualitative analysis

For qualitative analysis, such as phase determination without quantitative determination of phase fractions, simple search-match procedures are sufficient. This is based on the fact that the X-ray powder pattern is a unique "fingerprint" of a phase. The combination of unique d-spacings and intensities allow phase identification for powder patterns that approximate ideal diffraction conditions. If the pattern is severely affected by texture, then the measured intensity values will not be reliable, but reflection positions can still be used for ruling out the presence/absence of a phase. In the case of two phases having accidentally the same metric unit cell leading to a high degree of positional overlap, the intensity distribution in the pattern is expected to differ so that an unambiguous phase identification is possible. Furthermore, different thermal expansion is expected, so that changing the temperature would produce metrically different unit cells. Even a textured sample gives enough diffraction lines for an unambiguous identification. Databases of powder patterns are available from the *International Centre for Diffraction Data, ICDD* (www.icdd.com), where data of over 250,000 phases are compiled. Different data formats are available, and search-match algorithms have been implemented by the manufacturers of powder diffraction systems so multiphase samples can be analyzed qualitatively in a short time. Examples are phase analysis of mineralogical rock samples, where a large number of different minerals form the constituents of a rock.

9.2 Quantitative analysis

The constituent phases of a multiphase sample can be analyzed quantitatively. A volume or mass fraction of each constituent phase is desired, as for instance of a miner-

https://doi.org/10.1515/9783110610833-009

alogical rock sample containing several different minerals. Since each distinct phase present in a sample will give a distinct powder pattern that is not affected by the other phases present, diffraction techniques are uniquely suited for such tasks. As the intensities of the different powder patterns from the individual phase are proportional to their respective phase mass fractions, quantitative phase analysis based on diffraction methods is possible. With the Rietveld analysis, a method is available to quantify the fraction of each crystalline material present. In the following, a brief introduction to the Rietveld analysis/full pattern fitting is given.

9.3 Rietveld analysis

A quantitative analysis of a powder pattern needs to account for several different effects:
1. the intensities and the positions observed due to the structure;
2. the background in the pattern;
3. the shape of the profiles of the reflections.

Such a procedure was first described by H. M. Rietveld for neutron powder diffraction patterns [16]. While intensities and positions are given by the phase, the background and the peak profiles are a function of the diffractometer setup. Unless the phase itself affects the peak profiles due to strong shape and strain effects, the reflection profiles are mostly determined by the instrumental conditions. Calculation of the reflection positions and respective intensities has been discussed in the previous chapters, the details of the profile description and background modeling will be discussed in the following.

9.3.1 Background

Background subtraction
The background in a diffraction pattern may be due to a number of different effects. Scattering of the incoming X-ray radiation by air in the path will produce a background that has more intensity at a low angle. Diffuse scattering from the sample holder, such as an amorphous glass plate, will give a background that has a broad maximum. If the sample contains a possible noncrystalline component, its scattering will contribute to the background. For instance, the background for CuKα radiation from a glass slide usually has a broad maximum around $2\theta \approx 25°$. Furthermore, the detector system may have a dark count that is observed in the absence of X-ray radiation. Additional devices for the sample environment may contribute additional background scattering in certain angular ranges, that originate from heat shields, sample enclosures, etc. The

background intensity therefore may not be a simple function of 2θ: $I_{Background}(2\theta)$. Optimization of the diffractometer to reduce the background may include enclosed beam paths, adjustment of the X-ray beam footprint on the sample to reduce potential scattering from the sample holder, beam stops that control the incident beam and sample preparation with sample thicknesses optimized for absorption. For a simple pattern with a few reflections, the reflection peaks are resolved to the background baseline, so that the background in between the peaks is well-defined. The background can then either be subtracted, or fit with an empirical curve, such as a polynomial. For a background that does not vary strongly over the 2θ range, this approach will work if the pattern is not too complex. A background with more variability, for instance, from a sample environmental chamber that scatters intensity in specific angular ranges, may need a different approach, such as a separate background determination based on a reference pattern. This reference pattern can then be used to subtract the background due the diffractometer setup from the actual pattern. In contrast, in a complex pattern with many overlapping reflections, the peaks may not be resolved to the background and, therefore, the background itself cannot easily be estimated. An iterative procedure may therefore need to be applied for this case, where the background is reestimated and subtracted again.

An empirical background model can be constructed by manually picking background points and either interpolate a background curve, or fitting these background points to a continuous function, using for instance a spline interpolation. Such an empirical background curve can then be subtracted from the powder pattern. In case of a strongly angle dependent background that may include signals from the sample holder and environmental control chamber, an empirical background subtraction may be a good solution.

Background refinement
The background contributions to the powder pattern may be refined simultaneously with the structural parameters, using for instance a polynomial expression with a number of parameters. However, the polynomial expression of the background is mostly empirical, and the number of terms in the background function needs to be evaluated to ensure that the refinement of the background terms does not lead to insignificant values with large estimated standard deviations. In the case where the polynomial description fits the background well, then the overall refinement will be stable and the refined parameters will be significant. In contrast, when the background cannot be expressed by such functions, it will affect the quality of the refinement, and adding more terms to the polynomial expressions will not result in a better refinement. For such a case, a combination of both, background subtraction and background refinement may be applied.

9.3.2 Profile functions

The quantitative description of a reflection profile uses a *profile function* that handles the reflection peak top as well as the reflection tails. Different functions that can be mathematically handled are used to describe various diffraction setups, and a number of functions will be discussed.

Gauss function

A simple, well-behaved peak function to describe an X-ray reflection profile is the Gaussian function given in (9.1). For a given $2\theta_i$ in the powder pattern, a reflection at position $2\theta_K$ with width of H_K, with normalization factor proportional to H_K^{-1} contributes to $G(2\theta_i)$ as

$$G(2\theta_i) = \frac{\sqrt{4\ln 2}}{\sqrt{\pi}H_K} \exp\left(\frac{(2\theta_i - 2\theta_K)^2}{H_K^2} \right)$$ (9.1)

The Gaussian is shown in Figure 9.1. The prefactor normalizes the integrated area to 1.0.

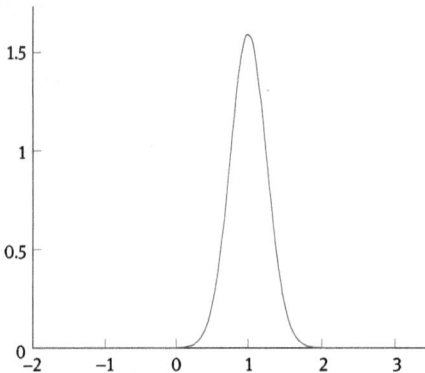

Figure 9.1: Gauss function.

The Gauss function fits the top of a reflection profile quite well, but the fast drop does not describe the extended tails of a reflection.

Lorentz function

The Lorentz function is a singly peaked function that is given by equation (9.2) for a reflection located at $2\theta_K$

$$L(2\theta) = \frac{4}{\pi H_K} \frac{1}{1 + 4(2\theta_i - 2\theta_K)^2/H_K^2}$$ (9.2)

The Lorentz function does not drop off as fast as the Gauss function, and describes the tails of a reflection better, but does not describe the peak top well. It is shown in Figure 9.2.

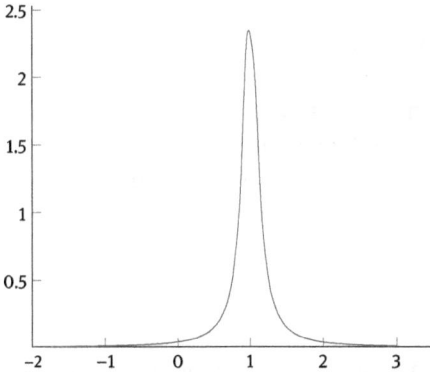

Figure 9.2: Lorentz function.

It should be noted that the Gauss and Lorentz functions both have the same FWHM of 0.25, and both functions are centered at $2\theta = 1$. While the Gauss function is close to 0 at a distance of 4× the FWHM, the Lorentz function still has value of 0.3. The contribution of a Lorentz function extends over a much larger range than the Gauss function, a fact that needs to be kept in mind when calculating intensities around a reflection position. If the Lorentz function does not drop off fast enough to describe the tails of a reflection, a squared Lorentz function may be considered. This function, also called a *modified Lorentzian,* may be a good alternative to fit a reflection profile.

Pearson-VII function

A reflection profile can be thought to combine the features of both, Gauss and Lorentz functions, and combinations of the two functions may provide a good approximation of reflection profiles. A function that combines the features of both the Gauss and Lorentz function is the *Pearson-VII* function. It is based on a Lorentz function raised to the power of m, and it is given by equation (9.3),

$$\text{PVII}(2\theta) = \frac{C_K}{H_K} \frac{1}{[1 + 4(2^{1/m} - 1)(2\theta_i - 2\theta_K)^2/H_K^2]^m} \tag{9.3}$$

with C_K a normalizing factor. For the parameter $m = 1$, the Pearson-VII function is a Lorentz function, and for $m \to \infty$, the Pearson-VII approaches a Gauss function. The parameter m can be refined to optimize the peak shape description. Furthermore, the parameter m is refined as a function of 2θ as

$$m = NA + \frac{NB}{2\theta} + \frac{NC}{(2\theta)^2}$$

with *NA*, *NB* and *NC* refinable parameters. The Pearson-VII function can be used for the description of peak shapes obtained from a laboratory diffractometer, which show profile shapes that are intermediate between Gaussian and Lorentzian. The Pearson-VII function has been further refined to deal with peak shape asymmetries by splitting the function into a left and right profile, the *split Pearson-VII* function. In the case of laboratory diffractometers where the peak asymmetries vary over the angular range, the split Pearson-VII function provides a good description of the peak shapes over an extended 2θ range. The split Pearson-VII function uses an asymmetry parameter *A* to split the function into a left and a right part with different FWHM H_K. The function also describes the angle dependent asymmetries in a Guinier powder pattern quite well, where peak asymmetries are observed due to the different angles of incidence of the diffracted intensities onto the detector.

Voigt and pseudo-Voigt function

The *Voigt* function is a convolution of a Lorentz and Gauss function. Unfortunately, this function is cumbersome to compute, and a different approach is chosen. The *pseudo-Voigt* function (equation (9.4)) combines the Lorentz and the Gauss function directly as

$$pV(2\theta) = \eta L(2\theta) + (1 - \eta)G(2\theta) \tag{9.4}$$

The mixing parameter η describes the relative contribution of the Lorentz and Gauss function to the reflection profile, and can be refined as a linear function in 2θ,

$$\eta = NA + NB \times 2\theta$$

with *NA* and *NB* refinable parameters. In Figure 9.3, the Lorentz function is shown in blue, the Gauss function in red, and the combined pseudo-Voigt function in green. The pseudo-Voigt function is often used, with additional corrections to account for reflection asymmetry.

The combination of Gauss and Lorentz functions into the pseudo-Voigt function allows fitting reflection profiles of X-ray powder patterns, and is used for both laboratory and synchrotron based systems. The extent of the tails of the reflection modeled predominantly by the Lorentz function indicates that, depending on the shape of the profiles, the profile calculations needs to be extended over a range of $\pm 6\times$ FWHM to capture the intensity contributions further away from the reflection intensity maximum.

9.3.3 The full width at half maximum

The Full Width at Half Maximum (FWHM) of reflection K, FWHM $= H_K$, is not constant over the angular range of the diffraction pattern and depends nonlinearly on the angle

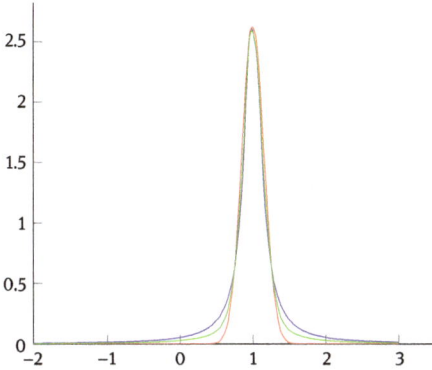

Figure 9.3: Pseudo-Voigt function (green), Gauss function (red) and Lorentz function (blue).

position of the reflection. Depending on the incident beam configuration using slits, monochromators, mirrors or combinations of these elements, a clear dependence of the FWHM $H_K(\theta)$ is observed. The FWHM generally increases with increasing diffraction angle for a laboratory Bragg–Brentano system, but different angular behaviors of the FWHM are found for different diffractometer designs. The angular dependence is generally modeled using the Caglioti function described as [17]

$$H_K^2 = U \tan^2 \theta + V \tan \theta + W \tag{9.5}$$

The parameters U, V and W in equation (9.5) can be refined, with W indicating the constant width part of the FWHM. In some programs, these parameters are referred to as GU, GV and GW, and are used to calculate the FWHM of the Gauss function. The change in the FWHM over the angular range of the diffraction pattern follows usually a monotonic behavior for a Bragg–Brentano diffractometer, with the smallest H_K^2 close to $2\theta = 0$.

The quantitative description of a the reflection profiles over the observed 2θ range uses a single profile function type (Gauss, Lorentz, Pearson-VII, Voigt or pseudo-Voigt function), and modifies the FWHM according to the Caglioti function. Additional parameters are introduced to address specific diffractometer setup effects on the reflection profiles.

9.3.4 Intensity calculation

The total intensity at position 2θ is given by a sum over all K structure factors $F_K(hkl)$ multiplied with the Lorentz and polarization $LP(2\theta)$ factors combined, the multiplicity factor $M(hkl)$, an absorption factor $A(2\theta)$, a profile function Φ, a scale factor S, a sample dependent factor for surface roughness s, plus the background intensity $I_{bkg}(2\theta)$.

This is expressed in equation (9.6),

$$I(2\theta_i) = s \times S(2\theta_i) \times A(2\theta_i) \times \left[\sum_K M(hkl) \times |F_K(hkl)|^2 \times LP(2\theta_i) \times \Phi(2\theta_i - 2\theta_K) \right] + I_{bkg}(2\theta_i) \quad (9.6)$$

With a known structure, knowledge of the diffractometer profile parameters and sample parameters, it is possible to calculate an ideal powder pattern, using equation (9.6).

9.3.5 Full pattern refinement

A quantitative fit of an observed powder pattern can be obtained by minimizing the difference squared between observed and calculated intensity values at each point i ($2\theta_i$) (least squares refinement),

$$\Delta = \sum_i \left[I(2\theta_i)_{obs} - I(2\theta_i)_{calc} \right]^2$$

This is at the heart of the Rietveld analysis that treats each point in the powder pattern, whether due to background or due to diffraction as an observation, and refines a model that explains each observation. Such a point-by-point profile fit of a powder pattern extracts the most information from the pattern. A given structural model can therefore be refined since the $|F(hkl)|^2$ together with the knowledge of the profile function and its changes over the angular range of the pattern can be calculated exactly by using equation (9.6). An ideal powder pattern has therefore a large number of nonoverlapping reflections, low background counts that do not vary strongly over the angular range of the pattern, high instrumental resolution for well-defined reflection profiles, together with an excellent sample of high crystallinity that gives well-defined reflection profiles obtained from an ideal diffractometer setup. Achieving such an ideal situation is difficult, but modern laboratory diffractometers can achieve excellent results. Synchrotron beamlines optimized for powder diffraction produce excellent patterns with high signal-to-noise ratio, high resolution and low background, closely approaching ideal conditions.

If a structural model is not available, but the pattern is successfully indexed, it is possible to use equation (9.6) to fit the pattern with the $|F(hkl)|^2$ values as refinable parameters, a *Pawley* fit. Such a full pattern fit will provide a good reference for a structural refinement using the pattern. However, for strongly overlapping reflections, correlations in the least squares refinement may produce negative $|F(hkl)|^2$ values. Therefore, extraction procedures for $|F(hkl)|^2$ values may need to be adjusted.

A different way of extracting $|F(hkl)|^2$ values is implemented in the LeBail fit. The unit cell parameters, a $2\theta_0$ offset, and profile parameters are refined, but the intensities are initially set to an arbitrary value. The algorithm does not refine the reflection peak heights, but apportions iteratively the intensity between overlapping reflections, producing a set of new intensities. These new estimated intensities are then used for the next iteration. In this way, correlations due to reflection overlap are eliminated, and negative $|F(hkl)|^2$ values are avoided.

Refinement

For a full pattern refinement, minimization of

$$\Delta = \sum_i w_i \left[I(2\theta_i)_{\text{obs}} - I(2\theta_i)_{\text{calc}} \right]^2$$

is carried out, where w_i is a weight function for the $I(2\theta_i)_{\text{obs}}$. An agreement factor, the *R-value* is defined as

$$R_{wp} = \left[\frac{\sum_i w_i \left[I(2\theta_i)_{\text{obs}} - I(2\theta_i)_{\text{calc}} \right]^2}{\sum_i w_i I(2\theta_i)^2_{\text{obs}}} \right]^{\frac{1}{2}}$$

This R-value is thus a measure of the quality of the overall pattern fit. It should be noted that for a pattern with high background, the observed intensities are large in areas of the pattern where no reflections are present. This will increase the denominator for R_{wp}, while not affecting the numerator, giving an artificially low agreement factor. Therefore, background subtraction may need to be considered to assess the quality of the structural model and to reduce the influence of the background itself. Other agreement factors have been devised, to address this issue. For instance, a R_{wp} for the structural model can be calculated. In this way, only the intensities where reflections are allowed by the structural model are considered.

While the R-value of a refined model may be low, it is not necessarily true that the refinement is fully converged. Since the number of parameters to be refined can be large, correlations between parameters can not be ruled out. Such correlations can produce a matrix that is pseudo-degenerate and the refinement will result in unreasonable values. Furthermore, a local minimum may have been found, and the refinement will not proceed to the global minimum.

A sequence of parameters to refine a powder pattern with a structural and a diffractometer model may be as follows:
- scale, and background
- $2\theta_0$, and lattice parameters
- peak profile function parameters
- atom positions
- displacement parameters
- preferred orientation

An unconstrained full matrix least squares refinement is desired that is stable against random perturbations. This indicates that a minimum in the agreement factor R_F is found. It remains to be proven that this minimum is also a global minimum.

Examples

Example: laboratory data of CaF_2

The powder pattern obtained with a Scintag PAD-V diffractometer equipped with a diffracted beam graphite analyzer was refined with GSAS-II (https://subversion.xray.aps.anl.gov/trac/pyGSAS) [18]. The X-ray source was a copper fine-focus tube, run at 45kV and 22 mA. The pattern start position is at $2\theta = 15°$, end position is at $2\theta = 120°$, with steps of $0.02°$, resulting in 5251 observations. The background was modeled using a Chebyshev polynomial with 5 terms, b_1 to b_5. The profile function chosen was a pseudo-Voigt function, with parameters for the Caglioti function U, V and W, and the Lorentz function FWHM given as $X \cos^{-1}(\theta) + Y \tan(\theta)$. The axial divergence is given by the SH/L parameter. The refinement used the structure described previously to start modeling the background and the reflection provides. The small number of observed reflections allows to define the background and reflection profiles, whereas the flat plate sample introduces some degree of texture. To address the texture, a spherical harmonic correction of order 4 ($C(4,1)$) was refined. The resulting parameters are given in Table 9.1.

Table 9.1: GSAS-II refinement of the powder pattern of CaF_2.

Space Group	$Fm\bar{3}m$
a	$5.47097(7)$ Å
Ca	$0, 0, 0$
U_{iso}(Ca)	$0.0010(5)$ Å^2
F	$\frac{1}{4}, \frac{1}{4}, \frac{1}{4}$
U_{iso}(F)	$0.0088(6)$ Å^2
$2\theta_{min}$	$15°$
$2\theta_{max}$	$120°$
Number of observations	5251
b_1	$59.1(4)$
b_2	$-20.0(9)$
b_3	$-7.8(25)$
b_4	$-19.8(14)$
b_5	$33.8(25)$
U	$3.3(7) \times 10^{-4}$ deg^2
V	$-8.7(9) \times 10^{-4}$ deg^2
W	$10.53(28) \times 10^{-4}$ deg^2
X	$0.18(11) \times 10^{-2}$ deg
Y	$0.5(3) \times 10^{-2}$ deg
SH/L	$0.0223(4)$
$C(4,1)$	$-0.002(19)$
R_F	0.0753
wR_F	0.054
Durbin–Watson	1.330

The refinement is acceptable, but indicates some of the issues with using laboratory data from a flat plate geometry to attempt a Rietveld refinement. In particular, the texture correction is needed, even though the sample is cubic, and the powder grains are quite isometric. The background model, for instance, could be reduced to 4 terms, since b_3 is not well-defined. The FWHM is adequately described, and the pseudo-Voigt function fits the reflection profiles well. Since the number of reflections is small due to the high symmetry of the CaF_2, the profiles and background can be modeled without difficulty. The structural model, with fixed atom positions, has only isotropic displacement parameters to refine, making CaF_2 a reasonable material to extract an instrument parameter file for future Rietveld studies. In Figure 9.4, the observed and calculated intensities are shown, with the difference curve shown with negative intensities.

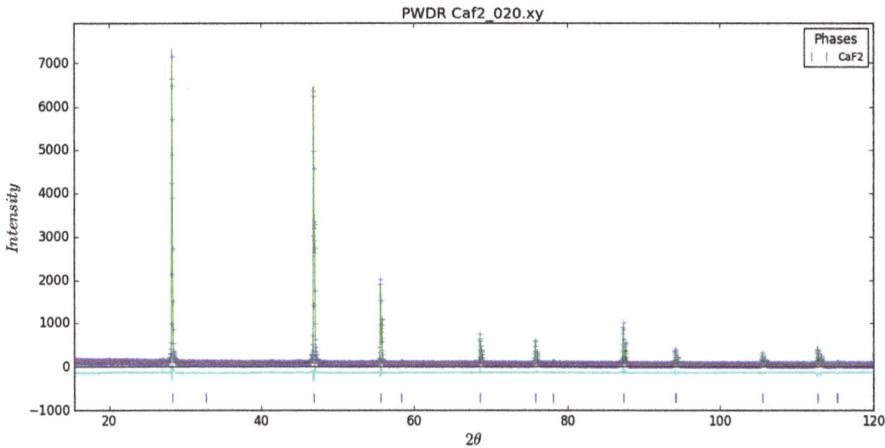

Figure 9.4: CaF_2 data and refined model. Data is given by the + sign, calculated intensities by the green line, and the difference curve in blue. Blue tick marks indicate reflection positions.

Example: synchrotron data of Barlowite

The quality of synchrotron data that can be obtained from a well-crystallized powder is well suited to the Rietveld analysis. The mineral barlowite, $Cu_4(OH)_6FBr$, has been synthesized in the laboratory using hydrothermal methods. The resulting powder has a deep blue color due to the presence of Cu^{2+}. This sample was sent to the Advanced Photon Source, Beamline 11-BM. The beam line has a silicon monochromator with narrow band pass characteristics, operating at a wavelength of 0.41454 Å, and almost 100 % polarization. The diffractometer at 11-BM is of the Debye–Scherrer type, with a 12-analyzer-detector system to reduce data acquisition time (https://11bm.xray.aps.anl.gov). The sample was contained in a kapton tube, and rotated during the data collection. The diffractometer measured an angular range from 0.425° to 50.0°, at steps

of 0.001°, giving 49575 observations. Rietveld refinement results of the room temperature structure of barlowite are given in Table 9.2, and the pattern is shown in Figure 9.5.
The structure contains a disordered copper atom (Cu 2) with an occupation factor of $\frac{1}{3}$
that is located in-between Kagome layers formed by Cu 1. The plot shows the excellent
resolution and dynamic range of the diffractometer and the narrow FWHM of all reflections. Reflections at higher angles have a high degree of overlap, but the well-resolved
low-angle reflections provide good values for the pseudo-Voigt profile function. The
quality of the data allows an unconstrained refinement of the all the parameters, instrumental and structural, resulting in refined values with small estimated standard
deviations.[1]

Table 9.2: Rietveld refinement of Barlowite, $Cu_4(OH)_6FBr$.

Symmetry	$P6_3/mmc$
a	6.675464(21) Å
c	9.298883(19) Å
Z	2
λ	0.41454 Å
$2\theta_{min}$	0.5°
$2\theta_{max}$	50.0°
Step width	0.001°
Observations	49575
Number of reflections	647
Number of parameters	34
R_F	0.082
wR_F	0.103

Atom	x	y	z	U_{iso} [Å^2]	occ
Cu 1	$\frac{1}{2}$	0	0	0.011	
Cu 2	0.37043(22)	0.74086	$\frac{3}{4}$	0.009	$\frac{1}{3}$
O	0.20158(7)	0.79842	0.90820(10)	0.007	
Br	$\frac{1}{3}$	$\frac{2}{3}$	$\frac{1}{4}$	0.014	
F	0	0	$\frac{1}{4}$	0.016	
H	0.12477	0.87523	0.87425	0.090	

Atom	u_{11}	u_{22}	u_{33}	u_{12}	u_{13}	u_{23}
Cu 1	0.00734(11)	0.00642(13)	0.01798(14)	0.0032	−0.0015(5)	−0.0030
Cu 2	0.0090(5)	0.0150(5)	0.0022(4)	0.0075	0.0000	0.0000
O	0.0050(4)	0.0050	0.0095(6)	0.0023(8)	−0.0006(4)	0.0006
Br	0.01582(18)	0.0158	0.01120(24)	0.0079	0.0000	0.0000
F	0.0125(9)	0.0125	0.0237(10)	0.0062	0.0000	0.0000

1 Barlowite Data taken at Beamline 11-BM at the Advanced Photon Source, Argonne National Laboratory, supported by the US Department of Energy, Office of Science, Office of Basic Energy Sciences,
under contract No. DE-AC02-06CH11357.

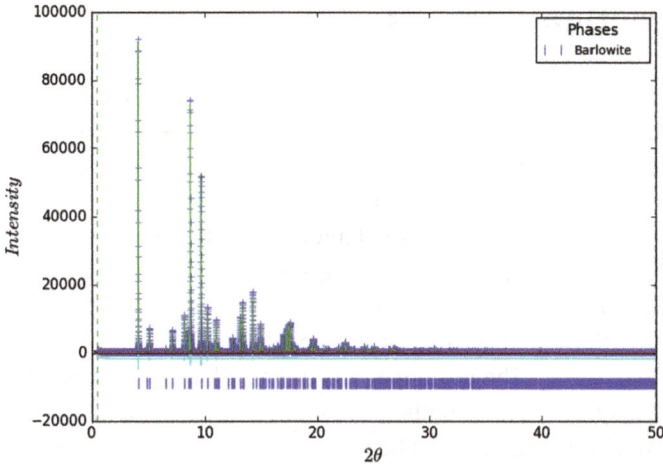

Figure 9.5: Synchrotron data and refined model of synthetic barlowite, $Cu_4(OH)_6FBr$. Data is given by the + sign, calculated intensities by the green line, and the difference curve in blue. Blue tick marks indicate reflection positions.

Bibliography

[16] H. M. Rietveld A Profile Refinement Method for Nuclear and Magnetic Structures. Journal of Applied Crystallography, 2(2):65–71, 1969 https://doi.org/10.1107/S0021889869006558.

[17] G. Caglioti, A. Paoletti, and F. P. Ricci. Nucl. Instrum. Methods, 35:223–228, 1958.

[18] B. H. Toby and R. B. Von Dreele. GSAS-II: the genesis of a modern open-source all purpose crystallography software package. Journal of Applied Crystallography, 46(2):544–549, 2013. https://doi.org/10.1107/S0021889813003531.

10 Structure solution methods

This chapter will discuss methods for the solution of a structure, based on intensity data and reasoning using knowledge of the chemistry about the atomic bonding between elements present in the structure. The chapter will be necessarily incomplete, but will touch on a number of methods that can be employed to determine the arrangement of atoms in a structure. There are several good tutorials on structure determinations available, either in the form of textbooks, or programs that are available freely. Usually, programs are set-up in suites that allow to produce a tentative solution, followed by further refining of the structure to arrive at a model of F_{calc} that matches the observed F_{obs}, or the squared F_{obs}^2. The positional and atomic displacement parameters will be refined using a least squares optimization, minimizing $(F_{obs} - F_{calc})^2$ or $(F_{obs}^2 - F_{calc}^2)$. Different weighting schemes can be introduced, giving a measurement of data reliability. In general, spherical atoms are assumed, but higher order electron distributions can be modeled by using modified electron densities that for instance reflect the fact that covalent bonding produces non-spherical electron distributions.

10.1 Example of a structure solution

An example of a solution of the structure of Cs_3CoCl_5 will be presented, using the publication by Powell and Wells [19]. The manuscript includes a description of the crystal habit, identifying the crystal class as tetragonal. Furthermore, the presence/absence of a pyroelectric effect was investigated, with a vanishing pyroelectric effect observed, concluding that the symmetry of the crystal is centrosymmetric. An X-ray measurements gave a tetragonal I-centered cell of $a = 9.18$ Å and $c = 14.47$ Å. Additionally, the stoichiometry was determined to be Cs_3CoCl_5, and a density measurement gave $\rho = 3.39 g/cm^3$ resulting in $Z = 4$. With the stoichiometry, density and the unit cell volume known, it becomes clear that the unit cell contains 4 cobalt atoms, 12 cesium atoms and 20 chlorine atoms. Analyzing the indexed diffraction pattern, consistent with I-centering, an additional systematic extinction is found: for $(0kl)$, $l = 2n$, indicating a c-glide plane perpendicular to the a and the b-axis. The possible space groups are therefore $I4cm$ and $I4/mcm$. The absence of a pyroelectric effect thus rules out $I4cm$, making $I4/mcm$ (#140) the actual space group symmetry of the crystal (see Figures 10.1 and 10.2).

With the symmetry determined and the number of the different atoms in the unit cell known, the possible distribution of atoms on particular space group positions (Wykoff positions) can be evaluated. Inspecting the special positions, there are four 4-fold positions, $4a$, $4b$, $4c$ and $4d$, four 8-fold positions $8e$, $8f$, $8g$ and $8h$, four 16-fold positions $16i$, $16j$, $16k$, $16l$ and one 32-fold position $32m$. It follows from the number of atoms in the unit cell that the $32m$ position is not occupied. It is now possible to constrain the number of distributions of atoms over the special positions. It is clear that

https://doi.org/10.1515/9783110610833-010

International Tables for Crystallography (2006). Vol. A, Space group 140, pp. 480–481.

$I4/mcm$ D_{4h}^{18} $4/mmm$ Tetragonal

No. 140 $I\ 4/m\ 2/c\ 2/m$ Patterson symmetry $I4/mmm$

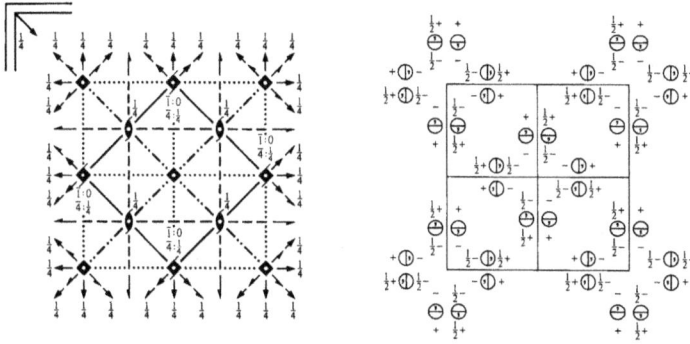

Origin at centre $(4/m)$ at $4/mc2_1/e$

Asymmetric unit $0 \le x \le \frac{1}{2};\ \ 0 \le y \le \frac{1}{2};\ \ 0 \le z \le \frac{1}{4};\ \ y \le \frac{1}{2}-x$

Symmetry operations

For $(0,0,0)+$ set

(1) 1	(2) 2 $0,0,z$	(3) 4^+ $0,0,z$	(4) 4^- $0,0,z$
(5) 2 $0,y,\frac{1}{4}$	(6) 2 $x,0,\frac{1}{4}$	(7) 2 $x,x,\frac{1}{4}$	(8) 2 $x,\bar{x},\frac{1}{4}$
(9) $\bar{1}$ $0,0,0$	(10) m $x,y,0$	(11) $\bar{4}^+$ $0,0,z;\ 0,0,0$	(12) $\bar{4}^-$ $0,0,z;\ 0,0,0$
(13) c $x,0,z$	(14) c $0,y,z$	(15) c $x,\bar{x},z$	(16) c x,x,z

For $(\frac{1}{2},\frac{1}{2},\frac{1}{2})+$ set

(1) $t(\frac{1}{2},\frac{1}{2},\frac{1}{2})$	(2) $2(0,0,\frac{1}{2})$ $\frac{1}{4},\frac{1}{4},z$	(3) $4^+(0,0,\frac{1}{2})$ $0,\frac{1}{2},z$	(4) $4^-(0,0,\frac{1}{2})$ $\frac{1}{2},0,z$
(5) $2(0,\frac{1}{2},0)$ $\frac{1}{4},y,0$	(6) $2(\frac{1}{2},0,0)$ $x,\frac{1}{4},0$	(7) $2(\frac{1}{2},\frac{1}{2},0)$ $x,x,0$	(8) 2 $x,\bar{x}+\frac{1}{2},0$
(9) $\bar{1}$ $\frac{1}{4},\frac{1}{4},\frac{1}{4}$	(10) $n(\frac{1}{2},\frac{1}{2},0)$ $x,y,\frac{1}{4}$	(11) $\bar{4}^+$ $\frac{1}{2},0,z;\ \frac{1}{2},0,\frac{1}{4}$	(12) $\bar{4}^-$ $0,\frac{1}{2},z;\ 0,\frac{1}{2},\frac{1}{4}$
(13) a $x,\frac{1}{4},z$	(14) b $\frac{1}{4},y,z$	(15) m $x+\frac{1}{2},\bar{x},z$	(16) $g(\frac{1}{2},\frac{1}{2},0)$ x,x,z

---

Maximal non-isomorphic subgroups *(continued)*

IIa	[2] $P4_2/ncm$ (138)	1; 2; 7; 8; 11; 12; 13; 14; (3; 4; 5; 6; 9; 10; 15; 16) $+ (\frac{1}{2},\frac{1}{2},\frac{1}{2})$
	[2] $P4_2/mbc$ (135)	1; 2; 7; 8; 9; 10; 15; 16; (3; 4; 5; 6; 11; 12; 13; 14) $+ (\frac{1}{2},\frac{1}{2},\frac{1}{2})$
	[2] $P4_2/nbc$ (133)	1; 2; 5; 6; 11; 12; 15; 16; (3; 4; 7; 8; 9; 10; 13; 14) $+ (\frac{1}{2},\frac{1}{2},\frac{1}{2})$
	[2] $P4_2/mcm$ (132)	1; 2; 5; 6; 9; 10; 13; 14; (3; 4; 7; 8; 11; 12; 15; 16) $+ (\frac{1}{2},\frac{1}{2},\frac{1}{2})$
	[2] $P4/ncc$ (130)	1; 2; 3; 4; 13; 14; 15; 16; (5; 6; 7; 8; 9; 10; 11; 12) $+ (\frac{1}{2},\frac{1}{2},\frac{1}{2})$
	[2] $P4/mbm$ (127)	1; 2; 3; 4; 9; 10; 11; 12; (5; 6; 7; 8; 13; 14; 15; 16) $+ (\frac{1}{2},\frac{1}{2},\frac{1}{2})$
	[2] $P4/nbm$ (125)	1; 2; 3; 4; 5; 6; 7; 8; (9; 10; 11; 12; 13; 14; 15; 16) $+ (\frac{1}{2},\frac{1}{2},\frac{1}{2})$
	[2] $P4/mcc$ (124)	1; 2; 3; 4; 5; 6; 7; 8; 9; 10; 11; 12; 13; 14; 15; 16
IIb	none	

Maximal isomorphic subgroups of lowest index

IIc [3] $I4/mcm$ ($\mathbf{c'} = 3\mathbf{c}$) (140); [9] $I4/mcm$ ($\mathbf{a'} = 3\mathbf{a}, \mathbf{b'} = 3\mathbf{b}$) (140)

Minimal non-isomorphic supergroups

I [3] $Fm\bar{3}c$ (226)

II [2] $C4/mmm$ ($\mathbf{c'} = \frac{1}{2}\mathbf{c}$) ($P4/mmm$, 123)

Figure 10.1: Space group I4/mcm, page 1.

CONTINUED

No. 140

I4/mcm

Generators selected   (1); $t(1,0,0)$; $t(0,1,0)$; $t(0,0,1)$; $t(\frac{1}{2},\frac{1}{2},\frac{1}{2})$; (2); (3); (5); (9)

Positions

Multiplicity, Wyckoff letter, Site symmetry		Coordinates $(0,0,0)+$ $(\frac{1}{2},\frac{1}{2},\frac{1}{2})+$				Reflection conditions
						General:
32	*m*	1	(1) x,y,z (2) $\bar{x},\bar{y},z$ (3) $\bar{y},x,z$ (4) $y,\bar{x},z$			$hkl : h+k+l = 2n$
			(5) $\bar{x},y,\bar{z}+\frac{1}{2}$ (6) $x,\bar{y},\bar{z}+\frac{1}{2}$ (7) $y,x,\bar{z}+\frac{1}{2}$ (8) $\bar{y},\bar{x},\bar{z}+\frac{1}{2}$			$hk0: h+k = 2n$
			(9) $\bar{x},\bar{y},\bar{z}$ (10) $x,y,\bar{z}$ (11) $y,\bar{x},\bar{z}$ (12) $\bar{y},x,\bar{z}$			$0kl : k,l = 2n$
			(13) $x,\bar{y},z+\frac{1}{2}$ (14) $\bar{x},y,z+\frac{1}{2}$ (15) $\bar{y},\bar{x},z+\frac{1}{2}$ (16) $y,x,z+\frac{1}{2}$			$hhl : l = 2n$
						$00l : l = 2n$
						$h00 : h = 2n$
						Special: as above, plus
16	*l*	..*m*	$x,x+\frac{1}{2},z$ $\bar{x},\bar{x}+\frac{1}{2},z$ $\bar{x}+\frac{1}{2},x,z$ $x+\frac{1}{2},\bar{x},z$			no extra conditions
			$\bar{x},x+\frac{1}{2},\bar{z}+\frac{1}{2}$ $x,\bar{x}+\frac{1}{2},\bar{z}+\frac{1}{2}$ $x+\frac{1}{2},x,\bar{z}+\frac{1}{2}$ $\bar{x}+\frac{1}{2},\bar{x},\bar{z}+\frac{1}{2}$			
16	*k*	*m*..	$x,y,0$ $\bar{x},\bar{y},0$ $\bar{y},x,0$ $y,\bar{x},0$			no extra conditions
			$\bar{x},y,\frac{1}{2}$ $x,\bar{y},\frac{1}{2}$ $y,x,\frac{1}{2}$ $\bar{y},\bar{x},\frac{1}{2}$			
16	*j*	.2.	$x,0,\frac{1}{4}$ $\bar{x},0,\frac{1}{4}$ $0,x,\frac{1}{4}$ $0,\bar{x},\frac{1}{4}$			$hkl : l = 2n$
			$\bar{x},0,\frac{3}{4}$ $x,0,\frac{3}{4}$ $0,\bar{x},\frac{3}{4}$ $0,x,\frac{3}{4}$			
16	*i*	..2	$x,x,\frac{1}{4}$ $\bar{x},\bar{x},\frac{1}{4}$ $\bar{x},x,\frac{1}{4}$ $x,\bar{x},\frac{1}{4}$			$hkl : l = 2n$
			$\bar{x},\bar{x},\frac{3}{4}$ $x,x,\frac{3}{4}$ $x,\bar{x},\frac{3}{4}$ $\bar{x},x,\frac{3}{4}$			
8	*h*	*m*.2*m*	$x,x+\frac{1}{2},0$ $\bar{x},\bar{x}+\frac{1}{2},0$ $\bar{x}+\frac{1}{2},x,0$ $x+\frac{1}{2},\bar{x},0$			no extra conditions
8	*g*	2.*mm*	$0,\frac{1}{2},z$ $\frac{1}{2},0,z$ $0,\frac{1}{2},\bar{z}+\frac{1}{2}$ $\frac{1}{2},0,\bar{z}+\frac{1}{2}$			$hkl : l = 2n$
8	*f*	4..	$0,0,z$ $0,0,\bar{z}+\frac{1}{2}$ $0,0,\bar{z}$ $0,0,z+\frac{1}{2}$			$hkl : l = 2n$
8	*e*	..2/*m*	$\frac{1}{4},\frac{1}{4},\frac{1}{4}$ $\frac{3}{4},\frac{1}{4},\frac{1}{4}$ $\frac{1}{4},\frac{3}{4},\frac{1}{4}$ $\frac{1}{4},\frac{3}{4},\frac{1}{4}$			$hkl : k,l = 2n$
4	*d*	*m*.*mm*	$0,\frac{1}{2},0$ $\frac{1}{2},0,0$			$hkl : l = 2n$
4	*c*	4/*m*..	$0,0,0$ $0,0,\frac{1}{2}$			$hkl : l = 2n$
4	*b*	$\bar{4}$2*m*	$0,\frac{1}{2},\frac{1}{4}$ $\frac{1}{2},0,\frac{1}{4}$			$hkl : l = 2n$
4	*a*	422	$0,0,\frac{1}{4}$ $0,0,\frac{3}{4}$			$hkl : l = 2n$

Symmetry of special projections

Along [001] $p4mm$	Along [100] $p2mm$	Along [110] $p2mm$
$\mathbf{a}' = \frac{1}{2}(\mathbf{a} - \mathbf{b})$ $\mathbf{b}' = \frac{1}{2}(\mathbf{a} + \mathbf{b})$	$\mathbf{a}' = \frac{1}{2}\mathbf{b}$ $\mathbf{b}' = \frac{1}{2}\mathbf{c}$	$\mathbf{a}' = \frac{1}{2}(-\mathbf{a} + \mathbf{b})$ $\mathbf{b}' = \frac{1}{2}\mathbf{c}$
Origin at $0,0,z$	Origin at $x,0,0$	Origin at $x,x,0$

Maximal non-isomorphic subgroups

I	[2]$I\bar{4}2m$ (121)	(1; 2; 5; 6; 11; 12; 15; 16)+
	[2]$I\bar{4}c2$ (120)	(1; 2; 7; 8; 11; 12; 13; 14)+
	[2]$I4cm$ (108)	(1; 2; 3; 4; 13; 14; 15; 16)+
	[2]$I422$ (97)	(1; 2; 3; 4; 5; 6; 7; 8)+
	[2]$I4/m11$ ($I4/m$, 87)	(1; 2; 3; 4; 9; 10; 11; 12)+
	[2]$I2/m2/c1$ ($Ibam$, 72)	(1; 2; 5; 6; 9; 10; 13; 14)+
	[2]$I2/m12/m$ ($Fmmm$, 69)	(1; 2; 7; 8; 9; 10; 15; 16)+

(Continued on preceding page)

Figure 10.2: Space group I4/mcm, page 2 (reproduced with permission of the IUCR).

the cobalt atoms are occupying one of the 4-fold positions. Since there are 12 cesium atoms, the distribution for cesium has to be 4 and 8. Therefore, one cesium atom is on a 4-fold position, and the other on an 8-fold position. The 20 chlorine atoms can be distributed either on one 4-fold position and one 16-fold position, or one 4-fold position and two 8-fold positions. It follows that the cobalt, cesium and chlorine each occupy one of the four 4-fold positions $4a$, $4b$. $4c$ or $4d$, while eight cesium atoms are located in one of the 8-fold positions $8e$, $8f$, $8g$, or $8h$. For chlorine, either one of the 16-fold positions $16i$, $16j$, $16k$, $16l$ is occupied or two of the 8-fold positions $8e$, $8f$, $8g$ or $8h$. However, the positions $8e$ and $8f$ can be ruled out due to very short distances to any of three 4-fold positions that must be occupied. For the next step, it is necessary to consult the observed intensities (see Table 10.1) and calculate possible distances between atoms to further narrow the possible atomic distributions over the space group positions.

The first intensity to consider is the value of the (400) reflection. Calculating the structure factors for $F(400)$ for the 4-fold positions shows that they are all identical:

$$4a; \ F(400) = 4 \times f \times e^{2\pi m} \quad n = \text{even}$$

and similar for $4b$, $4c$ and $4d$. Therefore, the contributions to the structure factor $F(400)$ of the atoms placed in the 4-fold positions do not depend on the particular 4-fold position they occupy, and it contains the respective sums of the form factors of the atoms in the 4-fold positions. Placing 8 cesium atoms in one of the 8-fold positions, at $8e$ and $8f$, is excluded, leaving only $8g$ and $8h$. Since the (400) reflection is not observed (weak), the cesium atoms have to be in destructive interference, which is only possible if the x-axis coordinate is variable, leaving $8h$ as the only possibility. Now, one of the 16-fold positions is occupied by chlorine. Inspecting the series of (00l) reflections, it is realized that the (00 10), and (00 14) reflections are weak. Since the atoms placed so far are either on fixed positions with $z = 0, 1/4, 1/2, 3/4$, the chlorine atoms in one of the 16-fold positions require a z-coordinate that is different. This excludes $16i$, $16j$ and $16k$, leaving $16l$ for the chlorine position. So far, the distribution of the atoms is therefore the following:

- 8 cesium atoms in $8h$: $x, x + \frac{1}{2}, 0; \dots$
- 16 chlorine atoms in $16l$: $x, x + \frac{1}{2}, z; \dots$
- 4 cesium, 4 cobalt and 4 chlorine atoms arranged in $4a$, $4b$, $4c$ or $4d$

It is further possible to restrict the ranges of the positional parameters of cesium in $8h$ and chlorine in $16l$ by investigating the ($hk0$) and (00l) reflection (Table 10.1), to obtain initial values for x in $8h$, and x, z in $16l$. The (00l) reflections allow an initial guess for the z-parameters in $16l$, $z = 1/12$ or $11/12$. This follows from the fact that the (00 12) reflection is strong, with the adjacent reflections weak. In addition, the cesium and cobalt atoms in the 4-fold positions need to be far from the cesium atoms in the 8-fold position due to local charge neutrality. To arrive at an atom distribution

Table 10.1: Visually estimated intensities for Cs_3CoCl_5: very strong (vs), strong (s), medium-strong (ms), medium (m), weak (w), very weak (vw) and not observed (nil). Data from [19], Table I.

(hkl)	$\sqrt{I}_{calc}$	I_{obs}	(hkl)	$\sqrt{I}_{calc}$	I_{obs}	(hkl)	$\sqrt{I}_{calc}$	I_{obs}	(hkl)	$\sqrt{I}_{calc}$	I_{obs}
200	110	w	206	461	s	554	35	nil	181	81	vw
400	30	nil	208	110	w	556	80	nil	183	201	m
600	580	vs	20 10	210	m	572	64	vw	185	93	w
800	67	nil	20 12	30	nil	352	70	w	581	82	w
10 00	35	nil	20 14	146	m	354	196	m	583	213	m
002	120	w	402	309	ms	356	93	nil	585	79	w
004	870	vs	404	166	m	372	20	nil	347	21	nil
006	350	s-	406	398	s	374	134	w	222	193	m
008	440	vs	408	113	w	376	146	w+	224	627	vs
00 10	90	nil	40 10	152	m-	392	154	m	226	94	vw
00 12	430	s-	40 12	8	nil	394	64	w+	228	283	ms
00 14	10	nil	602	40	nil	396	268	m+	22 10	79	vw
00 16	262	m	604	352	ms	121	300	ms	22 12	210	m-
110	40	nil	006	186	w	123	590	vs	242	138	vw
220	600	vs	608	290	ms	125	227	m+	244	339	ms+
330	390	vs	60 10	30	nil	127	162	w+	246	55	nil
440	370	vs	60 12	371	vs	129	270	ms	248	222	ms
550	29	nil	802	190	w	321	16	nil	442	156	m-
660	370	s	804	55	nil	323	45	vw	444	236	m+
770	38	nil	806	166	w	521	136	w	336	16	nil
310	540	vs	808	52	nil	523	305	ms	448	206	ms
350	230	m	80 10	183	m	525	134	w+	622	238	m
370	245	n	112	184	m-	527	116	w	624	93	w
390	150	w	114	166	w+	721	118	w	626	280	ms-
190	172	w+	116	197	m	723	192	m	628	72	nil
570	14	nil	132	110	w	725	118	w	642	182	m
590	170	w+	134	356	s-	141	200	m	644	96	vw
240	430	vs	136	150	w+	143	366	s	646	288	ms
260	40	nil	138	222	m	145	190	m	648	97	w
280	180	m-	332	270	s	147	148	m	662	34	w
2 10 0	222	m-	334	114	w	149	208	m	564	272	ms
460	0	nil	336	512	vs	541	124	m	666	138	w
480	165	w	338	77	w	543	251	m+	824	183	w+
4 10 9	230	m	33 10	172	w	545	123	w	844	182	m
202	566	s	33 12	187	w	547	127	w	846	70	vw
204	215	m	552	78	vw	653	41	vw			

that is consistent with the observed intensities, 4 cesium atoms are either in $4a$ or $4b$, 4 cobalt atoms in either $4a$ or $4b$ and 4 chlorine atoms in $4c$ or $4d$. The possible distribution of atoms in positions a, b, c and d fall into six groups, considering their effect on the (00l) intensities. However, five of these groups are incompatible with observations for any value of z for the chlorine in 16l. Calculating the intensities for the

($h00$) reflections allows to put further limits on the values for x for cesium in $8h$ and for x for the chlorine atoms in $16l$: For cesium, x or $\frac{1}{2}-x$ is between 0.13–0.19, for chlorine in $16l$, x or $\frac{1}{2}-x$ is between 0.11–0.22, leaving essentially only two combinations. However, one of these combinations results in short cesium to chlorine distances of the order of 2.7 Å, which are clearly too short. The trial solution for the structure therefore uses the values $x = 0.13 - 0.19$ for the cesium atom in $8h$, and $x = 0.11 - 0.22$ with $z = \frac{11}{12}$ for the chlorine atom in $16l$. It then becomes clear that 4 chlorine atoms are in $4c$, 4 cesium atoms in $4a$ and 4 cobalt atoms in $4b$. It is now necessary to consider a larger number of reflections to arrive at better values for the positional parameters for cesium in $8h$ and chlorine in $16l$. For this, the reflections ($h00$), ($hk0$) and ($hh0$) are calculated, since these reflections do not depend on the z-coordinate of chlorine. The positional values can be further constrained to $x = 0.15 - 0.18$ for cesium, and $x = 0.13 - 0.22$ for chlorine, and optimization of the calculated intensities results in $x = 0.167$ for cesium, and $x = 0.155$ for chlorine.

The structure is therefore described by the following parameters given in Table 10.2:

Table 10.2: Structural parameters Cs_3CoCl_5.

Space Group # 140, $I4/mcm$, $a = 9.18$ Å, $c = 14.47$ Å, $Z = 4$

atom	pos	x	y	z
Cs 1	4a	0	0	0
Cs 2	8h	0.167	0.667	$\frac{1}{4}$
Co	4b	0	$\frac{1}{2}$	0
Cl 1	4c	0	0	$\frac{1}{4}$
Cl 2	16l	0.155	0.655	0.917

Crystal structure determinations of Cs_3CoCl_5 in later years confirmed the structure, with only small adjustments to the parameters deduced from the film methods. The results of a recent study, carried out a temperature of 295 K, are given in Table 10.3. Note that these atomic positions include an origin shift of $(\frac{1}{2}, \frac{1}{2}, \frac{1}{4})$ [20].

Table 10.3: Structural data of Cs_3CoCl_5 (origin shift $(\frac{1}{2}, \frac{1}{2}, \frac{1}{4})$).

Space Group # 140, $I4/mcm$, $a = 9.2315(15)$ Å, $c = 14.45535(24)$ Å, $Z = 4$

atom	pos	x	y	z
Cs 1	4a	0	0	$\frac{1}{4}$
Cs 2	8h	0.665755(5)	0.16575	0
Co	4b	0	$\frac{1}{2}$	$\frac{1}{4}$
Cl 1	4c	0	0	0
Cl 2	16l	0.13915(10)	0.63915	0.15760(11)

A representation of the structure in Figure 10.3 shows the tetrahedrally coordinated cobalt ion, with the coordination tetrahedra shown in blue, cesium atom as green spheres and chlorine atoms as yellow spheres. The cesium atoms are 10- and 8-fold coordinated, respectively, since there are two crystallographically distinct cesium atoms. The $CoCl_4$ tetrahedra are well separated in the structure, resulting in small magnetic interactions between the Co^{2+} ions.

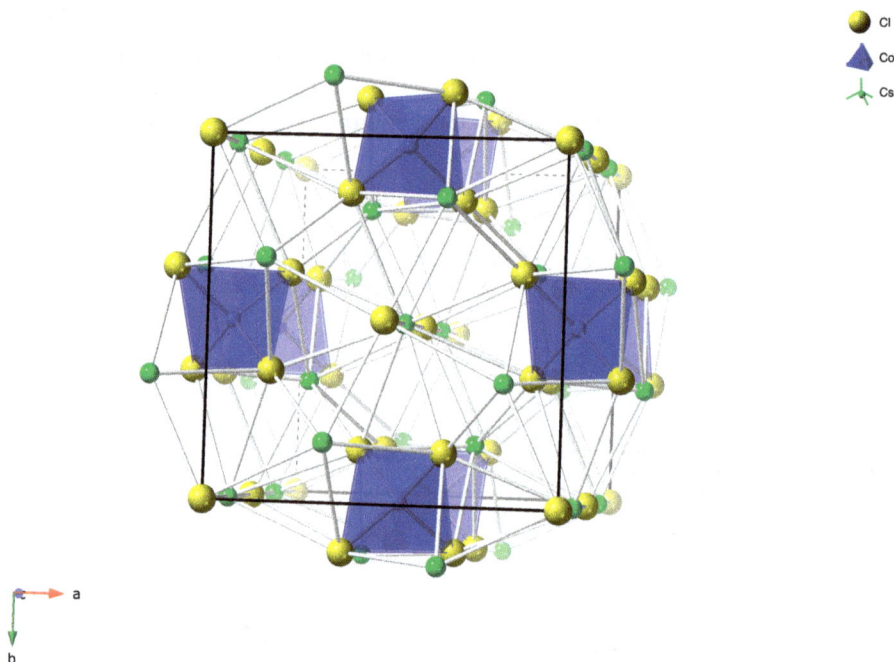

Figure 10.3: Structure of Cs_3CoCl_5: Blue tetrahedra represent the coordination polyhedron of Co, green spheres represent cesium atoms, yellow spheres chlorine atoms. The view is approximately along the c-axis.

This example of a structure solution relied on physical measurements, chemical knowledge, the space group symmetry and intensity data to obtain a trial structure. The calculations were minimal, and after a small number of possible solutions were identified, quantitative agreement was obtained. Even with intensity data obtained by visual estimation from film data, the structure could be determined unambiguously. Subsequent determination of the structure yielded more precise positions and, therefore, more precise bond distances, but did not change the overall structure.

In the following, a number of methods for the structure solution will be discussed. As shown in the example of Cs_3CoCl_5, the goal of a structure solution is the assignment of the correct phase value for each $F(hkl)$, so that a Fourier synthesis can be carried out. However, unless a holographic method is used to record the intensities and respective phases of the $F(hkl)$s, the phase information is not available in an intensity measurement. Therefore, methods have been developed that allow assigning the phase values. In the special case of a centrosymmetric structure with the inversion center at the origin, the phase values can only be 0 or π, and the complex exponential will be +1 or −1. If the number of symmetry independent reflections is of the order of a few hundred reflections N, then there are 2^N possibilities to assign the values +1 and −1. While a brute force approach has become possible due to the increased computing power available, such an approach is usually not taken. The diffracted intensities, however, carry additional information about the structure, and methods exploit this fact to arrive at trial structures for further refinements.

10.2 Patterson method

In a standard structure determination, the atoms can be considered point-like. This means that the autocorrelation function is finite only when the product of $\rho(\mathbf{r}) * \rho(\mathbf{x}-\mathbf{r})$ is nonzero. This is only the case if $\mathbf{x}$ is a vector between two atoms. The autocorrelation function, or Patterson function, is directly obtained from a Fourier synthesis using the observed $|F_{obs}(hkl)|^2$ values [21],

$$P(u, v, w) = \sum_{hkl} |F(hkl)|^2 e^{-2\pi i(hu+kv+lw)}$$

The function $P(u, v, w)$ therefore shows a 3-dimensional map containing all distances between atoms and their directions. The scattering densities are included, which implies that a heavy atom produces higher peaks than a lighter atom. The autocorrelation map for a structure with N atoms contains N^2 peaks, for the possible N^2 vectors between the atoms. Of these, N distances are from each atom to itself, producing a maximal peak at the origin of the map from their overlap. The rest, $N^2 - N$ peaks are distributed throughout the unit cell. The symmetry of the Patterson map is higher or equal to the symmetry of the actual structure. This is because the vectors from one atom to another can point in both directions, producing a center of symmetry. Patterson maps also do not contain translation symmetry from glide planes and screw axes. This loss of information reduces the symmetry of the Patterson maps from the 230 space groups to just 24 space groups, the Patterson groups, comprising the 11 Laue groups. If the space group has been determined (or is at least known to within a few possible groups), the coordinates of the vectors can be calculated. For these vectors between symmetry equivalent positions are computed. As an example, the space group $P2_1/c$ is considered, with four general symmetry equivalent positions (see Table 10.4).

The following 16 difference vectors are obtained by subtracting each position in Table 10.4 from each other (see Table 10.5).

A distinct heavy atom in a general 4-fold position can be located from the peaks that are found in the Patterson map, since the peak heights in the map scale with

Table 10.4: General symmetry elements of $P2_1/c$.

(1)	x, y, z
(2)	$\bar{x}, \bar{y}, \bar{z}$
(3)	$\bar{x}, \frac{1}{2} + y, \frac{1}{2} - z$
(4)	$x, \frac{1}{2} - y, \frac{1}{2} + z$

Table 10.5: Difference vectors for a general position in $P2_1/c$.

(i)	$4 \times (0, 0, 0)$
(ii)	$(2x, 2y, 2z), \ (-2x, -2y, -2z), \ (-2x, 2y, -2z), \ (2x, -2y, 2z)$
(iii)	$2 \times (2x, 1/2, 1/2 + 2z), \ 2 \times (-2x, 1/2, 1/2 - 2z)$
(iv)	$2 \times (0, 1/2 + 2y, 1/2), \ 2 \times (0, 1/2 - 2y, 1/2)$

the electron density. The difference vectors (i) are trivial and do not give any further information. The x and z coordinate can be deduced from peaks in the plane $y = \frac{1}{2}$ (difference vectors (iii)). Similarly, the y coordinate is found from peaks along the line given by the two planes $x = 0$ and $z = \frac{1}{2}$ (difference vectors (iv)). The peaks at locations given by the difference vectors (ii) then serve as cross check for the coordinates (xyz) of the atom. It can also be seen that even though the position is general, some of the vectors that are produced are located in special positions, on lines or planes. These special lines and planes are called *Harker lines* or *Harker planes*.

The analysis of Patterson maps may become rather tedious if there is a large number of heavy atoms in the unit cell, giving rise to strong peaks that are densely packed. The Patterson method works best for inorganic or metal-organic systems, where one atom is a distinctly stronger scatterer than the others, for instance in a metal oxide. The resolution of the Patterson function can be enhanced through modifications. For instance, the peak at the origin is not contributing to the overall information content (it describes distances of each atom to itself), and removing it removes a "bias" to the Patterson function. This requires that the intensities are properly scaled and the content of unit cell is known [22]. Further information can be found in the literature.

10.3 Direct methods

A majority of structures today are solved using direct methods [23, 24]. This is a statistical method that predicts phases with a certain probability. The development of direct methods has revolutionized the solution of crystal structures and was honored by the award of the 1985 Nobel Prize in Chemistry. The method is based on the fact that the scattering density is positive everywhere in direct space, since there cannot be a negative electron density. This condition is necessary, but not sufficient to determine the phases for the observed structure factors. However, the condition is strong enough to limit the choices for a phase of some of the structure factors. It is possible to build relations between structure factors, and to assigning more phases based in these relations. With a sufficient number of phases assigned, a trial structure can be obtained by a Fourier map calculation.

What are the known phases? First, there is the trivial result that the phase of $F(0,0,0)$ is always 0, meaning the forward scattered wave is always in phase with the incident wave. Then the phases for any pair of reflections (hkl) and $(\bar{h}\bar{k}\bar{l})$ have to have the opposite phase, so that the following relationship for the phase $\phi(hkl)$ holds:

$$\phi(hkl) + \phi(\bar{h}\bar{k}\bar{l}) = 0$$

Another trivial results is the following:

$$|F(hkl)|^2 \leq |F(000)|^2$$

This results is due to the Cauchy inequality, which states

$$\left|\sum a_j b_j\right|^2 \leq \left(\sum |a_i|^2\right)\left(\sum |b_j|^2\right)$$

This inequality also holds for functions a and b. Using this inequality in the expression for the structure factor and setting $a = \rho(\mathbf{r})^{1/2}$ and $b = \rho(\mathbf{r})^{1/2}e^{i\mathbf{qr}}$, one obtains

$$|F(hkl)|^2 \leq \int \rho(\mathbf{r})\,d\mathbf{r} \int \rho(\mathbf{r})|e^{i\mathbf{qr}}|^2\,d\mathbf{r}$$

For a positive density distribution, $|e^{i\mathbf{qr}}|^2 = 1$ and the right-hand side becomes $|F(000)|^2$. If the structure is centrosymmetric, then a different result is obtained due to the fact that $\rho(\mathbf{r}) = \rho(-\mathbf{r})$:

$$F(hkl) = \int \rho(\mathbf{r})\cos^2(\mathbf{qr})\,d\mathbf{r}$$

Using this expression for the centrosymmetric case, one obtains

$$|F(hkl)|^2 \leq \int \rho(\mathbf{r})d\mathbf{r} \int \rho(\mathbf{r})\cos^2(\mathbf{qr})\,d\mathbf{r}$$

This expressions reduces to

$$|F(hkl)|^2 \leq F(000)\left[\frac{1}{2}\int \rho(\mathbf{r})(1+\cos(2\mathbf{qr}))\,d\mathbf{r}\right]$$

and the result can be rewritten, using the expressions for the structure factors given earlier

$$F^2(hkl) \leq \frac{1}{2}F^2(000) + \frac{1}{2}F(000)F(2h,2k,2l)$$

Substituting by $u(hkl) = F(hkl)/F(000)$ gives

$$u^2(hkl) \leq \frac{1}{2} + \frac{1}{2}u(2h, 2k, 2l)$$

Since the magnitude and the phase for $u^2(hkl)$ is known (+), the only unknown is the phase of $u(2h, 2k, 2l)$, which can be (+) or (−). The relationship can now be written as

$$u^2(hkl) \leq \frac{1}{2}\left(\pm \frac{1}{2}|u(2h, 2k, 2l)|\right)$$

If the magnitudes of the u's are in the right range, this relation can force the unambiguous assignment of the phase for $u(2h, 2k, 2l)$. The usefulness of the above relation, unfortunately, is not very general. This is due to the fact that it requires the scattering amplitudes to be large, which means that most of the scattering has to occur in phase, which for complex structures with many atoms, is usually not the case. Also, the overall scattered intensity is exponentially dampened for increasing $\mathbf{q}$, which is due to the Debye–Waller factor. Additionally, the atomic form factor also produces a drop off of the scattering amplitude for higher $\mathbf{q}$. It would be convenient to work with ideal, point-like atoms, that are fixed in their positions. This will allow a uniform treatment of the structure factors, that are normalized, and consequently, phase relationships between different structure factors $F(hkl)$ can be better developed.

10.3.1 Unitary structure factor

The unitary structure factor is a normalized structure factor for a point-like atom, without any thermal motion. It therefore represents an object with the same number of electrons, multiplied by a δ-function. It is defined by

$$U(hkl) = \frac{F(hkl)_{point}}{F(000)}$$

with $F(hkl)_{point}$ the structure factor corrected for spatial extent of the atom and its atomic displacement. Since a point-like atom can be represented as δ function, the corresponding atomic form factor (Fourier transform) is a constant function of $\mathbf{q}$, and equal to the number of electrons of the atom, therefore, $f(\mathbf{q}) = Z$. $F(000)$ is the total number of electrons in the unit cell. The ratio of the idealized structure factor to the observed structure factor is then given as

$$\frac{F(hkl)_{point}}{F(000)} = \frac{Z}{f_0 e^{-(1/2)q^2 u^2}}$$

If there are different types of atoms in a structure, then it is not straightforward to obtain $F(hkl)_{point}$, since each atom has its own atomic displacement parameter, and usually a suitable average is used. The unitary structure factor $U(hkl)$ then has the same phase as its corresponding $F(hkl)$, but an amplitude between 0 and 1. The contribution of the ith atom in the structure to $U(hkl)$ is given as its fraction of the total electron count multiplied by its phase factor:

$$\frac{f_i}{F(000)} = n_i = \frac{f_i}{\sum f_i}$$

Since the atomic form factor depends on $\mathbf{q}$, this is somewhat tricky to evaluate. However, if all N atoms in the structure are the same, then the above expression is very simple:

$$n_i = 1/N$$

In the case of similar atoms, the expression is approximate. This is for instance the case for organic molecules that incorporate carbon, nitrogen, oxygen and hydrogen. Contributions to U by hydrogen can be neglected, and the contributions from carbon, nitrogen and oxygen are approximately equal. It is now possible to define an average value $U(hkl)$ for random phases, which are given by the geometric mean

$$\langle U^2(hkl) \rangle = \sum_i n_i^2$$

or for equal atoms

$$\langle U(hkl) \rangle_{rms} = \left(\sum_i n_i^2 \right)^{1/2} \approx \frac{1}{\sqrt{N}}$$

Using these unitary structure factors in inequality relations is, unfortunately, not very practical. The number of large $U(hkl)$'s drops quite fast for increasing number of atoms in a unit cell. Also the scaling is not quite uniform, and is only correct for a general reflection. Special reflections have to be normalized differently to assure a uniform scaling. For this purpose, a normalized structure factor $E(hkl)$ is introduced, defined by

$$E^2(hkl) = \frac{U^2(hkl)}{\langle U^2 \rangle} \tag{10.1}$$

The $E(hkl)$'s, the normalized structure factors (equation (10.1)) allow a statistical treatment, since their distribution in $\mathbf{q}$ is independent of the size and content of the unit cell. Only the presence or absence of a center of symmetry affects the distribution, giving a statistical test for a centrosymmetric or non-centrosymmetric atom distribution that can be used to distinguish between different space groups. The theoretical values for $E(hkl)$'s are given in Table 10.6.

Table 10.6: Distribution of E's.

	Centric	Noncentric
average $|E|^2$	1.000	1.000
average $|E^2 - 1|$	0.968	0.736
average $|E|$	0.798	0.886
$|E| > 1$	32.0 %	36.8 %
$|E| > 2$	5.0 %	1.8 %
$|E| > 3$	0.3 %	0.01 %

10.3.2 Structure invariants and semiinvariants

In the derivation of the expression of scattered intensity as the absolute square of the Fourier transform of the scattering density, the choice of the origin is not crucial, since the intensity cannot depend on the arbitrary choice of the origin of the unit cell. The phases, in contrast, do depend on the origin choice, and the assignment of phases therefore defines the origin of a structure. For a centrosymmetric structure, the choice of the origin is preferably one of the centers of inversion. For noncentrosymmetric structures, the choice is sometimes arbitrary in one or more directions. What is needed are therefore phase relationships that do not depend on the location of the origin. Such combinations of phases are called *structure invariants*. In addition, there are sets of phases that are invariant for shifts between equivalent origins (for instance, shifts between centers of inversion). These are called structure *semiinvariants*. The most important structure invariants have the form

$$\phi_n = \sum_i^n \phi(hkl)_i$$

with $(hkl)_1 + (hkl)_2 + \cdots + (hkl)_n = (0,0,0)$. This relationship does not depend on the origin choice as long as the sum of the phases of n reflections adds up to a multiple of 2π. Examples are the Friedel pairs (hkl) and $(\bar{h}\bar{k}\bar{l})$, where the sum of the indexes is $(0,0,0)$. Relationships involving three phases are no longer trivial, and can be set up in the following way:

$$\phi_3 = \phi(hkl)_1 + \phi(hkl)_2 + \phi((\bar{h}\bar{k}\bar{l})_1 + (\bar{h}\bar{k}\bar{l})_2)$$

Usually, one cannot specify the value of ϕ_3 a priori, but in special cases, the value is zero. In general, this relation defines reflections with constant relative phases, independent of the origin. For the centrosymmetric case, the relation simplifies, using the cosine function to give the sign (S) of the phase:

$$S_3 = S(hkl)_1 \cdot S(hkl)_2 \cdot S((\bar{h}\bar{k}\bar{l})_1 + (\bar{h}\bar{k}\bar{l})_2) = \pm 1$$

Unfortunately, there are usually not enough strong reflections to allow the assignment of phases based on inequalities alone. There are, however, a large set of intensities that are too small for inequalities, but still reasonable large, for which a set of equations can be set up that are probably true. As has been shown by D. Sayre [25] under a certain set of conditions, any structure factor $F(hkl)$ is determined by the products of all the pairs of structure factors whose indexes (hkl) add up to the (hkl) triplet. This means that

$$F(hkl) \propto \sum_{h'} F(hkl)F(h - h', k - k', l - l') \qquad (10.2)$$

For instance, $F(341)$ depends on the products of $F(420)$ and $F(\bar{1}11)$, $F(220)$ and $F(121)$, $F(252)$ and $F(1\bar{1}\bar{1})$, etc. Since the sum runs over all possible pairs, a large value of $F(hkl)$ can only be produced if other large $F(hkl)$'s add up in one direction. The main weight therefore stems from the pairs where both F's are large, and will trend toward +1 or −1. If the signs are considered, then the equation can be written in the centrosymmetric case for 3 strong reflections as

$$S(F(hkl)) \sim S(F(h'k'l')) \cdot S(F(h - h', k - k', l - l')) \qquad (10.3)$$

or

$$S(F(hkl)) \cdot S(F(h'k'l')) \cdot S(F(h - h', k - k', l - l')) \sim +1$$

Equation (10.3) gives the probabilities derived from equation (10.2) and is the basis for direct methods. To illustrate, the following relationship is found:

$$S(2h, 2k, 2l) \sim S(hkl) \cdot S(hkl)$$

which indicates that irrespective of the sign of $F(hkl)$, $F(2h, 2k, 2l)$ is likely positive if the reflections are strong. The probability P that this equation holds depends on the $F(hkl)$'s, or better, on the normalized structure factor $E(hkl)$, and is given by W. Cochran and M. M. Woolfson [26] as

$$P = \frac{1}{2} + \frac{1}{2} \tanh\left[\frac{1}{\sqrt{N}} |E(hkl)E(h'k'l')E(h - h', k - k', l - l')| \right]$$

with N being again the number of atoms (exact only for equal atoms). This probability function tends to be 1.0 for strong reflections, even for a large number of (equal) atoms in a unit cell. For medium to small $E(hkl)$'s, the probability that the relationship between the phases is correct, drops off. In a noncentrosymmetric structure, the above given relationships are no longer exact, since the phase is not just restricted to two values only, but may have any value. Fortunately, a strong reflection can only be produced if most of the scattering is in phase.

10.3.3 Phase assignment

The relationships derived now need to be applied to a real problem. Usually, one tries to build a tree, starting from the root, with a set of a few reflections with their phases known. Using the relationships outlined above, it is now possible to branch out and find phases of other reflections. A few different phase assignments may then be produced, and further tested. Obviously, the "tree building" is unstable, if one of the "root" reflections has an incorrect phase. One of the more common solutions due to an incorrect phase produces consistent phases for every reflection (0 or π) but a Fourier synthesis of such a solution will only show a single peak at the origin, and is therefore often called the "uranium solution." Since "tree building" has shown to be tedious, and ever so often is producing inconsistent phase assignments, other schemes have been devised to overcome those problems. The introduction of random phases and their subsequent refinement with the tangent formula has been incorporated into various programs, either as an option or as the default. It is therefore possible to produce a large number of trees based on different "root" phases and assess the resulting structures.

10.3.4 E-maps

The phases that were found using the direct methods are based on the $E(hkl)$'s, and not on the structure factor $F(hkl)$'s. Using a Fourier synthesis to obtain a starting model may not give a satisfactory answer, since the large $E(hkl)$'s tend to have large indexes and are therefore at medium to high angles. Their corresponding $F(hkl)$'s are often small, and in a Fourier synthesis, their contribution may be obscured by a few large terms. For this reason, an E-map is calculated by a Fourier transform. The $E(hkl)$'s, correspond to motionless point-like "atoms" and, therefore, an E-map will show very sharp peaks that can be associated with atoms in the structure. From these, it is possible to select a starting model for a subsequent structure refinement.

10.4 Charge flipping

A recent addition to ab-initio phase determination is the *charge flipping* algorithm, which is a remarkably simple way of exploring the possible phases of the structure factors. The algorithm is based on phase retrieval algorithms in optics, based on switching between direct and reciprocal space, and application of constraints in both spaces [27, 28], and has been developed for crystallography [29].

Constraints applied in direct space are not simultaneously present in reciprocal space, and similar, constraints in reciprocal space are not simultaneously present in direct space. To include constraints in both direct and reciprocal space, switching between direct and Fourier space repeatedly is required, making use a the *Fast Fourier*

Transformation (FFT). Furthermore, the object in real space is defined by a compara-tively small number of strong reflections in reciprocal space, and in direct space, the locations of high electron density are surrounded by large volumes where the den-sity is small or zero. The statistics of the $E(hkl)$'s reflects this, where the number of $|E| > 2$ is 5 % for the centrosymmetric case, and 1.8 % for the non-centrosymmetric case. Since the charge density in a material $\rho(\mathbf{r}) \geq 0$ everywhere, all structure factor terms (observed and unobserved) are necessary to achieve this in a Fourier transfor-mation. However, this is not possible, and small negative charge density regions are expected for data with finite resolution. It is this fact that the algorithm uses to explore the possible phases for each $|F_{obs}|$. With high resolution data $F_{obs}(hkl)$, the algorithm will attempt to find the plateaus where $|\rho(\mathbf{r})| \approx 0$.

10.4.1 The charge flipping algorithm

The structure factor $F(hkl) = |F_{obs}(hkl)|e^{i\phi(hkl)}$, with $F_{obs}(hkl)$ the observed structure factors and $\phi(hkl)$ the associated phases, which have to conform to Friedel's law $\phi(hkl) = -\phi(\bar{h}\bar{k}\bar{l})$. Symmetry is reduced to space group $P1$ so that all observed data is used in an unbiased way. The linear dimension of a voxel in the direct space is determined by the resolution, and is half the resolution value to ensure sufficient sampling of the charge density. For instance, if the maximum resolution is 0.8 Å, then the sampling interval is half of this value, 0.4 Å.

The charge flipping algorithm executes the following steps:

0. In the initial step, a random set of phases $\phi(hkl)$ are selected. Unobserved struc-ture factors are set to zero, as well as $F(000)$. All structure factors outside the reso-lution limit are also set to zero. A Fourier transform is executed to obtain an initial electron density $\rho(\mathbf{r})$.

1. The calculated charge density $\rho(\mathbf{r})$ is evaluated, and all density below a threshold δ is flipped, producing a modified charge density $g(\mathbf{r})$. A large negative charge density may also be flipped to ensure positive charge density.

2. The modified charge density is Fourier transformed using FFT to arrive at new structure factors $G(hkl) = |G(hkl)|e^{\phi(hkl)}$ within and outside the resolution sphere.

3. The phases $\phi(hkl)$ obtained in the previous step are now used to build new struc-ture factors $F(hkl) = |F_{obs}(hkl)|e^{i\phi(hkl)}$. Furthermore, $F(000) = G(000)$ is used, and all $F(hkl)$ beyond the resolution limit are set to zero.

4. The new structure factors $F(hkl) = |F_{obs}(hkl)|e^{i\phi(hkl)}$ are Fourier transformed to give a new charge density $\rho(\mathbf{r})$. The next iteration will then start at step 1.

While the algorithm is simple, it is not deterministic. A solution may be found or not be found within a sufficient number of cycles. On the other hand, running the algo-rithm for a second time may produce a different path due to the random assignment of phases in the initialization step, and a consistent phase assignment may result. What remains to be determined is a good value for the threshold δ. It was suggested that

$\delta = k\sigma$ with k a number of the order of $1.0 \ldots 1.2$ and

$$\sigma = (\langle \rho^2 \rangle - \langle \rho \rangle^2)^{\frac{1}{2}}$$

The algorithm steps are depicted in Figure 10.4. The efficient fast Fourier transform is applied repeatedly to switch between direct and reciprocal space to explore the space of possible phases that represent a structure.

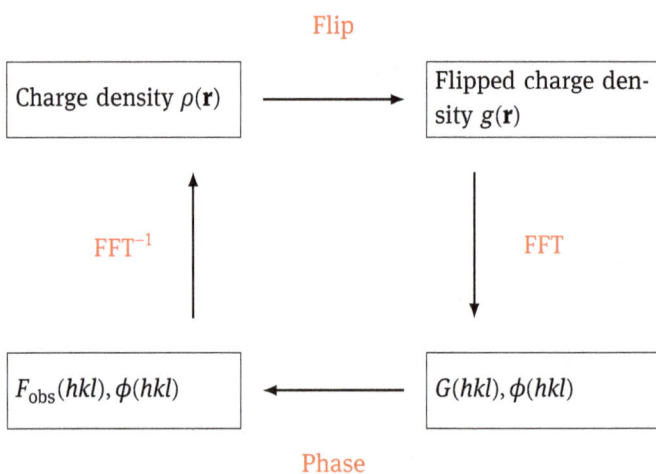

Flip

| Charge density $\rho(\mathbf{r})$ | $\longrightarrow$ | Flipped charge density $g(\mathbf{r})$ |

FFT^{-1} FFT

| $F_{obs}(hkl), \phi(hkl)$ | $\longleftarrow$ | $G(hkl), \phi(hkl)$ |

Phase

Figure 10.4: Charge flipping algorithm steps.

The exploration of the possible phases that define the regions with zero charge density proceeds in a chaotic way. Each charge flip step introduces perturbations that move the model along a path in the multidimensional space of possible phases. If a solution is found, these perturbations are not large enough to generate a new different charge density. Therefore, a stable charge density is a possible solution.

The charge flipping algorithm is remarkably simple and easy to implement. The calculation of the electron density on a grid is independent of atom types, chemical composition or even the total number of electrons in a unit cell. Symmetry information is not used, with the space group $P1$ allowing a general approach. This lets the charge density evolve without constraints until a stable solution is found. This solution is then analyzed, associating the charge density peaks with atoms and determination of the symmetry based on atomic positions. Therefore, charge flipping is truly *ab-initio*. This makes the method well suited for problems with pseudo-symmetry, disorder and unknown chemical composition.

A number of implementations of the charge flipping algorithm for crystallographic applications have been developed. These programs, however, are not yet as mature as direct method program suites. Their use is clearly growing, and the number of structures that are solved using the charge flipping algorithm is growing at a fast pace. Refinements of the charge flipping method are ongoing, and aspects of charge flipping have been incorporated into structure solution and refinement program suites, in parallel to direct methods.

10.5 Structure completion

Using one of the methods described in the previous sections, a set of self-consistent phases $\phi(hkl)$ are successfully assigned to the measured $|F_{obs}(hkl)|$. A Fourier transform using these phases will yield an electron density $\rho(\mathbf{r})$. This represents a starting model of the structure that may not yet be complete. For instance, this can be due to large differences in the form factor of the atoms in the structure, or the absence of weakly scattering atoms in the trial solution, placement of atoms in approximate positions, etc. To complete the structure, finding all atoms in the unit cell, Fourier maps and difference Fourier maps are calculated. Since an atom in the correct location allows an assignment of phases to the observed structure factors, the corresponding Fourier map should therefore reproduce this particular atom. However, with an incomplete starting model, not all of the calculated phases are correct. If the number of incorrect phases is too large, then the random phase noise is likely to obscure any information about additional atoms. To get information on the location of other atoms, a rejection criterion must be used. Usually, a reflection gets rejected, if the ratio between the calculated and observed F's is smaller than 0.1 to 0.3. The resulting Fourier map will show more peaks, of which a few will correspond to atoms in the structure. To find the correct ones, one may want to calculate distances between the new peaks and the ones from the starting model. It is then very often possible to choose a number of new correct locations and associate them with atoms. The new structure model is then used to calculate a better Fourier map. Within a few iterations, a starting model may be rejected, or it may have converged to a solution. However, in the case where a structure contains heavy as well as light atoms, it may be quite difficult to find the light atoms. In such a case, a *difference Fourier map* has a better chance of revealing the light atoms. In a difference Fourier map, the starting model is suppressed, allowing light atoms to show up. Such a map also shows all the errors of the model, since only the large deviations of the calculated $F_{calc}(hkl)$ from the observed $F_{obs}(hkl)$ are contributing. The structure factors that are not well explained by the model therefore contribute the most to the difference Fourier summation. In contrast to the Fourier map, which should have positive peaks only (corresponding to positive scattering density), a difference Fourier map also contains negative peaks, indicating regions with too much scattering density. Ideally, for a correct model, the difference Fourier map will show zero density. The difference Fourier map is therefore a versatile tool, not only to refine a structure, but also to assess the quality of the solution. The contours give information about the atomic displacement parameters, as well as the exact location of atoms. It may even be possible to locate hydrogen atoms in molecular compounds, which may be located using ΔF-maps.

To complete a structure starting from a model obtained in ways described previously, an iterative process is applied that includes

(i)   calculation of $\rho(\mathbf{r})$ from $F_{obs}(hkl)$ and the phases $\phi(hkl)$

(ii)   evaluation of the resulting Fourier map and searching for areas of high electron density not explained by the existing model

(iii)   calculation of distances from atoms to the new Fourier peaks

(iv)   inclusion of peaks and assignment of atom types consistent with chemistry, resulting in a new $\rho(\mathbf{r})$

(v)   calculation of new structure factors $F_{calc}(hkl)$ and phases $\phi(hkl)$

(vi)   repeat from step (i)

In this way, the structural model is completed and the differences between observed $F_{obs}(hkl)$ and calculated $F_{calc}(hkl)$ are diminshed.

10.6 Structure refinement

After a structure is completed, the next step minimizes the differences between the $F_{calc}(hkl)$s and the $F_{obs}(hkl)$s so that the structure model accurately describes the observed scattering factor amplitudes. The minimization uses a *least squares refinement*.

10.6.1 Linear least squares

An abbreviated introduction to the least squares refinement method is provided. Given is a linear scalar function f with n variables $\mathbf{x} = x_1, x_2, \ldots x_n$ and independent parameters $\mathbf{p} = p_1, p_2, \ldots p_n$, so that

$$f = \mathbf{p} \cdot \mathbf{x}$$

The function is evaluated by m measurements with $m > n$. The measurements are not exact, but contain errors that are assumed to follow a Gaussian distribution. The best set of parameters p_i that minimizes the deviations between the function and the measurements is now sought. This is achieved by minimizing the sums of the weighted squared differences between the calculated and observed function values for the m observations. In addition, proper weights for each of the m measurements, w_j, are needed. The quantity to be minimized is therefore

$$\Delta = \sum_{j=1}^{m} w_j (f_{obs,j} - f_{calc,j})^2$$

where $f_{obs,j}$ is one of the observed values of the function, and $f_{calc,j}$ the value calculated using the function. For an optimal fit, the parameters p_n are adjusted to minimize Δ.

Therefore, differentiating $\Delta(\mathbf{p})$ with respect to each of the n parameters p_n gives

$$\sum_{j=1}^{n} w_j(f_{obs,j} - f_{calc,j})\frac{\partial f_{calc,j}}{\partial p_i} = 0 \quad (i = 1, 2, 3, \ldots, n) \tag{10.4}$$

defining a set of n equations. Since the values of $\mathbf{x}$ are held fixed and the parameters $\mathbf{p}$ are adjusted, the order is reversed. The partial derivatives $\partial f_{calc,j}/\partial p_i$ for each of the m equations give a new set of equations

$$\sum_{j=1}^{m} w_j(f_{obs,j} - x_{l,1}p_1 - x_{j,2}p_2 - \cdots - x_{j,n}p_n)x_{j,1} = 0$$

$$\sum_{j=1}^{m} w_j(f_{obs,j} - x_{l,1}p_1 - x_{j,2}p_2 - \cdots - x_{j,n}p_n)x_{j,2} = 0$$

$$\vdots$$

$$\sum_{j=1}^{m} w_j(f_{obs,j} - x_{l,1}p_1 - x_{j,2}p_2 - \cdots - x_{j,n}p_n)x_{j,n} = 0$$

Rearranging these equations give

$$\sum_{j=1}^{m} w_j x_{j,1} x_{j,1} p_1 + \sum_{j=1}^{m} w_j x_{j,1} x_{j,2} p_2 \quad + \cdots + \sum_{j=1}^{m} w_j x_{j,1} x_{j,n} p_n \quad = \sum_{j=1}^{m} w_j f_{obs,j} x_{j,1}$$

$$\sum_{j=1}^{m} w_j x_{j,2} x_{j,1} p_1 + \sum_{j=1}^{m} w_j x_{j,2} x_{j,2} p_2 \quad + \cdots + \sum_{j=1}^{m} w_j x_{j,2} x_{j,n} p_n \quad = \sum_{j=1}^{m} w_j f_{obs,j} x_{j,2}$$

$$\sum_{j=1}^{m} w_j x_{j,3} x_{j,1} p_1 + \sum_{j=1}^{m} w_j x_{3,2} x_{j,2} p_2 \quad + \cdots + \sum_{j=1}^{m} w_j x_{j,3} x_{j,n} p_n \quad = \sum_{j=1}^{m} w_j f_{obs,j} x_{j,3}$$

$$\vdots$$

$$\sum_{j=1}^{m} w_j x_{j,n} x_{j,1} p_1 + \sum_{j=1}^{m} w_j x_{j,n} x_{j,2} p_2 \quad + \cdots + \sum_{j=1}^{m} w_j x_{j,n} x_{j,n} p_n \quad = \sum_{j=1}^{m} w_j f_{obs,j} x_{j,n} \tag{10.5}$$

Solving this set of equations (10.5) will then give the optimal values of the parameters p_i.

The structural model that was constructed needs to be refined so that the Fourier transform of the charge density $\rho(\mathbf{r})$ describes the observed structure factor amplitudes $|F_{obs}(hkl)|$. The least squares method is now applied to find optimal values for the structural parameters, such as the scale k and other overall parameters, as well as atom positions $\mathbf{x}, \mathbf{y}, \mathbf{z}$ and displacement parameters U_{ij} for each atom in the unit cell

$$\Delta = \sum_{hkl} w(hkl)(|F_{obs}(hkl)| - |kF_{calc}(hkl, \mathbf{p})|)^2 \tag{10.6}$$

with $w(hkl)$ the weights, k the scale factor, $\mathbf{p}$ a vector containing all parameters to be optimized, and sums over all n reflections (hkl) that were measured. It is again assumed that the errors of the measured structure factors $F_{obs}(hkl)$ follow a Gaussian distribution. The minimum is obtained by differentiating the above function, giving

$$\frac{\partial \Delta}{\partial p_l} = 0 = \sum_{hkl} w(hkl)(F_{obs}(hkl) - kF_{calc}(hkl, \mathbf{p}))\frac{\partial F_{calc}(hkl, \mathbf{p})}{\partial p_i} \quad (i = 1 \ldots n) \qquad (10.7)$$

Unfortunately, the structure factor does not depend linearly on the variables/parameters p_l. A linear approximation, a Taylor expansion terminating after the linear term, is therefore used:

$$F_{calc}(hkl, p_l) = F_{calc}(hkl) + \sum_n \frac{\partial F_{calc}(hkl)}{\partial p_i} dp_i$$

Using this expression in the minimization of equation (10.6) and neglecting the higher orders gives

$$\sum_{hkl} w(hkl)\left(|F_{obs}(hkl)| - |kF_{calc}(hkl, \mathbf{p}) - \frac{\partial |F_{calc}(hkl, \mathbf{p})|}{\partial p_1}dp_1 - \cdots - \frac{\partial |F_{calc}(hkl, \mathbf{p})|}{\partial p_n}dp_n\right)$$
$$\times \frac{\partial |kF_{calc}(hkl)|}{\partial p_i} = 0$$

with $(i = 1 \ldots n)$ for each refinable variable p_i. Since $|F_{obs}(hkl)| - |F_{calc}(hkl)| = \Delta F(hkl)$, it is substituted into the equation to give

$$\sum_{hkl} w(hkl)\left(\Delta F(hkl) - \frac{\partial |F_{calc}(hkl, \mathbf{p})|}{\partial p_1}dp_1 - \cdots - \frac{\partial |F_{calc}(hkl, \mathbf{p})|}{\partial p_n}dp_n\right)\frac{\partial |kF_{calc}(hkl)|}{\partial p_i}$$
$$= 0 \quad (i = 1 \ldots n)$$

This sum over all reflections (hkl) that includes the partial derivatives $\frac{\partial F_{calc}(hkl,\mathbf{p})}{\partial p_i}$ twice can be rearranged into a double summation giving a set of equations

$$\sum_{hkl} \sum_i w(hkl)\frac{\partial |F_{calc}(hkl)|}{\partial p_1}\frac{\partial |F_{calc}(hkl)|}{\partial p_i}dp_i = \sum_{hkl} w(hkl)\Delta F\frac{\partial |kF_{calc}(hkl)|}{\partial p_1}$$

$$\vdots$$

$$\sum_{hkl} \sum_i w(hkl)\frac{\partial |F_{calc}(hkl)|}{\partial p_n}\frac{\partial |F_{calc}(hkl)|}{\partial p_i}dp_n = \sum_{hkl} w(hkl)\Delta F\frac{\partial |kF_{calc}(hkl)|}{\partial p_n}$$

defining n equations for the changes dp_i.

These equations can be written out in matrix notation, with the matrix $\mathbf{A}$ given by

$$\mathbf{A} = A_{ij} = \sum_{hkl} w(hkl)\frac{\partial |F_{calc}(hkl)|}{\partial p_i}\frac{\partial |F_{calc}(hkl)|}{\partial p_j}$$

the vector $\mathbf{x} = dp_i$, and Δ' given as

$$\Delta' = \sum_{hkl} w(hkl)(|F_{obs}(hkl)| - |F_{calc}(hkl)|)\frac{\partial |F_{calc}(hkl)|}{\partial p_i}$$

resulting in

$$\mathbf{Ax} = \Delta'$$

Since the shifts to be applied to the parameters p_i are in the vector $\mathbf{x}$, the equation needs to be solved for $\mathbf{x}$ by

$$\mathbf{A}^{-1}\mathbf{Ax} = \mathbf{A}^{-1}\Delta'$$

and, therefore,

$$\mathbf{x} = \mathbf{A}^{-1}\Delta' \tag{10.8}$$

where the vector $\mathbf{x}$ in equation (10.8) represents the shifts to be applied to the original parameters p_i. The inversion of the matrix $\mathbf{A}$ is thus the basic operation in the refinement. Since this approach is based on the linearization of the complex equation for the structure factor, the shifts are approximate only, and the procedure has to be repeated until the calculated shifts approach zero. The refinement is considered converged once the calculated shifts are a small fraction of the estimated standard deviation. Numerical methods in dealing with these sometimes quite large matrices have been developed so that fast least squares refinements are achieved in a short time, even for systems with a large number of adjustable parameters.

The quantity Δ' defines a function in an n-dimensional space, and it will likely have local minima and maxima. It is quite possible that a least squares refinement will get "stuck" in a local minimum. However, such a refinement has not converged to a global minimum. In such a case, the starting model has to be modified to reset the refinement, since it can refine to a reasonable structure that may not be entirely correct. Methods to test the stability of a refinement are usually incorporated in the program suites available, such as the introduction of random shifts, followed by a number of refinement iterations.

If the parameters to be optimized are all uncorrelated, then the matrix $\mathbf{A}$ would be a diagonal matrix. This is generally not the case, and correlations between parameters do exist. For example, the scale and the displacement parameters are correlated. Also, if pseudo-symmetry is present, then larger correlations between atoms related by the pseudo-symmetry may affect the refinement. This can also happen, if a symmetry element has been missed, and atoms that are related by this symmetry element are treated as independent. In such a case, the lower symmetry space group does not give the proper constraints for the refinement.

The least squares method as described optimizes the structural parameters in respect to the measured structure factors $F_{obs}(hkl)$. These are obtained from the measured intensities, $I_{obs}(hkl)$, and with geometrical factors included, $F_{obs}(hkl) \propto \sqrt{I_{obs}(hkl)}$. The least squares refinement can also be applied to $F_{obs}^2(hkl) \propto I_{obs}(hkl)$, and a similar mathematical method is derived. Either refinement can be used, with the refinement on $F_{obs}^2(hkl)$ being preferred.

10.6.2 Residuals

The tests a refined model has to pass are quite stringent. First of all, the model should be refined to reasonably low residual value. The residual is defined as

$$R1 = \frac{\sum |F_{obs}(hkl) - F_{calc}(hkl)|}{\sum |F_{obs}(hkl)|} \tag{10.9}$$

In the case that the model describes the observations well, the numerator $\sum |F_{obs}(hkl) - F_{calc}(hkl)|$ becomes small, and the residual $R1$ becomes small. A similar expression with applied weights $w(hkl)$ is given by

$$wR2 = \left[\frac{\sum w(hkl)(F_{obs}^2(hkl) - F_{calc}^2(hkl))^2}{\sum (F_{obs}^2(hkl))^2} \right]^{\frac{1}{2}} \tag{10.10}$$

The residual in itself, however, is not the only quantity that has to be considered. The *goodness-of-fit*, which is defined as

$$\chi = GOF = \left[\frac{\sum w(hkl)(|F_{obs}(hkl)| - |F_{calc}(hkl)|)^2}{n - m} \right]^{\frac{1}{2}} \tag{10.11}$$

with n being the number of independent measurements and m the number of variables. Theoretically, the GOF value approaches 1 for a well-refined model, where the errors are random. So far, it was assumed that the measurements of the intensities were perfect, and that no errors had been introduced. This is usually not the case, and the treatment of errors, systematic and random, has to be carried out in detail to ensure that the estimated standard deviations of the refined variables are meaningful. There are several sources of errors, which are briefly discussed.

The intensity data depends very strongly on the quality of the crystal used for the measurement. Furthermore, the kinematical theory is deemed applicable, more precise, that after an incident wave is scattered, it does no longer interact with the crystal. However, the Ewald construction makes it clear that the scattered wave fulfills the conditions to be scattered again, back into the incident direction but with a phase shift of π. This effect is called *primary extinction*, since the scattered amplitude is reduced, and it is most often observed in the forward scattering region of crystals with large, perfect grains. Therefore, an *ideally imperfect crystal* is assumed, where the primary extinction is considered small. The *secondary extinction* can reduce the intensities in the forward scattering directions if there is a strong interaction with the incident radiation. In this case, the incident beam intensity is reduced due to diffraction as the incident wave propagates through the crystal. In both cases, the primary and secondary extinctions are path dependent, and an extinction coefficient is defined which can be refined. Secondly, absorption can seriously hamper a crystal structure determination, if the absorption coefficient is high. As the absorption coefficient is determined by the number of electrons per unit volume, it affects inorganic phases containing heavy elements more than organic molecular compounds. Absorption corrections, with the shape of the crystal measured, are usually the best corrections, and a series of absorption correction methods have been devised. Grinding a sphere is also recommended, but this may not be feasible for soft and for strongly anisotropic materials. With irregular crystals and absorption not too strong, a shape correction mapping the crystal onto a spherical shape is often included in the data reduction. A good test of the success of any absorption correction is the internal agreement between symmetry equivalent reflections, which are expected to have equal intensity. The merged residual (R_{int}) that determines the intensity deviations between symmetry equivalent reflections should be as low as possible. Data collected from spherical crystals can give internal agreement values R_{int} of 1 to 2%, with several sets of independent reflections merged. It is usually desirable to collect more than just the absolute minimum number of reflections, also to assess the quality of the data collection. Collecting a large set of redundant data is now routinely done with an area detector, with a reflection measured multiple times in different positions. In general, internal agreement between merged symmetry-equivalent intensities should be better than $R_{int} \leq 0.12$, giving a set of unique intensities for structure determination and refinement.

It has become customary to use a refinement on $F_{obs}^2(hkl) \propto I_{obs}(hkl)$, whereas earlier work usually refined a structure using $F_{obs}(hkl) \propto \sqrt{I_{obs}(hkl)}$. In both cases, the structure should refine to the same parameter values, whereas the calculated residuals will differ. For a refinement against $F_{obs}(hkl) \propto \sqrt{I_{obs}(hkl)}$, it is customary to set a limit for a reflection to be included in the refinement. An often used limit for accepting a reflection is $I_{obs}/\sigma(F_{obs}) \geq 2$. The transformation from $F_{obs}^2(hkl)$ to $F_{obs}(hkl)$ is non-linear, and reduces the dynamic range of the data. With unit weights, the refinement is stable and outliers have less of an influence. Another issue with refinement on $F_{obs}(hkl)$ is the possibility of a false minimum. For the final refinement, after the structure is completed, statistical weights are applied and the residuals are calculated.

A refinement on $F_{obs}^2(hkl) \propto I_{obs}(hkl)$ is now standard. The refinement requires proper weighting so that the strong reflections are not dominating the refinement. The advantage is that the standard deviations are well-defined, and quasi-unit weights of $w = 1/(4F_{obs}^2(hkl))$ deemphasize the influence of the strong reflections. If a measurement includes a very large number of weak/unobserved reflections, the residual can remain high, since the refinement is attempting to fit data that has considerable uncertainties due to low intensities. Adding a cut-off criterion to exclude weak reflections may therefore be added.

For non-centrosymmetric structures, the absolute configuration can be determined. A number of methods have been developed to compare the $F(hkl)$ and $F(\bar{h}\bar{k}\bar{l})$ or $F^2(hkl)$ and $F^2(\bar{h}\bar{k}\bar{l})$ (or intensity) values, which, due to the anomalous dispersion, are no longer equivalent. A way to quantify the absolute configuration of a non-centrosymmetric crystal is the *Flack parameter*, defined as

$$I_{obs}(hkl) \approx I_c = (1 - x)I_e(hkl) + xI_e(\bar{h}\bar{k}\bar{l})$$

where I_{obs} is the observed intensity, I_c the intensity of the crystal, $I_e(hkl)$ the calculated intensity of the pure enantiomorph, and with x a refinable parameter. For the case of an enantiopure crystal, x will refine to either $x = 0$ or $x = 1$. The case for $x = 0$ indicates that this is the correct absolute configuration, whereas for $x = 1$, the structure should be inverted, with atomic coordinates transformed from x, y, z to $-x, -y, -z$. For $x \approx 0.5$, the crystal is either twinned or racemic. Refined values outside this range are nonphysical [30, 31].

An excellent reference describing practical strategies for structure refinements is found in the paper by D. Watkins [32].

10.7 How does it all fit together: an example

As an example, a single crystal structure analysis will be presented. A crystal (Courtesy of Dr. M. Steigerwald, Columbia University), reported as containing cobalt, tellurium and tri-ethylphosphine groups, was measured using a κ-axis 4-circle diffractometer equipped with a CCD detector (Oxford Diffraction Xcalibur-2), using graphite monochromatized MoKα radiation. The data collection was extended to higher angles, up to $2\theta = 90°$, but the number of observed intensities above $2\theta = 60°$ was small. Therefore, the data was truncated at $2\theta = 60°$ ($\sin\theta/\lambda = 0.7035$). A short data collection was first run to determine the unit cell and the corresponding orientation matrix. Indexing a set of about 50 reflections gave an initial monoclinic primitive unit cell of $a = 7.52(1)$ Å, $b = 11.53(2)$ Å, $c = 13.89(2)$ Å and $\beta = 90.52(8)°$. Based on this unit cell, a strategy for the data collection was developed to ensure completeness and sufficient redundancy of the data. In a full data collection at a temperature of $200K$, a total of 31909 reflections were measured. The determination of the space group evaluated symmetry elements associated with the unique *b*-axis, with a systematic extinction condition in the $(h0l)$-plane of $h + l = 2n$, indicating an *n*-glide plane, and an extinction condition for $(0, k, 0)$ as $k = 2n$, indicating a 2_1 axis. Images of the $(hk0)$, $(hk1)$, $(h0l)$, $(h1l)$, $(0kl)$ and $(1kl)$ planes are given in Figure 10.5. The presence of the 2-fold axis is clearly observed in the reciprocal planes 10.5c and 10.5d. In addition, the extinction due to the *n*-glide plane removes half of the reflections (Figure 10.5c), and the extinction along the unique *b*-axis is clearly seen in Figures 10.5a and 10.5e. This combination is consistent for the space group $P2_1/n$, which is equivalent to $P2_1/c$ (space group number 14) with a unit cell transformation. The unit cell with space group $P2_1/n$ was retained.

Fitting all observed reflections to the unit cell parameters while optimizing the diffractometer model such as the sample-detector distance, the tilt and yaw as well as the intercept position of the incident beam on the detector, etc. yielded a unit cell of

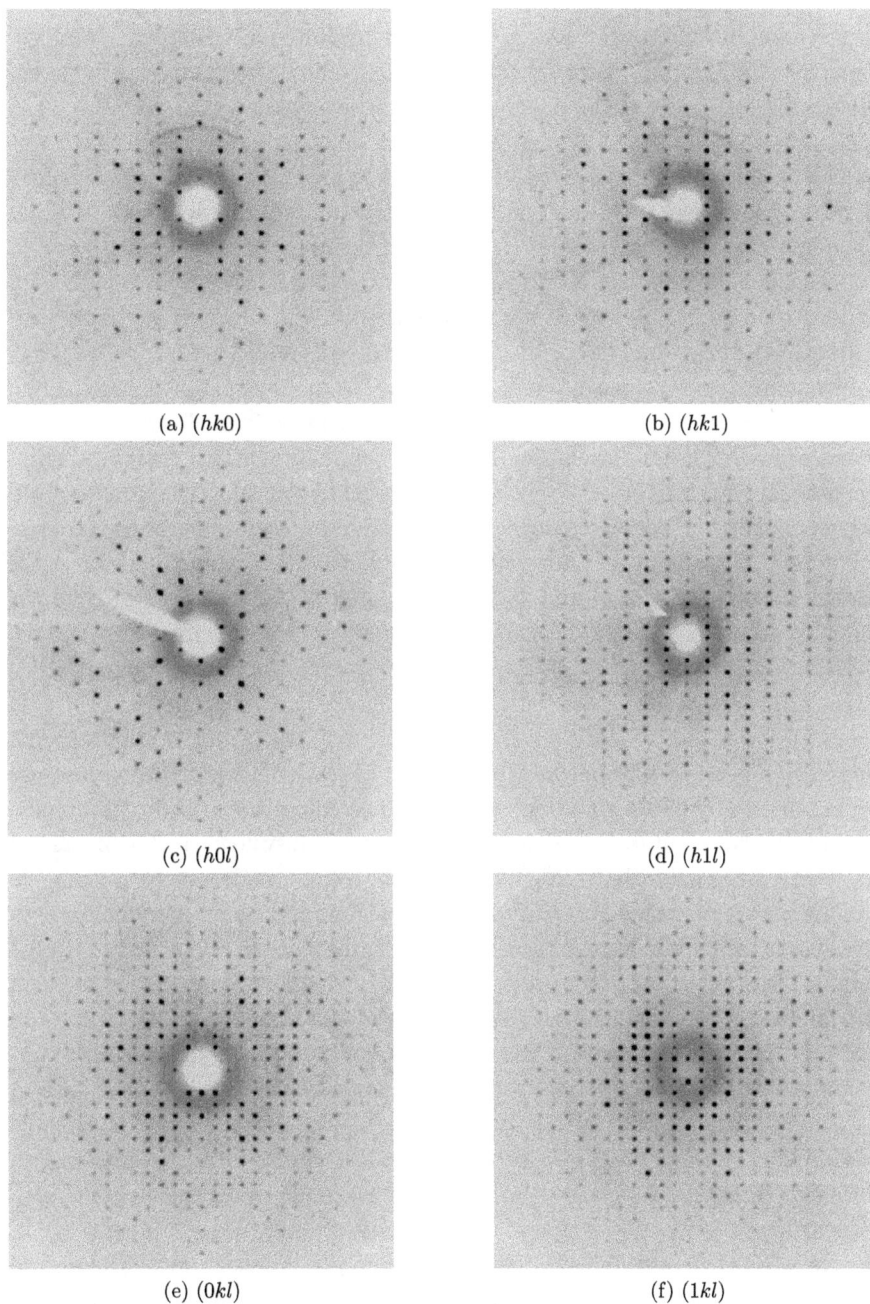

(a) $(hk0)$

(b) $(hk1)$

(c) $(h0l)$

(d) $(h1l)$

(e) $(0kl)$

(f) $(1kl)$

Figure 10.5: Reciprocal planes.

$a = 7.5345(5)$ Å, $b = 11.539(1)$ Å, $c = 13.898(1)$ Å and $\beta = 90.541(6)°$. With the orientation matrix well-defined via the unit cell refinement, the location of all the expected reflections can be calculated and corresponding integrated intensity data extracted (program suite CRYSALIS, Oxford Diffraction and Rigaku Corporation) For the latter, the integration volume is defined, but may be adjusted dynamically. The data collection is optimized to ensure that redundant data is collected, and that all reflections up to the desired resolution are measured multiple times. The high level of redundancy serves as a check on the stability of the sample and the diffractometer. A series of photographic sample images were taken for an absorption correction that is based on indexed faces to describe the crystal shape. The results of the integration of the intensities are stored in a file that contains the indexes h, k, l, observed intensity $I(hkl)$ and the estimated standard deviation $\sigma(I(hkl))$, or the values of $F^2(hkl)$ and $\sigma(F^2(hkl))$. Different manufacturers have different output formats, but in most cases, a reflection file ins ascii format with h, k, l, $F^2(hkl)$ and $\sigma(F^2(hkl))$ is written. Together with the unit cell and symmetry information, these data can then be imported into the program suite for the structure determination. In the following, the structure is solved and refined using the software package *CRYSTALS* developed in the Chemical Crystallography Laboratory at the University of Oxford [33]. Other program packages are available, such as *JANA2006*[1] and *SHELXL* [34], *SIR2014* [35], to mention just a few. A list of available crystallographic computing resources is given by the *International Union of Crystallography*, Commission for Crystallographic Computing.[2]

Using *CRYSTALS*, the data is imported and initial calculations are carried out to evaluate coverage (if all reflections are measured up to the desired resolution), the scale factor, and an overall Debye–Waller factor. The chemical formula of the compound is entered, or if not know exactly, approximated. As a rule of thumb, nonhydrogen atoms use about 18 Å^3 in reasonably dense packed systems. If a chemical formula gives unreasonable densities, adjustments should be made, since the chemical formula is used to calculate $F(000)$ and to estimate the scale factor. With the data imported, a structure solution can be attempted, using the tools described above. For the example, 31909 reflections were read and merged, giving 7126 reflections in the data set (redundancy of approximately 4), with a range of the indexes of $-11 \geq h \geq 11$, $-27 \geq k \geq 17$, and $-32 \geq l \geq 21$. However, the resolution range was adjusted to include reflections up to $\sin \theta / \lambda = 0.7035$. The expected stoichiometry was set to $Co_2Te_4P_2C_{12}H_{30}$, $Z = 2$, since details were not exactly known a priori. An initial attempt of solving the structure using direct methods gave two strong peaks in the Fourier map that were initially identified as tellurium, at positions (0.066 0.904 0.537) and (0.193 0.749 0.601). A refinement cycle gave an R-value of 0.47, but with strongly unequal displacement parameters. This indicates that the two positions may be correct but that the assignment of the atom types is incorrect. Furthermore, the distance between the two positions is calculated as 2.21 Å, too short for a tellurium-tellurium distance. Changing the

---

1 http://jana.fzu.cz
2 https://www.iucr.org/resources/other-directories/software

atom type of the atom with the high displacement parameter to cobalt and running another least squares refinement resulted in an R-value of 0.33. Therefore, these two positions are very likely positions of atoms in the structure, but the structure is obviously incomplete, and the atom assignment is inconsistent with the chemical bonding between cobalt and tellurium. Considering the calculated distance of 2.21 Å between these two positions, the distance is too short for a Co–Te bond. In addition, a longer bond of 2.66 Å between the strongest peaks in the structure is found. It is therefore very likely that tellurium is not present in the structure. A reassignment of atoms is therefore needed: cobalt is assigned to the strong peak, and the weaker peak is set to phosphorus, consistent with a Co–P distance. In this case, the electron ratio of tellurium to cobalt is approximately equal to the electron ratio between cobalt and phosphorus. The following backbone of the molecule is now identified: [P–Co–Co–P]. The distance between the two Co positions of 2.66 Å is a reasonable distance between two cobalt atoms, and the distance of cobalt to phosphorus of 2.1 Å is well within reported values. At this stage, the backbone of the molecule is likely correct. Another cycle of refinements with the new atom assignments, refining positional and isotropic displacement parameters gave a residual of 32 %. To complete the structure, further refinements and Fourier maps are needed. A Fourier map calculation using the calculated phases and the observed intensities identified a total of 17 peaks, 15 of which are new. Of these 15 peaks, 3 peaks are impossible atom positions due to very short distances to other peaks and are therefore not considered further. Of the remaining 12 peaks, there are 3 positions at 1.80 to 1.82 Å in a planar regular arrangement around the cobalt atom, with each of these new positions having another atom position at 1.12 Å (6 peaks total). Around the phosphorus atom, 6 atoms are found representing potential ethyl groups. These new atom positions are retained and included as carbon atoms. A series of least squares refinements of positions and isotropic displacement parameters brings the residual to 12.2 %, indicating that the basic arrangement of the molecule is correct. The structure is not yet correct, but the positions of nonhydrogen atoms have been identified. The atom type assignment, however, needs to be checked further. The short distances between the atoms around cobalt indicate that these are potential carbonyl groups. Furthermore, the furthermost atom position assigned as carbon showed smaller displacement parameter values than the positions closer to the cobalt atom. Therefore, changing the atom type from carbon to oxygen should improve the residual. Indeed, after a few cycles of least squares (positions and isotropic displacement parameters), the residual dropped to 10.2 %. The overall shape of the molecule is therefore well-defined. A subsequent Fourier map does not give additional atom positions and, therefore, the molecules can be identified as $Co_2(CO)_6P_2(C_2H_5)_6$ or $Co_2(CO)_6(PEt_3)_2$ with Et = ethyl. Switching from isotropic displacement parameters to anisotropic displacement parameters and running additional least squares cycles gave a residual of 7.94 %, indicating good agreement of the measured structure factors with the calculated structure factors. The structure of the molecule is now defined well enough to geometrically place the hydrogen atoms, assuming ideal angles and distances for carbon-hydrogen bonding. Since the hydrogen atoms are weakly scattering, their positions will not be freely refined, but tied to the respective carbon atoms. Additional refinement cycles with positional and displacement constraints on the hydrogen atoms lowered the residual to 6.97 % for the completed structure. The addition of an extinction parameter lowered the residual to 6.86 %, with a goodness-of-fit of 1.444, but gave nonphysical negative extinction values. Therefore, the secondary extinction was not refined. A number of additional least squares cycles brought the residual to 6.88 %, with the shifts applied to less than 0.0012, indicating that the refinement had converged. The refinement so far was on $F^2(hkl)$, with quasi-unit weights. This refinement used 127 parameters for 3581 observations, with about 28 reflections (observations) per refined parameter. The molecule is shown in Figure 10.6a using a *ball and stick* model, and the molecule with the refined displacement parameters represented as ellipsoids is shown in Figure 10.6b. The structure images of the molecule were generated using the program Mercury [36].

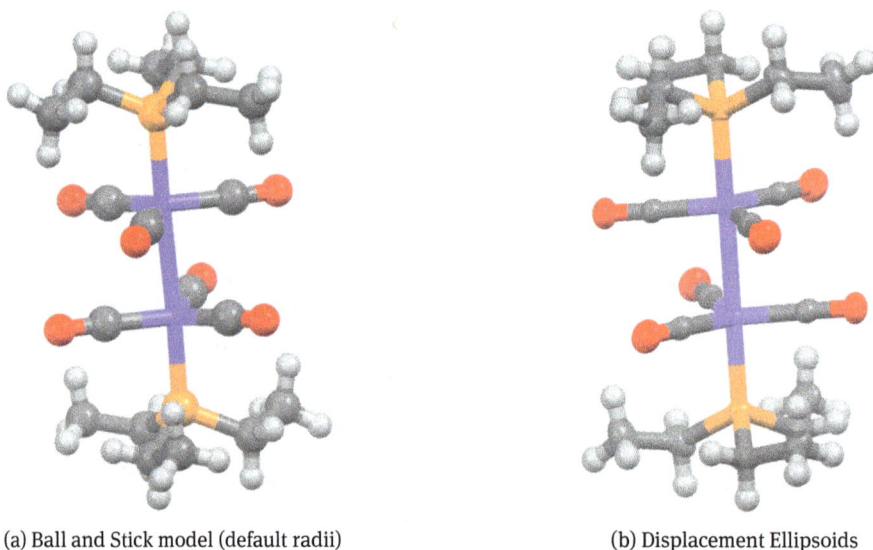

(a) Ball and Stick model (default radii)

(b) Displacement Ellipsoids

Figure 10.6: $Co_2(CO)_6(PEt_3)_2$ molecule.

In Figure 10.6b, the atomic displacement parameters are larger for the ethyl groups. This is an effect of the libration of the rigid molecule, where atoms farther from the molecular center of mass have less well-defined positions. In this case, the molecule has the center of mass located between the two cobalt atoms. The apparent displacement parameters of the ethyl groups are therefore a combination of the individual displacement parameters and the effects of the molecular librations. An analysis of the residuals for different structure factor values, shows that the strong structure factors are well explained by the model, whereas for the weak structure factors, the residuals are, as expected, larger. Therefore, the weighting scheme of the least squares can be optimized to take into account that the strong reflections are usually better determined than weak reflections. With *autostatistical* weights, the residuals are $R = 6.90\%$, and $Rw = 11.89\%$, with a Goodness-of-fit of 0.959. The final shifts are less than 0.0007, and therefore, the refinement can be considered as converged. Changing the weighting scheme to Chebyshev weights gives $R = 6.92\%$, $Rw = 7.65\%$, and Goodness-of-fit as 1.069. The position of the atoms in the structure, however, have not changed significantly by the change in the weighting scheme. The structure can therefore be considered as solved, with an acceptable residual. The molecular packing is shown in Figure 10.7, using the *capped stick* model, with hydrogen atoms omitted (rendered using the program Mercury, Cambridge Crystallographic Data Centre)

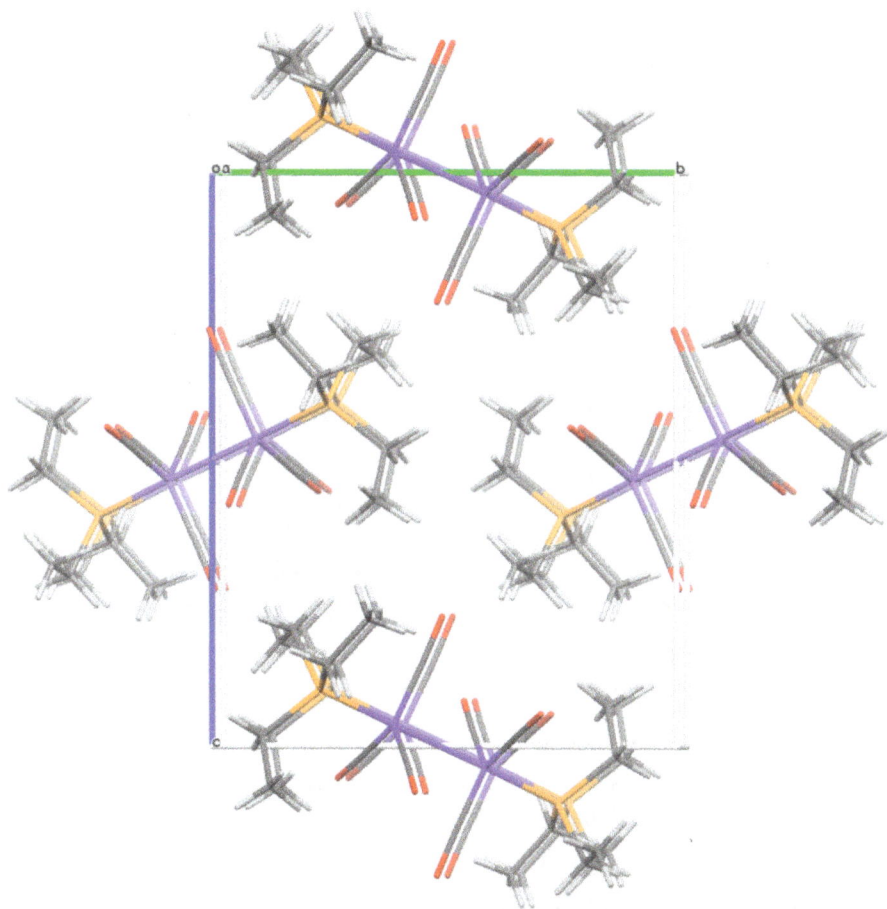

Figure 10.7: Capped Stick Model for the Molecular Packing of $Co_2(CO)_6(PEt_3)_2$ with the view approximately parallel to the a-axis.

It is instructive to plot the extracted $F_{obs}(hkl)$ (×) and refined $F_{calc}(hkl)$ (∘) versus $1/d = (2\sin\theta/\lambda)$, given in Figure 10.8, with the magnitude shown in color (red: strong; blue: weak). As expected the strong reflections are found at low angles, and the observed intensities drop off towards higher angles. Since the 3-dimensional data is collapsed onto a linear scale $(1/d)$, the density of reflections increases with larger $1/d$-values. The agreement between $F_{obs}(hkl)$ and $F_{calc}(hkl)$ is reflected how far apart their respective symbols are located. Figure 10.8 also shows that at short d-spacing (large 2θ), relatively few very strong reflections are observed, and truncating a measurement around $2\theta = 60°$ for MoKα radiation gives sufficient data for a structure determination, with the number of observations per refined parameter larger than 10.

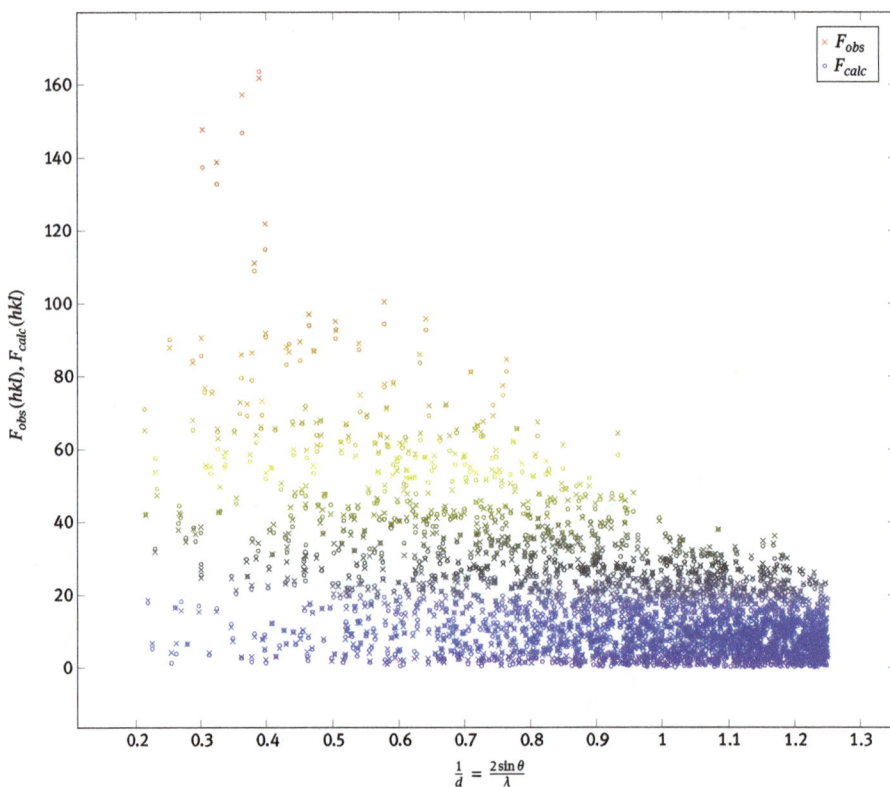

Figure 10.8: Plot of $F_{obs}(hkl)$ (×) and $F_{calc}(hkl)$ (∘) of $Co_2(CO)_6(PEt_3)_2$ vs ($1/d = 2\sin\theta/\lambda$). The color indicates the magnitude of $F_{obs}(hkl)$ and $F_{calc}(hkl)$.

Plotting the calculated $F_{obs}(hkl)$ versus the observed $F_{calc}(hkl)$ is expected to follow a linear relationship for a well refined structure, where the residuals are small. The deviations are larger for the weak intensities, however, no obvious outliers are visible in

Figure 10.9, where the large magnitude reflections are indicated in red, and the weak reflections in blue. Analysis of the refinement should be done to identify areas in reciprocal space and intensity distribution where the model and the measurement differ to identify potential problems with the structure. For instance, shielding of strong low angle reflections by the diffractometer beam stop will result in a large F_{calc}, whereas the F_{obs} will be too low. Such outliers can be identified, and the source of the discrepancy investigated.

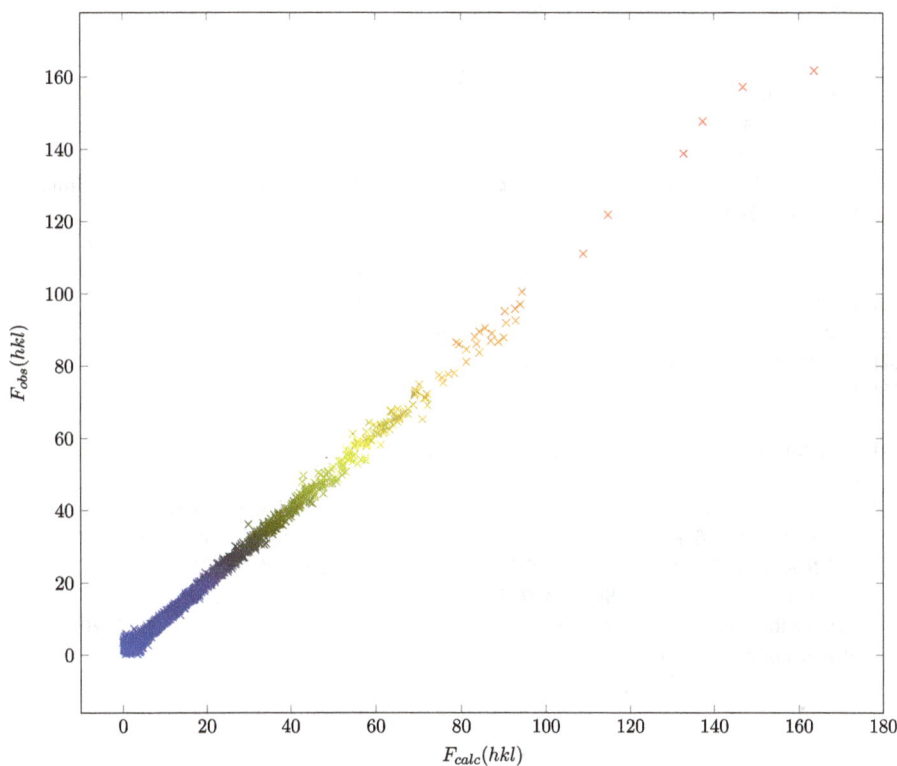

Figure 10.9: Plot of $F_{obs}(hkl)$ vs $F_{calc}(hkl)$ for $Co_2(CO)_6(PEt_3)_2$.

Single crystal structural analysis of materials has evolved into a powerful method for chemical analysis. The large searchable crystallographic data base systems allow identification of materials by their unit cell parameters, a distinctive finger print for any crystalline material. In combination with X-ray spectroscopy to identify elemental composition, diffraction methods can provide chemical information of new phases within hours. Further advances in diffraction methods and algorithm development are expected to shorten this time even further.

Bibliography

[19] H. Powell and A. Wells. The Structure of Caesium Cobalt Chloride, Cs_3CoCl_5. J. Chem. Soc., 359–362, 1935.

[20] P. A. Reynolds, B. N. Figgis, and A. H. White. Acta Crystallogr. B, 37:508–513, 1981.

[21] A. L. Patterson. A direct method for the determination of the components of interatomic distances in crystals. Z. Kristallogr., 90:517, 1935. https://doi.org/10.1524/zkri.1935.90.1.517.

[22] H. Lipson and W. Cochran. The Determination of Crystal Structures. G. Bell and Sons Ltd, London, 1953.

[23] M. M. Wolfson. Acta Crystallogr., A43:593–612, 1987. https://doi.org/10.1107/S0108767387098854.

[24] J. Karle and H. Hauptman. Acta Crystallogr., 3:181, 1950. https://doi.org/10.1107/S0365110X50000446.

[25] D. Sayre. The squaring method: a new method for phase determination. Acta Crystallogr., 5:60–65, 1952. https://doi.org/10.1107/S0365110X52000137.

[26] W. Cochran and M. M. Woolfson. Acta Crystallogr., 8:1, 1955.

[27] R. W. Gerchberg and W. O. Saxton. A practical Algorithm for the Determination of Phase from Image and Diffraction Plane Pictures. Optik, 39:237–249, 1972.

[28] J. R. Fienup. Phase retrieval algorithms: a comparison. Appl. Opt., 21:2758–2769, 1982.

[29] G. Oslányi and A. Sütő. The charge flipping algorithm. Acta Crystallogr. A, 64:123–134, 2008.

[30] D. J. Watkin and R. I. Cooper. Chemistry, 2:796–804, 2020.

[31] H. D. Flack. Acta Crystallogr. A, 39:876–881, 1983.

[32] D. Watkins. J. Appl. Crystallogr., 41:491–522, 2008.

[33] P. W. Betteridge, J. R. Carruthers, R. I. Cooper, K. Prout, and D. J. Watkin. J. Appl. Crystallogr., 36:1487, 2003. http://www.xtl.ox.ac.uk/crystals.1.html.

[34] G. M. Sheldrick. Acta Crystallogr. C, 71(1):3–8, 2015. https://doi.org/10.1107/S2053229614024218.

[35] M. C. Burla, R. Caliandro, B. Carrozzini, G. L. Cascarano, C. Cuocci, C. Giacovazzo, M. Mallamo, A. Mazzone, and G. Polidori. J. Appl. Crystallogr., 48:306–309, 2015. http://www.ic.cnr.it.

[36] C. F. Macrae, I. Sovago, S. J. Cottrell, P. T. A. Galek, P. McCabe, E. Pidcock, M. Platings, G. P. Shields, J. S. Stevens, M. Towler, and P. A. Wood. Mercury 4.0: from visualization to analysis, design and prediction. J. Appl. Crystallogr., 53:226–235, 2020. https://doi.org/10.1107/S1600576719014092.

11 Crystallographic information file

Crystallographic data used to be published, for instance, as a list of (hkl) and measured F_{obs} and calculated F_{calc}. This is no longer feasible, and electronic storage is now the rule. To facilitate the data exchange, a standardized format to encode crystallographic data was developed in the early 1990s, the *Crystallographic Information File, CIF*. The *CIF* format is not static, and is further under development into the *Crystallographic Information Framework*, and a *CIF2* file format has been described.[1] The original *CIF1* format will be supported by the International Union of Crystallography in perpetuity, and is a convenient way of depositing crystallographic information in database systems, such as the *Cambridge Crystallographic Data Centre* (CCDC),[2] the *Inorganic Crystal Structure Database* (ICSD)[3] and the *Crystallography Open Database* (COD).[4] A detailed description of the CIF1 standard is available at http://ww1.iucr.org/iucr-top/cif/standard/cifstd1.html and [37].

A partial description to the CIF1 format will be illustrated by an example. The file structure is ASCII free-format and can be directly edited using any text editor. This ensures that the data structure is independent of any particular computer architecture, and allows data to be read by humans as well as by machines. Specialized programs that can edit a CIF and validate the syntax are available, and are often part of program suites.

The CIF contains data blocks, with each block composed of individual data items. The data blocks group data that is associated with data sets. Individual data items are identified with a unique data name that is self-explanatory. In addition, multiple data items can be combined in a loop structure. An excerpt of the CIF of barlowite is given below. The data identifiers are preceded by a "_" and are shown in teletype font. The first block of data contains the unit cell and symmetry information.

```
data_Barlowite                                          Start of a data block
_computing_structure_refinement    GSAS                 program package used
_refine_ls_number_parameters        52                  # of refined parameters
_refine_ls_goodness_of_fit_all      1.80
_refine_ls_number_of_restraints     0
_refine_ls_matrix_type              full

_chemical_formula_sum               "Br Cu4.00 F H6 O6"  Formula
_chemical_formula_weight            455.13
_cell_formula_units_Z               2
```

1 https://www.iucr.org/resources/cif/cif2
2 https://www.ccdc.cam.ac.uk
3 https://icsd.products.fiz-karlsruhe.de
4 http://www.crystallography.net/cod

https://doi.org/10.1515/9783110610833-011

```
# Unit Cell Information                            Comment line
_cell_length_a                    6.675463(15)   unit cell length a in Å and
_cell_length_b                    6.675463       ESD
_cell_length_c                    9.298900(23)   unit cell length c
_cell_angle_alpha                 90.0
_cell_angle_beta                  90.0
_cell_angle_gamma                 120.0
_cell_volume                      358.8600(20)
_symmetry_cell_setting            hexagonal
_symmetry_space_group_name_H-M    "P 6/3m m c"   Space group symbol

loop_                                             Start of a loop
_symmetry_equivalent_pos_site_id                  first field of the loop, a number
_symmetry_equiv_pos_as_xyz                        second field of the loop, the (x, y, z)'s
1 +x,+y,+z
2 x-y,+x,+z+1/2
3 -y,x-y,+z
4 -x,-y,+z+1/2
5 y-x,-x,+z
6 +y,y-x,+z+1/2
7 y-x,+y,+z
8 -x,y-x,+z+1/2
9 -y,-x,+z
10 x-y,-y,+z+1/2
11 +x,x-y,+z
12 +y,+x,+z+1/2
-1 -x,-y,-z
-2 y-x,-x,-z+1/2
-3 +y,y-x,-z
-4 +x,+y,-z+1/2
-5 x-y,+x,-z
-6 -y,x-y,-z+1/2
-7 x-y,-y,-z
-8 +x,x-y,-z+1/2
-9 +y,+x,-z
-10 y-x,+y,-z+1/2
-11 -x,y-x,-z
-12 -y,-x,-z+1/2
```

The atom positions are described in a loop, giving the atom type, the atom label, the position (x, y, z), occupation parameters (usually 1.0 unless refined), a mnemonic for the displacement type (Uiso, Uani) and the multiplicity of the site symmetry. Estimated standard deviations are given for the parameters where refined, and omitted for symmetry related values. It is followed by a loop for the anistropic displacement parameters that were refined.

```
# Atomic Coordinates and Displacement Parameters
loop_                                                          Coordinate loop
_atom_site_type_symbol
_atom_site_label
_atom_site_fract_x
_atom_site_fract_y
_atom_site_fract_z
_atom_site_occupancy
_atom_site_thermal_displace_type
_atom_site_U_iso_or_equiv
_atom_site_symmetry_multiplicity
Cu Cu1 0.5 0.0 0.0 1.0 Uani 0.00979 6
Cu Cu2 0.37044(5) 0.74089(10) 0.75 0.33333 Uani 0.00693 6
O O1 0.20134(7) 0.79866(7) 0.90848(10) 1.0 Uani 0.00734 12
Br Br 0.33333 0.66667 0.25 1.0 Uani 0.01319 2
F F 0.0 0.0 0.25 1.0 Uani 0.01629 2
H H 0.1248(10) 0.8752(10) 0.8743(15) 1.0 Uiso 0.090(7) 12    last atom
;                                                             end of the loop

loop_                                                         loop for U_ijs
_atom_site_aniso_label
_atom_site_aniso_U_11
_atom_site_aniso_U_12
_atom_site_aniso_U_13
_atom_site_aniso_U_22
_atom_site_aniso_U_23
_atom_site_aniso_U_33
Cu1 0.00752(9) 0.00317(6) -0.00140(7) 0.00635(12) -0.00280(13) 0.01512(13)
Cu2 0.00835(29) 0.00671(21) 0.0 0.0134(4) 0.0 0.0007(4)
O1 0.00642(35 0.0031(4) -0.00040(19) 0.00642(35) 0.00040(19) 0.0091(6)
Br 0.01567(12) 0.00783(6) 0.0 0.01567(12) 0.0 0.00823(23
F 0.0140(6) 0.00698(30) 0.0 0.0140(6) 0.0 0.0209(11)
```

This is an abbreviated CIF that contains the necessary structural information to render a structure. Measured and calculated F-values can be added, as well as author information. Additional information can be embedded into the CIF for publication submission to *Acta Crystallographica C*. Data name categories include

_audit	contains data about the creation and updates of the CIF
_atom	describes atom types and atom sites
_cell	unit cell data
_chemical	composition and chemical properties
_computing	programs used in the structure analysis
_diffrn	diffraction experiment details
_exptl	crystal measurements, such as size, shape, color, etc.
_refine	structure refinement parameters
_symmetry	space group

There are a number of programs available to check for self-consistency of the data, with *CheckCIF* (https://checkcif.iucr.org) a web based interface that will give feedback on a structure determination. Submission of a CIF for publications that include a structure determination is now standard. This reflects the success of the CIF format as a means for data exchange across platforms. It facilitates data storage, and is adaptable for future developments.

Bibliography

[37] S. R. Hall, F. H. Allen, and I. D. Brown. The crystallographic information file (CIF): a new standard archive file for crystallography. Acta Crystallogr., A47:655–685, 1991.

Index

www.ingramcontent.com/pod-product-compliance
Lightning Source LLC
Chambersburg PA
CBHW061405210326
41598CB00035B/6104